Studies in Computational Intelligence

Volume 517

Series editor

J. Kacprzyk, Warsaw, Poland
email: kacprzyk@ibspan.waw.pl

For further volumes:
http://www.springer.com/series/7092

About this Series

The series "Studies in Computational Intelligence" (SCI) publishes new developments and advances in the various areas of computational intelligence—quickly and with a high quality. The intent is to cover the theory, applications, and design methods of computational intelligence, as embedded in the fields of engineering, computer science, physics and life sciences, as well as the methodologies behind them. The series contains monographs, lecture notes and edited volumes in computational intelligence spanning the areas of neural networks, connectionist systems, genetic algorithms, evolutionary computation, artificial intelligence, cellular automata, self-organizing systems, soft computing, fuzzy systems, and hybrid intelligent systems. Of particular value to both the contributors and the readership are the short publication timeframe and the world-wide distribution, which enable both wide and rapid dissemination of research output.

Nazmul Siddique

Intelligent Control

A Hybrid Approach Based on Fuzzy Logic, Neural Networks and Genetic Algorithms

Foreword by Bernard Widrow

Springer

Nazmul Siddique
School of Computing and Intelligent Systems
University of Ulster
Londonderry
UK

ISSN 1860-949X ISSN 1860-9503 (electronic)
ISBN 978-3-319-02134-8 ISBN 978-3-319-02135-5 (eBook)
DOI 10.1007/978-3-319-02135-5
Springer Cham Heidelberg New York Dordrecht London

Library of Congress Control Number: 2013949695

Printed on acid-free paper

Springer is part of Springer Science+Business Media (www.springer.com)

To the loving memory of my mother who struggled to educate her children and faced all sort of hardships but never gave up hope. It was only the cancer—she could not cope up with and it didn't even give me a chance to thank her for all she did for me

Foreword

Intelligent Control is a hybrid approach to control systems based on Fuzzy Logic, Neural Networks and Genetic Algorithms. Intelligent Control (IC) breaks new ground on many levels and demonstrates the effectiveness of an alternative approach to traditional mathematical model-based control approaches. Control systems have evolved over the centuries. The 1950s and 1960s have seen the development of optimal control, modern control, and the space race unraveling the most multifaceted control problems. Despite all its successes, dissatisfaction was growing among control system scientists and intelligent control was becoming popular in the 1970s and 1980s. IC brought fresh interdisciplinary approaches into the field addressing a higher level of complexity. In recent years, the IC approach has been growing rapidly in visibility and importance.

The concept of intelligent control began to crystallize in the nineteen eighties, at a time when AI was undergoing an identity crisis, moving from logic to probability theory. There were many competing methodologies, among them the traditional AI, fuzzy logic, neurocomputing and evolutionary computing. Each of these methodologies had a community, with each community claiming superiority over others. In this monograph, different combinations of these techniques are presented in a very reader-friendly way. A brief discussion on an exemplary non-linear system, a flexible robotic arm, is presented which demonstrates the amount of effort required in developing a mathematical model of a nonlinear system. A reader can easily understand the complexities of the traditional model based control and the difference between intelligent control and model-based control.

The guiding principle of intelligent control is that, in general, superior performance is achieved when fuzzy logic, neurocomputing, and genetic algorithms are used in combination rather than in stand-alone mode. The three paradigms have their own advantages and disadvantages. The advantage of a fuzzy system is that it works using approximate information and it represents knowledge in a linguistic form, which resembles human-like reasoning. The NN has the advantage of learning from experience. GA optimizes or learns through adaptation of a system's structure or parameters by evolving a random population while exploring the search space in several directions. The fusion of the three paradigms creates a system with a certain degree of autonomy and enhanced adaptive performance.

This monograph on intelligent control is a valuable contribution to the literature—it shows an alternative approach to control problems using fuzzy logic, neural networks and genetic algorithms. It is insightful and reader-friendly. The book will be useful for advanced students, researchers and practitioners who are interested in the conception, design and utilization of intelligent control systems. Dr Nazmul Siddique and the Springer Verlag deserve our compliments.

California, October 10, 2013 Bernard Widrow

Preface

Traditional mathematical model-based control has a history of over hundred years. The term traditional or conventional control is used to refer to theories and methods that were developed in the past decades to control dynamical systems, the behaviour of which is primarily described by differential and difference equations.

In fact, it is well known that there are control problems that cannot be adequately formulated and studied in the form of differential or difference equation mathematical framework. To address these problems in a systematic way led researchers to develop a number of methods that are collectively known as intelligent control methodologies.

In this context, the term intelligent control has come to mean some form of control using fuzzy logic and/or neural network methodologies. However, intelligent control does not restrict itself only to those methodologies. Research into intelligent control incorporates and integrates different techniques and concepts from different disciplines including control theory, computer science, fuzzy logic, neural networks and genetic algorithms. Considerable research is currently being devoted to intelligent control techniques for systems that are ill-defined, poorly understood or highly nonlinear such as flexible-link robot arm. However, application of intelligent control to flexible-link robot arm is not widespread.

Modelling and control of flexible robot arm for both space and industrial applications is a research area that has recently aroused considerable interest. For robots to meet the demands of industry, lightweight arms are needed so that they move faster without requiring high-powered bulky actuators. As manipulator arms are made lighter, their deformation under stress increases. Conventional control methods of flexible manipulators require fast and accurate models for dynamic performance. The demand for such an accurate mathematical model for the system under a variety of different operating conditions complicates the design of control systems. A non-conventional control strategy is sought without requiring expensive computing machinery.

The aim of this research monograph is to develop intelligent control schemes. Application of those control techniques was verified on a flexible robotic arm. To allow this, first, investigations into modelling and simulation of flexible arm are carried out. A simulation environment characterizing the dynamic behaviour of an arm is initially developed for test and verification of controller designs. Second, investigations into different types of fuzzy controller such as PD-, PI- and PID-type controllers and their performances are carried out. Third, a rule reduction scheme for a fuzzy PID-type controller is developed by implementing a switching PD-PI-type controller where a PD-type controller is executed first to attain fast rise time and smaller overshoot and then switched to a PI-type controller to gain the advantage of minimized steady-state error.

To optimise and tune the membership functions of the fuzzy controller, evolutionary algorithms and neural learning systems are applied. Efforts are made in developing systems that are capable of learning in a real-time manner by utilizing intelligent methodologies such as fuzzy logic, neural networks, genetic algorithms and a combination of those paradigms. The flexible robot arm is utilized in this work as a practical level test and verification platform for controller design.

Acknowledgments

As a final remark, it is necessary to thank a number of people who have helped in many ways unknown to them to the endeavour of this book. I would like to thank Dr. Osman Tokhi for his guidance and help that made it possible to carry out the research work at the Department of Automatic Control and Systems Engineering.

It is my great pleasure to thank Dr. Edin Begic and Dr. Takatoshi Okuno whose friendship enlivened both the research and my stay at Sheffield. Particularly, the wonderful experience with Dr. Begic in part taking on the discussions on different aspects of this research work is simply unforgettable. Further, my sincere thanks go to the entire staff of the Department of Automatic Control and Systems Engineering, whose help was indispensable for pursuing this research work.

I gratefully acknowledge the financial support of the Commonwealth Scholarship Commission in the United Kingdom, and other necessary supports of Khulna University, Bangladesh during 1998–2001.

The author would like to thank many of the collaborators, Dr. Bala Amavasai, Dr. Richard Mitchell, Dr. Michael O'Grady, Dr. Mourad Oussalah, Dr. John St. Quinton, Prof. Atta Badii, Prof. Alamgir Hossain, Prof. Ali Hessami, Dr. Takatoshi Okuno, Dr. Faraz Hasan, Prof. Akira Ikuta, Prof. Hydeuki Takagi, Dr. Filip Ponulak, Dr. David Fogel, Prof. Hojjat Adeli and Prof. Bernard Widrow. The author would like to thank all staff of Springer Verlag associated with the publication of this book, especially the project editors Oliver Jackson and Charlotte Cross for their kind support and help throughout the publication process from the manuscript to print.

The author would like to express his gratitude to his brothers, sisters, brothers-in-law and sisters-in-law for their constant inspiration, support and love. The author would like to thank his wife Kaniz for her love and patience during the entire endeavour of the book without which this book would have never been published, and his daughters Oyndrilla, Opala and Orla for not making any complaints during this time.

Contents

Chapter 1
Introduction

1.1 Intelligent Control

The history of feedback control dates back to the period from about 300 BC to about 1200 AD with the float valve regulators for keeping accurate track of time of the Hellenic and Arab worlds (Meyer 1970). The known Greek engineers Ktsebios and Philon of the third century BC and Heron of the first century AD developed float regulator for water clock, regulator for keeping a constant level of oil in lamps, automatic dispensing of wine, siphon for maintaining constant water level and various other applications. The water clock tradition was continued by the Arab engineers from 800 AD through 1200 AD. Among them are Banu Musa borthers (c. 873), Al-Jazari (c. 1203) and Ibn al-Sa-ati (c. 1206). Float regulators in the tradition of Heron were also constructed by the three brothers Banu Musa in Baghdad in the ninth century AD (Bissell 2009; Hill 1979). During this period the important feedback principle of on/off control was used, which came up again in connection with minimum-time problems in the 1950s.

The first independent European feedback control system was the temperature regulator developed by Cornelius Drebbel in the 17th century, which was an alcohol thermometer used to operate a valve controlling furnace fuel. The device included screws to alter what is called the set point in today's control systems (Bissell 2009). The float valve regulator does not appear to have spread to Europe rather it seems to have been reinvented during the industrial revolution in the 18th century started with the invention of James Watt's steam engine in 1769. The Industrial Revolution followed many inventions initiating the renewed interest in the development of a variety of control systems including float regulators, temperature regulators, pressure regulators, and speed control devices. The problem associated with the rotary steam engine was controlling the speed of revolution. It was not until 1788 when James Watt completed the design of the centrifugal flyball governor for regulating the speed of the rotary steam engine. The operation of the flyball governor was clearly visible and its principle of feedback mechanism gave an exotic flavour to new industrial age.

N. Siddique, *Intelligent Control*, Studies in Computational Intelligence 517,
DOI: 10.1007/978-3-319-02135-5_1,

The field of control engineering had to wait another century up until 1868 since the invention of steam engine in 1769 for the first mathematical description of system by J.C. Maxwell using differential equations. It is J.C. Maxwell, who demonstrated the mathematical and analytical approach to developing system models and methods in understanding and analyzing behaviour of systems and pioneered the mathematical system and control theory for theoretical analysis of stability problems encountered with flyball governor over the century. Since Maxwell the control theory made significant strides in the next 100 years up until 1960s. The classical control theory was naturally developed using frequency domain methods and Laplace transforms in the 1930s and 1940s. The classical control design approach mainly relied on transform methods, mathematical system description and the transfer function. An exact description of the internal system dynamics is not needed. Only the input–output behaviour of the system is of importance.

Optimal control methods and state space analysis were developed in the 1950s and 1960s representing them as the modern control theory, which have made it possible to meet the demanding control specification for increasingly complex dynamic systems (Antsaklis and Passino 1995). Modern control theory is fundamentally a time-domain technique. An exact state-space model of the system to be controlled is required. This is a first-order vector differential equation of the form

$$\frac{dx}{dt} = Ax + Bu \tag{1.1}$$

$$y = Cx + Du \tag{1.2}$$

where $x(t)$ is a vector of internal variables or system states, $u(t)$ is a vector of control inputs, and $y(t)$ is a vector of measured outputs. It is possible to add noise terms to represent process and measurement noises. A, B, C and D are matrices describing the interconnections of the dynamical system. The advantageous feature of the state-space model is that it can also be used to represent a MIMO system in terms of SISO systems. Modern controls techniques are well established for linear systems, which can be extended to nonlinear systems using the Lyapunov approach, dynamic programming, and other techniques.

In fact it is well known that there are control problems where the system cannot be adequately described in a differential equations framework. It is fact that there are unmodelled dynamics in systems that cannot be described using traditional mathematical methods, which makes the control design a difficult task and do not let them control to projected accuracy or intended satisfaction. Deliberately modern control theories were then followed by the progress in stochastic, robust and adaptive control methods in the 1960s. Adaptive control mainly estimates the control parameters from the short-term (or recent) data, which does not require any memory. Learning control came as an expected consequence when a robust control design necessitates the learning of parameters from long-term history demanding storage of previous states and appropriate responses.

All of these control methods discussed hitherto demand exact model of the system. On the contrary, it has been demonstrated that the use of highly complex mathematical models can seriously inhibit the ability to develop control algorithms. Furthermore, it is required to cope with significant unmodeled and unanticipated changes in the system, in the environment and in the control objectives. This will involve the use of advanced decision-making processes to generate control actions so that a certain performance level is maintained even though there are drastic changes in the operating conditions. Thus, the dissatisfaction with conventional control started growing with increasing complexity of dynamical systems and necessitates the use of more human expertise and knowledge in controlling such processes. Intelligent control is thus a manifestation of the crucial time when human knowledge is becoming more and more important in control systems design as an alternative to traditional model-based control whose structure and consequent outputs in response to external commands are determined by experimental evidence, i.e., the observed input–output behaviour of the system, rather than by reference to a mathematical or model-based description of the controller. The controller is then a so-called intelligent controller. The term intelligent control was originally coined by Fu (1971) and was defined as an approach to generate control actions by employing aspects of natural or human intelligence, operation research and automatic control systems. Intelligent control systems are evolutionary rather than revolutionary and evolve from existing controllers in a natural way meeting and competing demanding challenges of the time starting from James Watt's flyball governor.

Intelligent control approaches are suitably aimed at processes that are ill-defined, complex, nonlinear, time varying and stochastic. Intelligent control systems are not defined in terms of specific algorithms. They employ techniques that can sense and reason without much a priori knowledge about the environment and produce control actions in a flexible, adaptive and robust manner (Harris 1994). Central to intelligent control is the construction of the process model. Many real world processes are not amenable to mathematical modeling because

(i) the processes are too complex to represent mathematically
(ii) process models are difficult and expensive to evaluate
(iii) there are uncertainties in process operation
(iv) the process is nonlinear, distributed, incomplete and stochastic in nature.

In the minds of many people, particularly outside the control area, the term *intelligent control* has come to mean some form of control using fuzzy logic and/or neural network methodologies. This perception has been reinforced by researchers in the 1980s, 1990s and to date. The area of intelligent control is in fact interdisciplinary and it attempts to combine and extend theories and methods from other disciplines including artificial intelligence, modern adaptive control, optimal control, learning theory, reinforcement learning, fuzzy logic and neural networks. Each discipline is approaching intelligent control from a different perspective, using different methodologies and toolsets towards a common goal. The interrelationship between these disciplines is illustrated in Fig. 1.1.

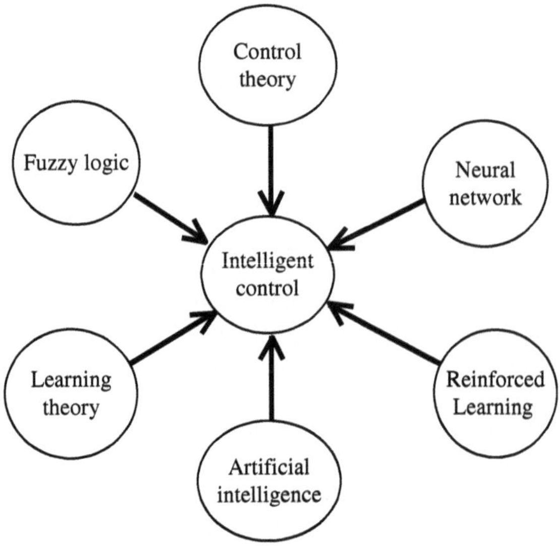

Fig. 1.1 Tools of intelligent control

Ozguner gives an outline of decentralized control-theoretic view of intelligent control in Ozguner (1989). Functional and structural hierarchies are studied in Acar and Ozguner (1990). Fundamentals of intelligent systems such as the principle of increasing intelligence with decreasing precision are discussed in Meystel (1985) and Saridis (1985, 1987). Albus introduces the theory of intelligent control that has received considerable attention since his publication of cerebellum model articulation control (CMAC) (Albus 1981, 1991). Harris et al. introduce intelligent control as an aspect of fuzzy logic and neural networks in Harris et al. (1993) and Harris (1994). Zilouchian and Jamshidi introduce intelligent control using soft computing, where the inherent components of soft computing are fuzzy logic, neural networks, evolutionary computing and probabilistic and chaos theory, in Zilouchian and Jamshidi (2001).

1.2 Intelligent Control Architecture

The engineering is in such a vital stage with the advent of so many new technologies when dissatisfaction with conventional control is growing with increasing complexity of dynamical systems and necessitates the use of more human expertise and knowledge in controlling processes. As systems become more complex, uncertainty in modelling increases and human intervention is more likely. The human operator can only respond to complex set of observations and constraints to satisfy multiple objective performance criteria. It is not known how a human operator controls such a complex process. Hence a clear-cut architecture of intelligent control cannot be given. Many researchers give the general framework

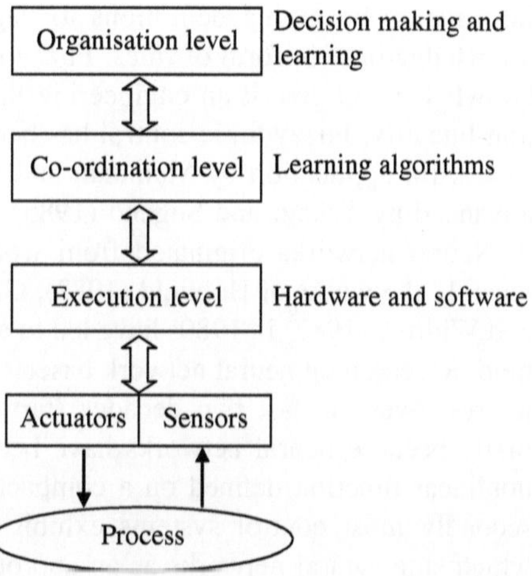

Fig. 1.2 Three-level intelligent control architecture

or the architecture that support and integrate intelligent control system in practice. A variety of architectures have emerged. Among these the hierarchical architecture has appeared as a natural one in the literature. Figure 1.2 shows the three-level intelligent control architecture. Attempts to increase precision will result in decreasing intelligence. Such architecture is not confined to three levels and in practice it may involve more than three levels.

1.3 Approaches to Intelligent Control

Intelligent controllers use experiential knowledge about the process that generally produces a model in terms of input–output behaviour. The question is how to model this human knowledge and represent it in such a manner to be computationally efficient. According to Harris et al. (1993) there are three basic approaches to intelligent control. These are

(i) Knowledge-based systems
(ii) Fuzzy logic
(iii) Neural networks.

Knowledge-based systems utilize a collection of information of different forms such as facts, heuristics, common sense and other forms of knowledge and use reasoning methods to inferences.

It has been challenged by researchers that measurements, process modelling and control can never be exact for real and complex processes. Also there are uncertainties such as incompleteness, randomness and ignorance of data in the process model. The seminal work by Zadeh (1965, 1968, 1973) introduced the concept of fuzzy logic to model human reasoning from imprecise and incomplete information by giving definitions to vague terms and allowing construction of knowledge-base in form of rules. Fuzzy logic can incorporate human experiential knowledge and give it an engineering flavour to control ill-defined systems with non-linearity. Fuzzy logic control has been very popular control method since the successful application by Mamdani and Assilian (1974) in the 1970s and further advanced by Takagi and Sugeno (1985) in the 1980s.

Neural networks originated from works of Hebb (1949) in 1940s and more recently the works of Hopfield (1982), Grossberg (1982), Rumellhart et al. (1986) and Widrow (1987) in 1980s have led to a resurgence of the research interest in the field. Research on neural network based control systems has received considerable interest over the last two decades starting from the renewed interest in 1980s, firstly because neural networks have been shown to be able to approximate any nonlinear function defined on a compact set of data to a specified accuracy and secondly most control systems exhibit certain types of unknown nonlinearity, which suits neural networks as an appropriate control technology.

Optimisation of control parameters and tuning of scaling factors in terms of control objectives and system performance is a natural extension to any control design. Although fuzzy logic controllers and neural network controllers have been successfully applied to many complex industrial processes, but they experience a deficiency in knowledge acquisition and rely to a great extent on empirical and heuristic knowledge, which, in many cases, cannot be objectively elicited. The inherent problem is that the traditional optimisation techniques cannot be applied to fuzzy logic or neural network based control systems. As an alternative to these, evolutionary algorithms, specifically genetic algorithms, appear to be the more suitable method for searching optimal control performance. Genetic algorithms are exploratory search and optimisation methods that were devised on the principles of natural evolution introduced by Holland (1975). Efforts have been made to automate the construction of rule-bases, define the membership functions, finding the optimal neural network architecture and weights in various ways using genetic algorithms.

1.4 Experimental Rig of Flexible Arm

The demand for the employment of robots in various applications has increased in line with the increasing demand for system automation. The dominant factor that contributes largely to performance limitations of the robot is the limited capability of its control system especially in applications requiring high-speed and/or large payloads. The need for lightweight elastic robot arms has increased, as they are capable of improving the speed of operation and handling larger payloads in comparison to rigid arms with the same actuator capabilities. However, their structural flexibility results in oscillatory behaviour in the system. The problem of oscillatory behaviour due to the arm flexibility has traditionally been solved by mechanically stiffening the arm. However, this leads to an increase in the weight of the arm. Thus, a conventional robot arm does not achieve the objective of the lighter weight requirement of the flexible arm.

There are other potential advantages arising from the use of flexible arms. These include

1. Lower energy consumption: lighter arms have lower inertia and therefore require less power to produce the same acceleration as rigid arms with the same payload capacity.
2. Smaller actuator required: the reduced power requirement means that smaller and, generally, cheaper actuators can be used.
3. Safer operation due to reduced inertia: in the event of a collision less damage would be caused.
4. Compliant structure: flexible arms introduce mechanical compliance into the robot structure, which is useful for delicate assembly operations.
5. Possible elimination of gearing: increasingly relevant with the development of motors with high power/weight ratios, leading to a possible reduction in costs and backlash as well as improvement in actuator linearity.
6. Less bulky design.

The properties discussed above on the flexible arm make it an interesting bench-mark problem for the verification and application of intelligent control. The intelligent control approaches developed in this monograph will be tested on this experimental rig to verify the performances of the different proposed techniques.

1.5 Overview of the Book

The aim of the book is to demonstrate some of the features and application of fuzzy logic, neural networks and genetic algorithms in the development of intelligent control strategies. The book comprises of ten chapters covering the different aspects of intelligent control schemes with verification to an experimental rig. A brief outline of the book contents is as follows:

Chapter 2: Presents a brief introduction to an exemplar non-linear system of flexible robotic arm. The flexible robotic arm will be used as the experimental rig for different control strategies developed throughout the book. The chapter describes the flexible robotic arm with necessary instrumentation and measurement. An experimental investigation aimed at estimating the model of the hub-angle, hub-velocity and end-point acceleration is also provided.
Chapter 3: Describes different control strategies for the nonlinear model of the flexible arm. A simulation environment is developed accordingly and its performance in characterizing the behaviour of the system investigated. The environment, thus developed is used for testing and verification of control strategies proposed in this book.

Chapter 4: Presents the mathematical background for fuzzy modelling and control in general. These include analysis of different membership functions, fuzzification, different inferencing mechanisms such as Mamdani-type, Sugeno-type and Tsukomoto-type fuzzy inferencing, rule-base construction, different defuzzification methods and their analysis.
Chapter 5: Presents an investigation into different types of fuzzy controllers namely PD-, PI-, PID-type and describes the development of a PD-PI-type fuzzy logic controller. A rule reduction method is proposed by introducing a switching PD-PI-type controller without compromising the performance.
Chapter 6: Presents the parameter optimization of the PD-PI-type fuzzy controller, mainly parameters of the membership functions. Triangular membership functions are parameterized and a derivative free optimization procedure, genetic algorithm, is applied to optimize the parameters of membership functions. The feature of the genetic algorithm used in this chapter is its small population size. The problems of genetic algorithm with a small population size are analyzed. The essence of dynamic mutation rate is investigated in overcoming the decrease of population diversity over generations in genetic algorithms.
Chapter 7: Presents tuning of the parameters of the PD-PI-type fuzzy controller using neural networks. Replacement of multi-layer neural network with a single neuron network with linear and non-linear activation function is described. The effect of linear and non-linear activation function on the performance is investigated.
Chapter 8: Investigates genetic algorithm-based learning of the parameters of the neural network used to tune the scaling factors of the fuzzy controller. Optimisation of the shape of sigmoidal function is included in the genetic algorithm based learning of the weights and bias of the neural network.
Chapter 9: Presents a general overview on the different methods of stability analysis and robustness issues in fuzzy control. Especially, stability analysis is very important for a fuzzy controller, where a mechanism is used to switch from a PD-type to PI-type controller.
Chapter 10: Presents concluding remarks and future extension of the research. There are a number of ideas, which could be exploited in this research, but time constraints did not permit to do so. These ideas can be implemented as an extension of this research work.

References

Acar L, Ozguner U (1990) Design of knowledge-rich hierarchical controllers for large functional systems. IEEE Trans Syst Man Cybern 20(4):791–803

Albus JS (1981) Brains, behavior, and robotics. Peterborough, McGraw-Hill/NH, BYTE

Albus JS (1991) Outline for a theory of intelligence. IEEE Trans Syst Man Cybern 21(3):473–509

Antsaklis PJ, Passino KM (1995) Introduction to intelligent control systems with high degrees of autonomy, in intelligent control. Kluwer Academic Publishers, edt

Bissell CC (2009) A history of automatic control, Chap. 4.In: Springer Handbook of Automation, Edt. S. Y. Nof, Springer, Berlin

Fu KS (1971) Learning control systems and intelligent control systems: an intersection of artificial intelligence and automatic control. IEEE Trans Autom Control 6:70–72

Grossberg S (1982) Studies of the mind and brain. Reidel Press, Drodrecht, Holland

Harris CJ, Moore CG, and Brown M (1993) Intelligent control: Aspects of fuzzy logic and neural nets, World Scientific (World Scientific series vol. 6), Singapore, NJ, London, Hong Kong

Harris CJ (1994) Introduction to intelligent control in 'Advances in intelligent control', Edt. CJ Harris, Taylor and Francis

Hebb DO (1949) The organisation of behaviour. John Wiley, New York

Hill DR (1979) The book of ingenious devices, Translated from Kitab al-Hiyal by Banu Musa bin Shakir (d. 873), D. Reidel Publishing Company, Dordrecht, Boston, London

Hopfield JJ (1982) Neural networks and physical systems with emergent computational abilities. Proc Natl Acad Sci 79:2554–2558

Holland JH (1975) Adaptation in natural and artificial systems. University Michigan Press, Ann Arbor

Mamdani EH, Assilian S (1974) Application of fuzzy algorithms for control of simple dynamic plant. Proc IEE 121:1585–1588

Mayer O (1970) The origins of feedback control. The MIT Press, Cambridge

Meystel A (1985) Intelligent control: issues and perspectives. In: Proceedings of IEEE workshop on intelligent control, pp 1–15

Ozguner U (1989) Decentralized and distributed control approaches and algorithms. In: Proceedings of the 28th IEEE conference on decision and control, Tampa, FL, pp 1289–1295

Rumelhart DE, Hilton GE, Williams RJ (1986) Learning representations by back-propagating errors. Nature 323:533–536

Saridis GN (1985) Foundations of the theory of intelligent controls. Proceedings of IEEE workshop on intelligent control, pp 23–28

Saridis GN (1987) Knowledge implementation: structure of intelligent control systems. Proceedings of IEEE international symposium on intelligent control, pp 9–17

Takagi T, Sugeno M (1985) Fuzzy identification of systems and its application to modeling and control. IEEE Trans Syst Man Cybern 15:116–132

Widrow B (1987) "ADALINE and MADALINE—1963", Plenary speech. In: Proceedings of 1st IEEE international conference on neural networks, San Diego, CA, pp 143–158

Zadeh LA (1965) Fuzzy sets. Inf Control 8:338–353

Zadeh LA (1968) Fuzzy algorithms. Inf Control 12:94–102

Zadeh LA (1973) Outline of a new approach to the analysis of complex systems and decision process. IEEE Trans Syst Man Cybern 3:28–44

Zilouchian A, Jamshidi Mo (2001) Intelligent control using soft computing methodologies, edt. A. Zilouchian and Mo Jamshidi CRC Press, London

Chapter 2
Dynamical Systems

2.1 Introduction

Anything that is changing with time is called dynamic. A system whose states are changing with respect to time is called a dynamic system. In general, a dynamic system is described by differential equations. The precise control of high speed motion demands a realistic dynamic model of the robot arm. The dynamic equations of motion of a general robotic arm is rather complex. There are different approaches used by researchers to derive the equations of motion for robotic systems. The first is the Lagrange-Euler formulation (Schilling 1990) based on the concept of generalised coordinates, energy, and generalised force. An alternative approach is the recursive Newton-Euler formulation, which is computationally more efficient (Asada and Stoline 1986; Fu et al. 1987).

Using the D' Alembert's principle of the fundamental classical laws of motion, dynamics of arms can be derived by summing all of the forces acting on the coupled rigid bodies that form the robot arm. But the Lagrangian derivation of the dynamics has the advantage of requiring only the kinetic and potential energies of the system. Therefore, Lagrangian analysis is employed to compute the equations of motion.

The Lagrangian, L, is defined based on the notion of generalised coordinates, energy and generalised forces, as the difference between the kinetic energy and potential energy of the system:

$$L(q,\dot{q}) = K(q,\dot{q}) - V(q) \tag{2.1}$$

where K is the kinetic energy and V is the potential energy of the forces of the system, q represents the joint variables, $\dot{q} = dq/dt$ represents the joint velocities. The general equations of motion of a robotic arm can be defined in terms of Lagrangian function as follows:

$$\frac{d}{dt}\frac{\partial}{\partial \dot{q}_i}L(q,\dot{q}) - \frac{\partial}{\partial \dot{q}_i}L(q,\dot{q}) = F_i \tag{2.2}$$

N. Siddique, *Intelligent Control*, Studies in Computational Intelligence 517,
DOI: 10.1007/978-3-319-02135-5_2, © Springer International Publishing Switzerland 2014

where F_i, $i = 1, 2, \cdots, n$, is the generalized forces acting on the ith joint. The Lagrangian equation of robot dynamics consists of n-second order nonlinear differential equations in the vector of joint-space of q. To derive any dynamic equation for any robotic system, we need to formulate expressions for kinetic energy K, potential energy V, and generalised forces F. The computation of the total kinetic energy of a robotic arm $K(q, \dot{q})$ in Lagrangian function is the most complicated term. The potential energy $V(q)$ is the gravitational forces. The generalized forces are the residual forces acting on the robot arm.

2.2 Dynamics of Robot Manipulator

The dynamic equation of an n-axis robot arm can be derived from Lagrangian function in (2.2) when the detailed expressions for the kinetic energy, potential energy and generalised forces are available. The general dynamic model of the robot arm can be described by the following equation with the joint variables q and actuator torques τ:

$$\sum_{j=1}^{n} D_{ij}(q)\ddot{q}_j + \sum_{k=1}^{n}\sum_{j=1}^{n} C^i_{kj}(q)\dot{q}_k\dot{q}_j + h_i(q) + b_i(\dot{q}) = \tau_i, \quad i = 1, \cdots, n \tag{2.3}$$

The first term $D_{ij}(q)\ddot{q}_j$ is an acceleration term that represents the inertial forces and torques generated by the motion of the links of the arm. The second term $C^i_{kj}(q)\dot{q}_k\dot{q}_j$ is a product velocity term associated with Coriolis and centrifugal forces. The third term $h_i(q)$ is the position term representing loading due to gravity. The fourth term $b_i(\dot{q})$ is a velocity term representing the friction acting opposite to the motion of the arm. The n separate scalar equations in (2.3) can be written as a single vector equation as follows:

$$D(q)\ddot{q} + c(q, \dot{q}) + h(q) + b(\dot{q}) = \tau_i \tag{2.4}$$

The term $c(q, \dot{q})$ is called the velocity coupling vector. There are two distinct types of inter-axis velocity coupling arising from here, which can be expanded into two terms as follows:

$$c(q, \dot{q}) = \sum_{k=1}^{n} C^i_{kk}(q)\dot{q}_k^2 + \sum_{k=1}^{n}\sum_{j \neq k} C^i_{kj}(q)\dot{q}_k\dot{q}_j \tag{2.5}$$

The first summation corresponds to squared velocity terms associated with centrifugal forces acting on joint i due to motion of joint k. The second summation corresponds to product velocity terms associated with Coriolis forces acting on joint i due to the combined motions of joints k and j such that $j \neq k$.

2.3 Dynamics of Flexible-Arm

To design a flexible arm efficiently, whether intended for a specific application or for a range of applications, several factors need to be considered. These include the strength and flexibility of the arm, fast speed and acceleration capability, good payload requirements, choice of suitable actuator and sensing equipment for the control mechanisms intended to be employed. The problem of oscillatory behaviour due to arm flexibility has traditionally been solved by mechanically stiffening the arm. However, this leads to an increase in the weight of the arm. Thus, a conventional industrial robot arm does not achieve the objective of the lighter weight requirement of the flexible arm.

The issue of flexible arm design and control thus primarily caters for the design of controllers to either compensate for the structural flexibility or to be robust in the presence of structural flexibility. It has been shown that using joint position and velocity sensors in a feedback control scheme for a rigid robot is adequate to ensure satisfactory performance (Khosla and Kanade 1988; Seraji and Moya 1987). However, these sensors may not be sufficient to provide the necessary information for the control of the elastic behaviour of a flexible arm. In addition to measuring joint position and velocity, it is desirable to obtain the state of the end-point as well. Although the deflection information of the arm can be theoretically determined if the dynamic model of the system is available, this will require high computing power and speed for on-line computation in addition to the uncertainties usually associated with formulating the dynamic model of the flexible arm. This argument for control purposes leads to the requirement of a suitable measuring system for the flexible arm's end-point.

A description and dynamic characterisation of the flexible arm utilised to verify different intelligent control strategies is presented in this chapter. Similar experimental arms have been constructed in the past (Cannon and Schmitz 1984; Hastings and Ravishankar 1988). The principal originality of the arrangement presented here is that the deflection of the flexible arm is measured and controlled using an accelerometer at the end-point. The following sections are intended to recall, in an abbreviated form, the design procedures of a laboratory facility for experimental study of a single-link flexible arm using end-point acceleration feedback (Azad 1994; Tokhi and Azad 1997). The main purpose of the design procedure is to relate a set of criteria which are useful in the design procedure, such as accuracy of end-point, allowable payload mass, maximum joint velocity, maximum joint acceleration and operating bandwidth of the arm. This will lead to preliminary results of the arm parameters. The actuating system is also studied and incorporated into the design procedure. Moreover, the design procedure will also indicate the significance of flexibility for a range of specifications.

2.3.1 Strength and Stiffness

Arm strength is the ability to withstand loads, which create stress in the system. These loadings arise from attempts to move and stop the arm and to maintain a position in gravity or other force fields. These loadings limit the capability of the arm to perform its specified functions by limiting the speed with which it can move. This speed limitation can yield a basis for comparison against stiffness limitations. The stiffness itself is the tendency of the arm to resist deflections, which may take the form of vibrations. For a distributed system such as a flexible arm, these may take place at an infinite number of frequencies. Strength is seemingly a more compelling requirement since inadequate provisions for strength can result in catastrophic failures if a component fractures.

Arm strength and stiffness are qualitatively affected in the same way by a number of parameters including payload mass, arm's material density and length and dimensions of the cross section. Varying one of these parameters to make the arm stronger will result in a stiffer arm. Other parameters directly affecting only strength include the maximum allowable stress for the arm material, inertial loading from accelerating the arm and its payload, and the gravity or other constant body forces present in the environment. Stiffness is characterised by the natural frequencies, and its magnitude is additionally affected by the value of the Young's modulus for the material used.

Two other factors influencing characterisation of an arm are rigidity and flexibility. The rigidity of the system can be checked by using a rule of thumb (Book 1984) which can be summarized as: if the servo frequency ω_S of a controller is less than a third of the first resonance frequency of the arm $[\omega_1]$, then the arm can be considered as rigid. If the servo frequency is between $\omega_1/3$ and $\omega_1/2$ then vibration will occur but will be well damped. If the servo frequency is greater than $\omega_1/2$, oscillations with insufficient damping will result. This will help to indicate the significance of flexibility and rigidity for a range of specifications.

For the single-link arm considered, there are only two parameters, namely thickness and width that can be altered to increase the strength of the arm. The minimum thickness, as discussed later, will limit the stability of the arm. The arm can be considered as a pinned-free flexible beam, with a lumped inertia at the hub, which can bend freely in the horizontal plane but is stiff in vertical bending and torsion. To avoid the difficulties arising in the case of a beam with time-varying length, the length l of the arm is assumed to be fixed. Moreover, shear deformation, rotary inertia and effect of axial force are also neglected. A schematic representation of the single-link flexible arm system under consideration is shown in Fig. 2.1, where E, I, ρ, M_p and I_h are the Young's modulus, the area moment of inertia, the mass density per-unit length, payload mass and hub inertia respectively.

For a given value of payload mass at the end point of the arm and a desired linear acceleration at the centre of gravity of the payload mass, the desired moment at the joint (hub) can be calculated as (Book 1984)

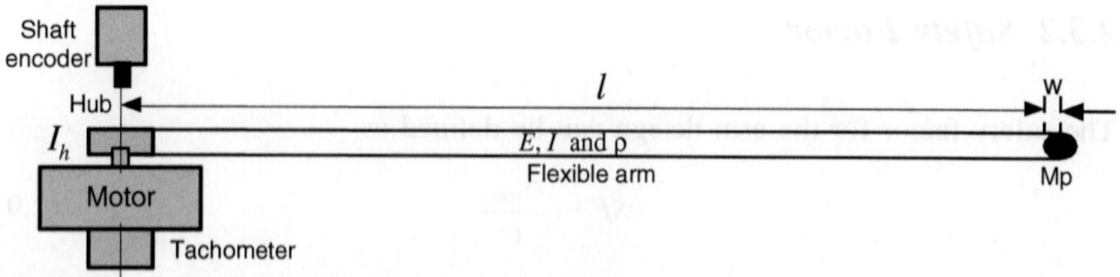

Fig. 2.1 Outline of the flexible arm system

$$\sigma = \alpha M_p w + \alpha M_p l + \frac{\rho}{2}\left[\frac{\alpha - g}{l + w} + g\right] l^2 - \frac{\rho(\alpha - g)}{6(l + w)} l^3 + \frac{I_h \alpha}{l + w} \tag{2.6}$$

where,

σ = required moment at the hub of the flexible arm,
α = linear acceleration at the centre of gravity of the payload mass,
w = offset between the centre of gravity of the payload mass and the end-point of the arm,
g = acceleration due to gravity.

For a given flexible arm, the maximum torque, which can be applied to the joint is given by

$$\tau_{max} = \frac{\sigma_{max} I}{C} \tag{2.7}$$

where,

τ_{max} = maximum applied torque,
σ_{max} = maximum tensile stress (without changing the shape). This depends upon the material used for the flexible arm construction;
I = second moment of inertia of the arm,
C = half of the thickness of the arm (thickness/2).

For the arm to be strong enough to handle the desired torque, the following criterion must be satisfied

$$\sigma_{max} \geq \sigma \tag{2.8}$$

where equality applies to the optimum structure of the arm. It follows from Eq. (2.7) that the minimum thickness of the flexible arm depends upon the amount of torque required to be applied at the joint, i.e. upon the linear acceleration at the centre of gravity of the payload mass.

2.3.2 Safety Factor

The safety factor for the arm design can be defined as

$$SF = \frac{\sigma_{max}}{\sigma} \tag{2.9}$$

For optimum design, the safety factor is to be unity. Figure 2.2 shows the relation between end-point acceleration α and thickness T of the arm with various values of safety factor for a given material (e.g. Aluminium alloy) with a payload mass of 10 g, length of 0.96 m, and hub inertia of 5.86×10^{-4} kgm^2. For a given end-point acceleration and safety factor, the thickness of the flexible arm can be found from this diagram for a specific configuration.

The design and development of a flexible arm includes three areas (Tokhi and Azad 1997): (a) design and construction of the mechanical structure; (b) choice of suitable transducers for the specific application; (c) development of required amplifiers and processing circuits and their calibration.

For the flexible arm utilized in this research as an experimental rig, a printed circuit armature type motor is used as the drive actuator due to its low inertia, low inductance and physical structure, which allows to be connected to the flexible arm. The system incorporates an LA5600 amplifier/controller for controlling the required dc-current driving the motor. This is due to its several features such as motor clamping, directional clamping and output current monitoring provision. The details of LA5600 are presented in Sect. 2.3.5.

For measurement of angular movement, the developed processing circuit is able to produce both digital and analogue outputs at the same time. This enables the use of analogue output from a 16-bit D/A as a feedback signal to the amplifier/controller and digital output to the computer. This circuit can be modified to control a microprocessor where the 8-bit output of the THCT2000 could be directly connected with the computer's data bus. The system uses a new type of velocity measurement transducer instead of the conventional tachometer. A special

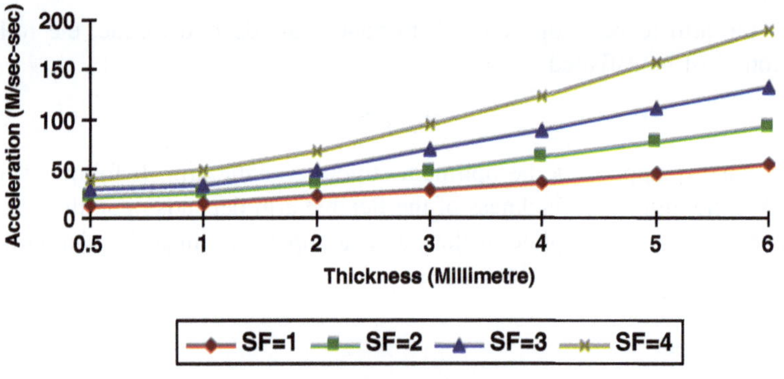

Fig. 2.2 Relation between end-point acceleration and required thickness of the flexible arm

feature of this transducer is that the output is totally free from any noise induced from commutator friction, which is very important for low-level feedback signals. Moreover, the inertia for this system is very small. In selecting the type of accelerometer and strain gauge, size, weight and frequency range constraints are considered strictly. The accelerometer utilised includes a built-in FET source follower, which allows for a lower output impedance level.

Due to the irregular shape of the flexible arm hub, the hub inertia is measured experimentally. This is important for modelling and simulation of the system. Motor friction is also measured experimentally to verify the supplier's data-sheet, which has been found to be 0.011 Nm. However, this parameter is not considered in the model because the effect of friction is not significant as compared to the applied torque.

Noise and interference have been a serious problem than others during testing and experiments. To overcome this, a linear amplifier is used and all the signal amplifiers are powered from a battery instead of from the mains power supply.

Pentium I microcomputer in conjunction with an ADC-44d I/O board is used with the flexible arm system. The experimental setup requires one analogue output to the motor drive amplifier, four analogue inputs from the hub-angle and velocity transducer, accelerometer and motor current sensor. The interface board is used with a conversion time of 3 μs for A/D conversion and settling time of 20 μs for D/A conversion, which are satisfactory for the system under consideration. The details of the ADC-44d I/O board are presented in Sect. 2.3.7.

A more detailed description of the flexible arm system is given in the next section.

2.3.3 Experimental Flexible Arm

The experimental rig constituting the flexible arm system consists of two main parts: a flexible arm and measuring devices. The flexible arm contains a flexible link driven by a printed armature motor at the hub. The measuring devices are shaft encoder, tachometer, accelerometer and strain gauges along the length of the arm. The shaft encoder, tachometer and accelerometer are utilised in this work. The experimental rig is shown in Fig. 2.3a and the schematic diagram of the rig is shown in Fig. 2.3b.

The shaft encoder is used for measuring the hub angle of the arm. A tachometer is used for measurement of the hub velocity. An accelerometer is located at the end-point of the flexible arm measuring the end-point acceleration. The flexible arm is constructed using a piece of thin aluminium alloy. The parameters of the flexible arm are given in Table 2.1.

(a)

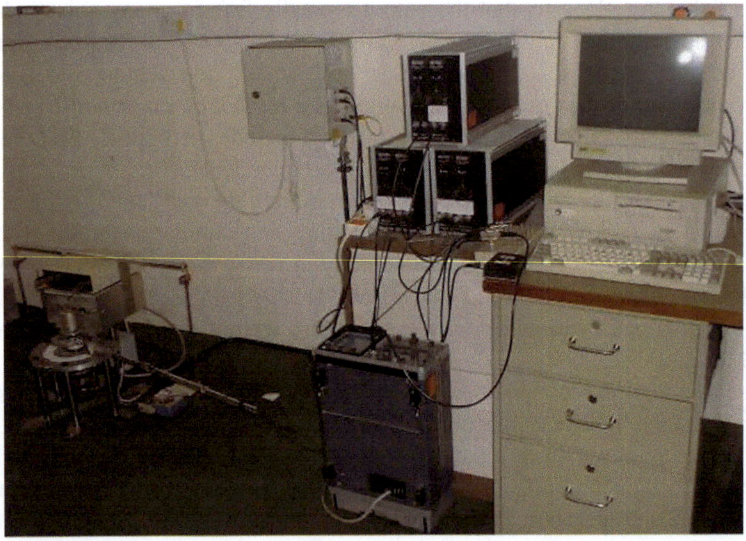

(b)

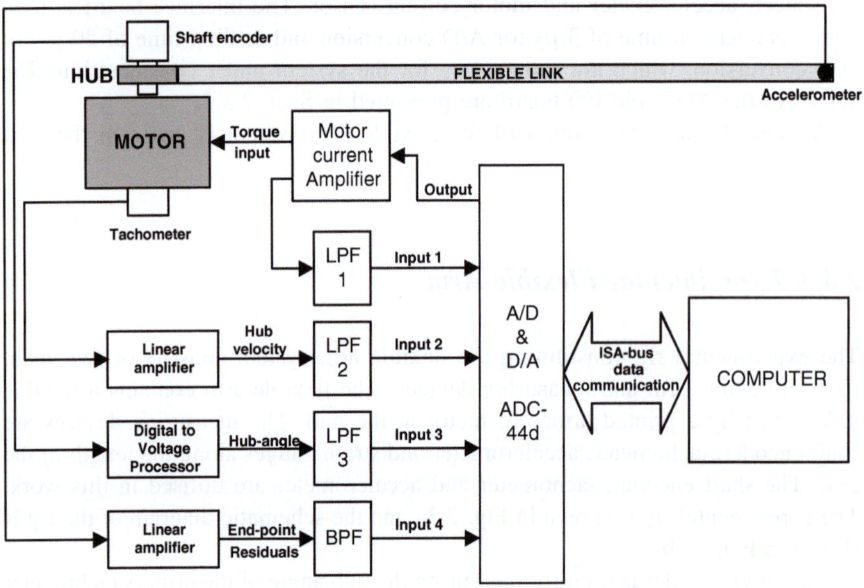

Fig. 2.3 Experimental rig Flexible arm (**a**); Schematic diagram of the experimental rig (**b**)

2.3.4 Printed Armature Motor

The experimental rig is equipped with a U9M4AT type printed circuit motor driving the flexible arm. The specifications of the motor are given in Table 2.2.

Table 2.1 Parameters and characteristics of the flexible arm system

Parameter	Value
Length	960.0 mm
Width	19.008 mm
Thickness	3.2004 mm
Mass density per unit volume	2,710 kgm^{-3}
Second moment of inertia, I	5.1924×10^{-11} m^4
Young modulus, E	71×10^9 Nm^{-2}
Moment of inertia, I_b	0.04862 kgm^2
Hub inertia, I_h	5.86×10^{-4} kgm^2

Table 2.2 Specifications of the U9M4AT motor

Parameters	Unit	Value
Peak torque	Nm	3.075
Peak current	A	72.0
Peak acceleration	$Krad.s^{-2}$	52.3
Continuous torque	Nm	0.325
Continuous current	A	8.68
Maximum terminal voltage	V	22.9
No load speed	RPM	4593
Power output	W	101.7
Torque constant	Nm/A	0.043
Back e.m.f. constant	V/Krpm	4.5
Armature resistance	Ω	0.66
Moments of inertia	$Nm.s^2$	5.88×10^{-5}
Armature inductance	μH	<100
Mechanical time constant	μs	20.52
Electrical time constant	μs	<0.15
Voltage constant	V/Krpm	2.25

This type of motor gives significant performance advantages for motion control applications, which can be listed as follows (Azad 1994):

- It gives high acceleration since it is able to produce high torque combined with low armature inertia. This means shorter cycle times, more displacement per second and higher throughput.
- It has a very low inductance, which leads to a negligible electrical time constant and a short mechanical time constant (less than one millisecond). This implies almost instantaneous application of full torque. This is a key to fast motion and accurate tracking.
- It does not have armature associated torque loss due to its construction and as a result delivers more torque over its entire speed range. Moreover, the torque is almost constant throughout its speed range. These properties provide a non-varying transfer function over the entire operating range of the motor.
- Due to the absence of any iron in the rotor and a large number of commutator bars and slots, extremely smooth torque with no "cogging" is achieved.

- With the non-magnetic printed circuit armature and axial magnetic field, the problem of demagnetisation of the permanent magnet is virtually eliminated. Most printed armature motors are rated for a peak current of 10 times the continuous rating. This property enables them to produce rapid acceleration and deceleration when higher than the rated torque is usually required.
- As the inductance is nearly zero, there is no stored energy in the armature to be dissipated during commutation. This eliminates arcing, which is the major cause of brush wear. This increases the reliability in operation and life expectancy.

2.3.5 Motor Drive Amplifier

Linear amplifiers may be classified as either unidirectional with dynamic breaking capability or bi-directional. For a unidirectional motor, deceleration is only dependent on system friction and viscous damping, which takes a longer time to stop the motor. This can be modified by introducing a circuit which shorts the motor terminals when the motor current becomes zero, allowing a negative current to circulate in the armature, thus stopping the motor. However, the drive amplifier used is a bi-directional one because the motor needs to be driven in both directions to control the arm vibration. The motor driver used is the LA5600 manufactured by Electro-Craft Corporation. This motor drive amplifier (current amplifier) delivers a current proportional to the input voltage. It serves as a velocity/position controller as well as a motor driver. The reasons for using this amplifier/controller can be listed as follows:

1. The main objective is to control the flexible arm by applying a controlled amount of torque at the arm hub. This amplifier/controller can be used in torque-controlled mode. That is, for a given amount of input voltage it can produce a proportional current output to the motor.
2. For off-line identification of the system, it is needed to operate the system in a joint based position and velocity feedback system. With this amplifier/controller, such a system can be implemented by feeding back the tachometer and shaft encoder output to the amplifier.
3. This amplifier has a four clamp facility, which can be used to limit or restrict the function of the amplifier and consequently of the entire system to accommodate various application requirements. The inhibit (INH) clamp is used to disable the amplifier by turning all transistors in the output stage OFF when the clamp is activated. The motor hold (MHO) clamp is used to stop the motor by effectively driving the amplifier input signal to zero. It decelerates the motor and load to zero speed as fast as possible and then holds the motor when stopped. The system resists external torque applied to the motor shaft. These two clamps could be used to stop the motor at any extreme operational situation by disconnecting them from the ground (COM). The forward direction (FAC) clamp and reverse direction (RAC) clamp affect the specific direction of motor movement. When one of these clamps modes is activated, it prevents movements of the motor in

that direction. These forward and reverse direction clamps can be used by wiring through normally closed travel limit micro switches. This will stop the flexible arm when moved beyond its range of movement.

4. There are three other output signals from this amplifier/controller, which could be used as input to the controller or for monitoring purposes. The first one is the motor velocity output (MVO) signal, which is an analogue voltage signal and is proportional to the actual instantaneous motor speed. The second one is the motor current output (MCO), which is a voltage proportional to the actual motor current. Knowing the amount of current output from the amplifier, the applied torque can be calculated by multiplying the current (I_a) with motor torque constant (K_t). The third output is the system status output (SSO), which is a logic level signal.

2.3.6 Accelerometer

There are two types of accelerometer commonly in use: (a) strain gauge type accelerometers; and (b) piezoelectric accelerometers. Due to weight and size constraints, the flexible arm system used in this work incorporates a piezoelectric type transducer. Table 2.3 shows the specifications of the accelerometer used.

In this type of accelerometer, the crystal-sensing element is isolated from the case and compressed or sheared between the accelerometer base and a seismic mass. A dynamic force applied to the accelerometer in either direction along its sensitive axis causes the crystal to generate a charge proportional to the acceleration. The generated voltage can be represented as

$$v = \frac{DM}{C} A \tag{2.10}$$

where D, M, C, A are the piezoelectric constants of the material, the mass of the seismic mass, the capacitance and the acceleration respectively.

Table 2.3 Specifications of the 303A03 accelerometer

Parameters	Unit	Value
Range	g	±500
Resolution	g	0.01
Sensitivity	mV/g	10
Frequency range	Hz	1–10,000
Output impedance	Ω	100
Linearity	%	1
Transverse sensitivity (max)	%	5
Excitation voltage	V	18–24
Excitation current	mA	2–20
Weight	g	2

2.3.7 Computer Interfacing

The computer used for this experimental rig was an IBM-PC compatible Pentium I 100 MHz CPU. Data acquisition and control was accomplished through the utilization of an ADC-44d I/O board. This board provides a direct interface between the microcomputer and the actuators and transducers..

The ADC-44d board contains an NEC μPD71055 device, which is equivalent to an Intel i8255 PIO. This device produces 24 programmable digital I/O channels. It is suitable for sensing the presence of TTL driving connections. It contains a 12-bit A/D converter with a conversion time of 3μs. Various configurations of its CMOS multiplexer enable it to receive different numbers of input channels with different voltage ranges. A/D conversion can be initiated by one of three possible means: using a software convert command, applying an external TTL logic level signal or by programming the NEC μPD71055 chip. The ADC44d is able to generate an interrupt when one of three independent conditions has occurred: when an A/D conversion is completed, when an overrun condition has occurred or when a given number of counts have finished. The board contains two independent voltage output channels, each with its own 12-bit D/A converter, which can produce an output from ±10 V to ±50 mV. For time related digital I/O applications, the NEC μPD71055/NEC μPD71054 counter/timer chip provides the ADC-44d with three independent 16-bit channels that can be used for such counter/timer functions as event counting, frequency measurement, single pulse output and time proportional output. The board is mapped into the microcomputer's I/O channel structure as a block of 16 consecutive bytes, addressable on any unoccupied 16-byte boundary from address 200H to 3FFH.

2.3.8 Operating Characteristics

Many systems and signal sources have non-linear characteristics associated with actuation and sensing. Possible sources and types of input (actuation) non-linearity and output (sensor) non-linearity are illustrated in Fig. 2.4.

A typical input non-linearity is the dead-zone $\pm U_d$. This represents a signal level below which no actuation signal $u_{act}(t)$ is sent to the system. The signal $u(t)$ must be designed to ensure that it has a minimum amplitude in this region.

The characteristics of frictional forces between two contacting surfaces often depend on several factors including the composition of the surfaces, the pressure between two surfaces and their relative velocity. An exact mathematical description of the frictional forces is thus difficult to obtain (Kuo and Tal 1978). The amount of friction associated with the motor has previously been determined experimentally (Azad 1994). The measured friction (viscous coefficient B) was found to be 0.029 mNm/rad/second.

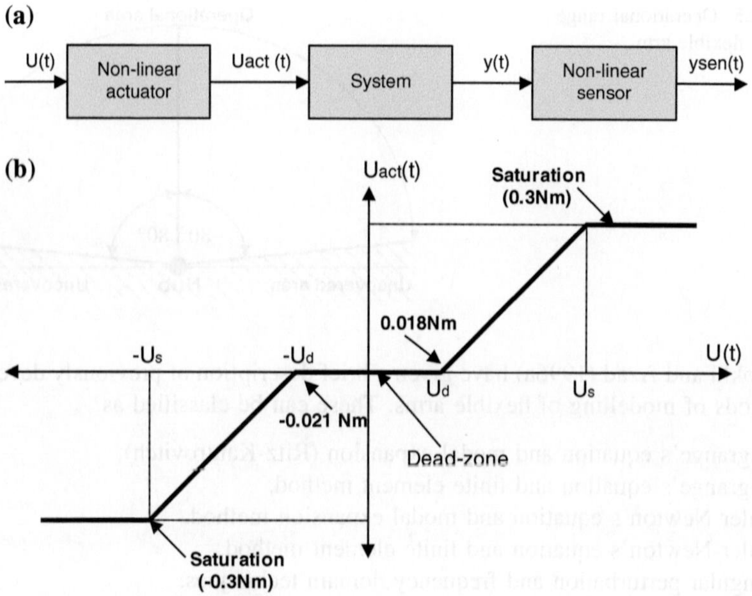

Fig. 2.4 Non-linear characteristics associated with actuation and sensing. Actuation and sensing (**a**); Input non-linearity with saturation limits (**b**)

The torque required to eliminate the *dead-zone* has been found to be at least ±10.0 mNm. The presence of the flexible beam, including the clamp, would obviously affect the *dead-zone*. This implies that a *dead-zone* compensator would be required. Thus, if the torque has low amplitude components, which are within the *dead-zone*, a minimum amount of torque should be maintained outside the *dead-zone*. It is noted in Fig. 2.4b that the characteristic of the *dead-zone* of the flexible arm lies between +18 and −21 mNm. By considering the error between both, the angle of demand and position and the hub-velocity, the compensation torque could be started.

The flexible arm hub-movements is defined to be in the range of ±80° of angle only, as illustrated in Fig. 2.5. The uncovered angle here is the undesired angles, which are beyond the defined range.

2.4 Previous Research and Developments

Research in the area of flexible arm systems has been carried out for over 20 years. These range from a single-link arm rotating about a fixed axis to three-dimensional multi-link arms. However, most of the experimental work is limited to single-link arm systems. This is due to the complexity of multi-link arm systems.

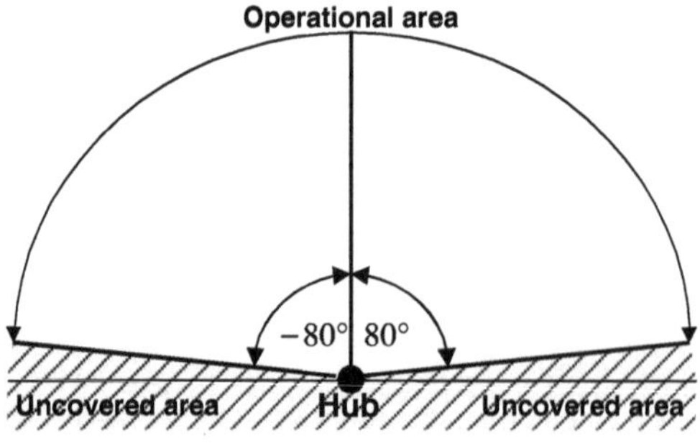

Fig. 2.5 Operational range of the flexible arm

Tokhi and Azad (1996a) have given a brief description of previously developed methods of modelling of flexible arms. These can be classified as

- Lagrange's equation and modal expansion (Ritz-Kantrovitch),
- Lagrange's equation and finite element method,
- Euler-Newton's equation and modal expansion method,
- Euler-Newton's equation and finite element method,
- Singular perturbation and frequency domain techniques.

In Lagrange's equation and modal expansion method, the model is in the form of a summation of modes. Each mode is a product of two functions: one is a function of distance along the length of the arm, the other is a function of generalised co-ordinates dependent upon time. The model contains an infinite number of modes. In practice, a finite number of modes is used. The Lagrange's equation and finite element method is similar to the Lagrange's equation and modal expansion method. In Lagrange's equation and finite element method, the generalised co-ordinates are the displacement and/or slope at specific points along the arm.

The Euler-Newton's method is a more direct method of calculating system dynamics. The derivatives of linear and angular momentum are derived explicitly. Newton's second law is applied to equate these terms to the applied forces and torques. The linear and angular momentum of arm is the unknown. These are expressed in terms of assumed modes of finite elements. This leads to a dynamic model containing time dependent elements, which relate to the applied forces and torques. The basic approach in the Euler-Newton's equation and assumed mode method is to divide the system into a number of finite elements and balance each element dynamically. This method can be tedious if the total number of elements is large. The advantage of this method is that it is easier to include non-linear effects without complicating the basic linear model.

In the singular perturbation technique, the characteristic modes of the system are separated into two groups. These include a set of low-frequency, or slow modes, and a set of high frequency, or fast modes. In a flexible arm system, the rigid body modes are the low frequency modes and the vibration modes are the high frequency modes. The dynamics of the system can then be divided into two sub-systems. The low frequency sub-system has the same order as that of the rigid body sub-system. The low frequency variables are considered as constant parameters for the high frequency sub-system.

An alternative method for modeling the arm system is the frequency domain approach. This method can be used to develop a transfer matrix model using the Euler-Bernoulli's beam equation for a uniform beam. The disadvantage of this method is that it is not possible to include the interaction between the gross motion and vibration of the arm in the model. Hence, the model can only be considered as an approximate method (Azad 1994; Tokhi and Azad 1996a).

An analytical model based on the Lagrange's equation and modal expansion method has previously been proposed (Tokhi and Azad 1996a). This model contains an infinite number of natural modes. It is used to develop a state space and equivalent frequency domain model of the system. These models are used as background knowledge for the identification and modeling process in this study.

The finite difference (FD) method has been used to obtain an efficient numerical method of solving the governing dynamic equation, partial differential equation (PDE), of the flexible arm system through discretisation, both, in time and space (distance) coordinates of the system. The algorithm proposed in this study allows the inclusion of distributed actuator and sensor terms in the PDE and modification of boundary conditions. The development of such an algorithm for a system without the inclusion of structural damping has previously been reported (Tokhi and Azad 1995). However, to provide a more realistic characterization of a flexible arm, the development of a simulation environment incorporating a mode frequency dependent structural damping can be pursued.

Both open-loop and closed-loop control strategies have been considered for control of flexible arms (Tokhi and Azad 1996b). Open-loop control methods have included bang-bang torque input, low-pass filtered and *Gaussian* shaped torque inputs. Closed-loop control, on the other hand, has included joint based collocated control and end point feedback (non-collocated control).

An important aspect of the flexible arm control that has received little attention is the interaction between the rigid and flexible dynamics of the links. As the arm configuration changes, the spatial boundary conditions of the links change, thereby modifying their characteristic frequencies and modes. Carrying a load leads to a change in the natural modes. This has important implications for control design, where the performance of a controller designed on the basis of a fixed linear model of the dynamics may be seriously degraded. There are several possible options to solve this problem: a controller based on a more general model of the system, perhaps even a non-linear model, a *robust control scheme* which can maintain satisfactory level of performance despite changes in the arm dynamics, an *adaptive control scheme* which modifies the controller in line with changes in the system. Model based control can work best for a particular model of the system. The robust fixed-parameter control scheme is limited to a very narrow range of system changes. There is always a trade-off between the range of system changes and the quality of the final control. Adaptive control schemes, combining plant model and controller adaptation, require necessary computing capability at the speed necessitated by the dynamics of the flexible arm.

It is still not clear which of the various control schemes available represents the best all-round solution for flexible arm control or even whether there is a single best solution. Unfortunately, trends towards intelligent control of flexible arm to date are not satisfactory and researches reported are very few. This area of research demands further attention from the research community.

2.5 Dynamic Equations of Flexible Robotic Arm

A schematic representation of the flexible arm system considered in this work is shown in Fig. 2.6, where X_0OY_0 and XOY represent the stationary and moving coordinates respectively, $\tau(t)$ represents the applied torque at the hub. E, I, ρ, V, I_h and m_P represent the Young modulus, area moment of inertia, mass density per unit volume, cross sectional area, hub inertia and payload of the arm respectively. In this work, the motion of the arm is confined to the X_0OY_0 plane.

The flexible arm system can be modelled as a pinned-free flexible beam, incorporating an inertia at the hub and payload mass at the end-point. The model is developed through the utilisation of Lagrange equation and modal expansion method (Hasting and Book 1987). For an angular displacement θ and an elastic deflection u, the total displacement $y(x,t)$ of a point along the arm at a distance x from the hub can be described as a function of both the rigid body motion $\theta(t)$ and elastic deflection $u(x,t)$ measured from the line OX. The dynamic equations of motion of the arm can be obtained using the Hamilton's extended principle (Meirovitch 1967) with the associated kinetic, potential and dissipated energies of the system. Ignoring the effects of the rotary inertia and shear deformation, a fourth order PDE representing the arm motion can be obtained as (Azad 1994)

$$EI\,\frac{\partial^4 u(x,t)}{\partial x^4} + \rho\,\frac{\partial^2 u(x,t)}{\partial t^2} = -\,\rho x\ddot{\theta} \tag{2.11}$$

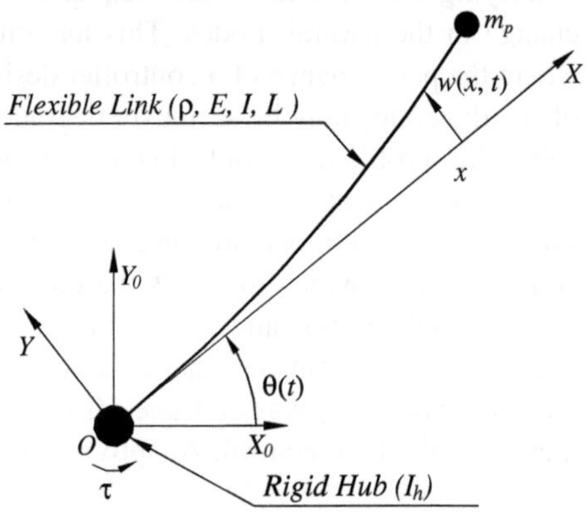

Fig. 2.6 Schematic representation of the single-link flexible arm

where $u(x,t)$ is the deflection of the link, ρ is density of the material, EI is the flexural stiffness. This equation is very difficult to solve because of the third term on the left-hand side.

To obtain the corresponding boundary conditions, the following must hold

- The displacement at the hub $\{u(0,t)\}$ must be zero,
- The total forces at the hub must be the same with the applied torque,
- The shear force at the end-point must be equal to $M_P \frac{\partial^2 u(x,t)}{\partial t^2}$ (Tse et al. 1980).
- The stress at the end-point must be zero, that is, no force should be present at the free end;

$$\begin{aligned} u(0,t) &= 0 \\ I_h \frac{\partial^3 u(0,t)}{\partial t^2 \partial x} - EI \frac{\partial^2 u(0,t)}{\partial x^2} &= \tau(t) \\ M_p \frac{\partial^2 u(l,t)}{\partial t^2} - EI \frac{\partial^3 u(l,t)}{\partial x^3} &= 0 \\ EI \frac{\partial^2 u(l,t)}{\partial x^2} &= 0 \end{aligned} \tag{2.12}$$

where l is the length of the arm.

Equation (2.11) with the corresponding boundary conditions in Eq. (2.12) represents the dynamic equation of motion of the flexible arm system assuming no damping in the system. In practice, however, such an effect is always present in the system. There are several possible forms of damping within the system. To incorporate damping into the governing dynamic equation of the system, a mode frequency dependent damping term proportional to $\frac{\partial^3 u(x,t)}{\partial x^2 \partial t}$ can be introduced (Davis and Hirschorn 1988). Equation (2.11) can thus be modified to yield

$$EI \frac{\partial^4 u(x,t)}{\partial x^4} + \rho \frac{\partial^2 u(x,t)}{\partial t^2} - D_S \frac{\partial^3 u(x,t)}{\partial x^2 \partial t} = -\rho x \ddot{\theta} \tag{2.13}$$

where D_S is the resistance to strain velocity, that is, rate of change of strain and $D_S \frac{\partial^3 u(x,t)}{\partial x^2 \partial t}$ represents the resulting damping moment dissipated in the arm structure during its dynamic operation. The corresponding boundary conditions are:

$$\begin{aligned} u(0,t) &= 0 \\ I_h \frac{\partial^3 u(0,t)}{\partial t^2 \partial x} - EI \frac{\partial^2 u(0,t)}{\partial x^2} &= \tau(t) \\ M_p \frac{\partial^2 u(l,t)}{\partial t^2} - EI \frac{\partial^3 u(l,t)}{\partial x^3} &= 0 \\ EI \frac{\partial^2 u(l,t)}{\partial x^2} &= 0 \end{aligned} \tag{2.14}$$

Substituting for $u(x,t)$ using total displacement $y(x,t) = x\theta(t) + u(x,t)$, Eq. (2.13), the boundary conditions, manipulating and simplifying yields the governing equation of motion of the arm in terms of $y(x,t)$ as

$$EI\frac{\partial^4 y(x,t)}{\partial x^4} + \rho\frac{\partial^2 y(x,t)}{\partial t^2} - D_S\frac{\partial^3 y(x,t)}{\partial x^2 \partial t} = 0 \tag{2.15}$$

Equation (2.15) gives the fourth-order PDE which represents the dynamic equation describing the motion of the flexible arm. To solve this equation and develop a suitable simulation environment characterising the behaviour of the system, the Finite difference (FD) method can be used. Thus, a set of equivalent difference equations defined by the central finite difference quotients of the FD method are obtained by discretising the PDE in Eq. (2.15) with its associated boundary and initial conditions. The process involves dividing the arm into N sections each of length Δx and considering the deflection of each section at sample times Δt. In this manner, a solution of the PDE is obtained by generating the central difference formulae for the partial derivative terms of the response $y(x,t)$ of the arm at points $x = i.\Delta x$, $t = j.\Delta t$ (Azad 1994; Burden and Faires 1989; Lapidus 1982):

$$\begin{aligned}
\frac{\partial^2 y(x,t)}{\partial t^2} &= \frac{y_{i,j+1} - 2y_{i,j} + y_{i,j-1}}{\Delta t^2} \\
\frac{\partial^2 y(x,t)}{\partial x^2} &= \frac{y_{i+1,j} - 2y_{i,j} + y_{i-1,j}}{\Delta x^2} \\
\frac{\partial^3 y(x,t)}{\partial x^3} &= \frac{y_{i+2,j} - 2y_{i+1,j} + 2y_{i-1,j} - y_{i-2,j}}{2\Delta x^3} \\
\frac{\partial^4 y(x,t)}{\partial x^4} &= \frac{y_{i+2,j} - 4y_{i+1,j} + 6y_{i,j} - 4y_{i-1,j} + y_{i-2,j}}{\Delta x^4} \\
\frac{\partial^3 y(x,t)}{\partial t^2 \partial x} &= \frac{y_{i,j+1} - 2y_{i,j} + y_{i,j-1} - y_{i-1,j+1} + 2y_{i-1,j} - y_{i-1,j-1}}{\Delta x \Delta t^2} \\
\frac{\partial^3 y(x,t)}{\partial x^2 \partial t} &= \frac{y_{i+1,j} - 2y_{i,j} + y_{i-1,j} - y_{i+1,j-1} + 2y_{i,j-1} - y_{i-1,j-1}}{\Delta t \Delta x^2}
\end{aligned} \tag{2.16}$$

where, $y_{i,j}$ represents the response $y(x,t)$ at $x = i\Delta x$ and $t = j\Delta t$ or $y(x_i, t_j)$. A time-space discretisation is adopted in the evaluation of the response of the arm.

2.5.1 Development of the Simulation Algorithm

A solution of the PDE in Eq. (2.15) can be obtained by substituting for $\frac{\partial^2 y}{\partial t^2}$, $\frac{\partial^4 y}{\partial x^4}$ and $\frac{\partial^3 y}{\partial x^2 \partial t}$ from Eq. (2.16) and simplifying to yield

$$\begin{aligned}
&\frac{EI}{\Delta x^4}\left[y_{i+2,j} - 4y_{i+1,j} + 6y_{i,j} - 4y_{i-1,j} + y_{i-2,j}\right] + \frac{\rho}{\Delta t^2}\left[y_{i,j+1} - 2y_{i,j} + y_{i,j-1}\right] \\
&\quad - \frac{D_S}{\Delta x^2 \Delta t}\left[y_{i+1,j} - 2y_{i,j} + y_{i-1,j} - y_{i+1,j-1} + 2y_{i,j-1} - y_{i-1,j-1}\right] = 0
\end{aligned}$$

or

$$\begin{aligned} y_{i,j+1} = &-c\left[y_{i+2,j} + y_{i-2,j}\right] + b\left[y_{i+1,j} + y_{i-1,j}\right] + a\, y_{i,j} - y_{i,j-1} \\ &+ d\left[y_{i+1,j} - 2y_{i,j} + y_{i-1,j} - y_{i+1,j-1} + 2y_{i,j-1} - y_{i-1,j-1}\right] \end{aligned} \tag{2.17}$$

where $a = 2 - \frac{6EI\Delta t^2}{\rho\Delta x^4}$; $b = \frac{4EI\Delta t^2}{\rho\Delta x^4}$; $c = \frac{EI\Delta t^2}{\rho\Delta x^4}$; $d = \frac{D_S\Delta t}{\rho\Delta x^2}$.

Equation (2.17) gives the displacement of section i of the arm at time step $j+1$. It follows from this equation that, to obtain the displacements $y_{n-1,j+1}$ and $y_{n,j+1}$, the displacements of the fictitious points $y_{n+2,j}, y_{n+1,j}$ and $y_{n+1,j-1}$ are required. These can be obtained using the boundary conditions related to the dynamic equation of the flexible arm. The discrete form of the corresponding boundary conditions are:

$$y_{0,j} = 0 \tag{2.18}$$

$$y_{-1,j} = y_{1,j} + \frac{\Delta x I_h}{EI\Delta t^2}\left[y_{1,j+1} - 2y_{1,j} + y_{1,j-1}\right] + \frac{\Delta x^2}{EI}\tau(j) \tag{2.19}$$

$$y_{n+2,j} = 2y_{n+1,j} - 2y_{n-1,j} + y_{n-2,j} + \frac{2\Delta x^3 M_p}{\Delta t^2 EI}\left[y_{n,j+1} - 2y_{n,j} + y_{n,j-1}\right] \tag{2.20}$$

$$y_{n+1,j} = 2y_{n,j} - y_{n-1,j} \tag{2.21}$$

2.5.2 Hub Displacement

Note that the torque is applied at the hub of the flexible arm. Thus, $\tau(i,j) = 0$ for $i \geq 1$. Using Eqs. (2.17) and (2.18), the displacement $y_{1,j+1}$ can be obtained as

$$y_{1,j+1} = -c\left[y_{3,j} + y_{-1,j}\right] + by_{2,j} + ay_{1,j} - y_{1,j-1} + d\left[y_{2,j} - 2y_{1,j} - y_{2,j-1} + 2y_{1,j-1}\right] \tag{2.22}$$

Substituting for $y_{-1,j}$ from Eq. (2.19) into Eq. (2.22) and simplifying yields

$$y_{1,j+1} = K_1\, y_{1,j} + K_2\, y_{2,j} + K_3\, y_{3,j} + K_4\, y_{1,j-1} + K_5\, y_{2,j-1} + K_6\tau(j) \tag{2.23}$$

where

$$K_1 = \frac{c\Delta t^2 EI + 2c\Delta x I_h + (a-2d)\Delta t^2 EI}{\Delta t^2 EI + c\Delta x I_h} \qquad K_4 = -\frac{c\Delta x I_h + (1-2d)\Delta t^2 EI}{\Delta t^2 EI + c\Delta x I_h}$$

$$K_2 = \frac{(b+d)\Delta t^2 EI}{\Delta t^2 EI + c\Delta x I_h} \qquad K_5 = -\frac{d\Delta t^2 EI}{\Delta t^2 EI + c\Delta x I_h}$$

$$K_3 = -\frac{c\Delta t^2 EI}{\Delta t^2 EI + c\Delta x I_h} \qquad K_6 = \frac{c\Delta x^2\Delta t^2}{\Delta t^2 EI + c\Delta x I_h}$$

2.5.3 End-Point Displacement

Using Eq. (2.17) for $i = n - 1$, yields the displacement $y_{n-1,j-1}$ as

$$\begin{aligned} y_{n-1,j+1} = & -c\left[y_{n-3,j} + y_{n+1,j}\right] + b\left[y_{n,j} + y_{n-2,j}\right] + a\,y_{n-1,j} - y_{n-1,j-1} \\ & + d\left[y_{n,j} - 2y_{n-1,j} + y_{n-2,j} - y_{n,j-1} + 2y_{n-1,j-1} - y_{n-2,j-1}\right] \end{aligned} \tag{2.24}$$

Similarly, using Eq. (2.17) for $i = n$, yields the displacement $y_{n,j+1}$ as

$$\begin{aligned} y_{n,j+1} = & -c\left[y_{n-2,j} + y_{n+2,j}\right] + b\left[y_{n+1,j} + y_{n-1,j}\right] + a\,y_{n,j} - y_{n,j-1} \\ & + d\left[y_{n+1,j} - 2y_{n,j} + y_{n-1,j} - y_{n+1,j-1} + 2y_{n,j-1} - y_{n-1,j-1}\right] \end{aligned} \tag{2.25}$$

The fictitious displacements $y_{n+1,j}$ and $y_{n+2,j}$, appearing in Eqs. (2.24) and (2.25), can be obtained using the boundary conditions in Eqs. (2.20) and (2.21). $y_{n+1,j-1}$ can easily be obtained by shifting $y_{n+1,j}$ from time step j to time step $j - 1$. Substituting for $y_{n+1,j}$ from Eq. (2.21) into Eq. (2.24) yields the displacement $y_{n-1,j+1}$ as

$$\begin{aligned} y_{n-1,j+1} = & K_7 y_{n-3,j} + K_8 y_{n-2,j} + K_9 y_{n-1,j} + K_{10} y_{n,j} + K_{11} y_{n-2,j-1} + K_{12} y_{n-1,j-1} \\ & + K_{13} y_{n,j-1} \end{aligned} \tag{2.26}$$

where

$$\begin{aligned} K_7 &= -c & K_{11} &= -d \\ K_8 &= (b + d) & K_{12} &= -(1 - 2d) \\ K_9 &= (a + c - 2d) & K_{13} &= -d \\ K_{10} &= -(2c - b - d) & & \end{aligned}$$

Similarly, substituting for $y_{n+2,j}$ and $y_{n+1,j}$ from Eqs. (2.20) and (2.21) into Eq. (2.25), and simplifying yields the displacement $y_{n,j+1}$ as

$$y_{n,j+1} = K_{14} y_{n-2,j} + K_{15} y_{n-1,j} + K_{16} y_{n,j} + K_{17} y_{n,j-1} \tag{2.27}$$

where

$$\begin{aligned} K_{14} &= \frac{-2c\Delta t^2\,EI}{\Delta t^2\,EI + 2c\Delta x^3 M_P} \\ K_{15} &= \frac{4c\Delta t^2\,EI}{\Delta t^2\,EI + 2c\Delta x^3 M_P} \\ K_{16} &= \frac{\Delta t^2\,EI}{\Delta t^2\,EI + 2c\Delta x^3 M_P}\left\{a + 2b - 4c + \frac{4c\Delta x^3 M_P}{\Delta t^2\,EI}\right\} \\ K_{17} &= \frac{-\Delta t^2\,EI}{\Delta t^2\,EI + 2c\Delta x^3 M_P}\left\{\frac{2c\Delta x^3 M_P}{\Delta t^2\,EI} + 1\right\} \end{aligned}$$

Equations (2.17), (2.23), (2.26) and (2.27) represent the dynamic equation of the arm for all the grid points (stations) at specified instants of time t in the presence of hub inertia and payload.

$$\begin{aligned} y_{i,j+1} = & - c\left[y_{i+2,j} + y_{i-2,j}\right] + b\left[y_{i+1,j} + y_{i-1,j}\right] + a y_{i,j} - y_{i,j-1} \\ & + d\left[y_{i+1,j} - 2y_{i,j} + y_{i-1,j} - y_{i+1,j-1} + 2y_{i,j-1} - y_{i-1,j-1}\right] \end{aligned} \tag{2.28}$$

$$y_{1,j+1} = K_1 y_{1,j} + K_2 y_{2,j} + K_3 y_{3,j} + K_4 y_{1,j-1} + K_5 y_{2,j-1} + K_6 \tau(j) \tag{2.29}$$

$$\begin{aligned} y_{n-1,j+1} = & K_7 y_{n-3,j} + K_8 y_{n-2,j} + K_9 y_{n-1,j} + K_{10} y_{n,j} + K_{11} y_{n-2,j-1} + K_{12} y_{n-1,j-1} \\ & + K_{13} y_{n,j-1} \end{aligned} \tag{2.30}$$

$$y_{n,j+1} = K_{14} y_{n-2,j} + K_{15} y_{n-1,j} + K_{16} y_{n,j} + K_{17} y_{n,j-1} \tag{2.31}$$

2.5.4 Matrix Formulation

Using matrix notation, Eqs. (2.28)–(2.31) can be written in a compact form as

$$\mathbf{Y}_{i,j+1} = \mathbf{A}\mathbf{Y}_{i,j} + \mathbf{B}\mathbf{Y}_{i,j-1} + \mathbf{C}\mathbf{F} \tag{2.32}$$

where $\mathbf{Y}_{i,j+1}$ is the displacement of grid points $i = 1, 2, \cdots, n$ of the arm at time step $j + 1$, $\mathbf{Y}_{i,j}$ and $\mathbf{Y}_{i,j-1}$ are the corresponding displacements at time steps j and $j - 1$ respectively. **A** and **B** are constant $n \times n$ matrices whose entries depend on the flexible arm specification and the number of sections the arm is divided into, **C** is a constant matrix related to the given input torque and **F** is an $n \times 1$ matrix related to the time step Δt and mass per unit length of the flexible arm;

$$\mathbf{Y}_{i,j+1} = \begin{bmatrix} y_{1,j+1} \\ y_{2,j+1} \\ \vdots \\ y_{n,j+1} \end{bmatrix}, \mathbf{Y}_{i,j} = \begin{bmatrix} y_{1,j} \\ y_{2,j} \\ \vdots \\ y_{n,j} \end{bmatrix}, \mathbf{Y}_{i,j-1} = \begin{bmatrix} y_{1,j-1} \\ y_{2,j-1} \\ \vdots \\ y_{n,j-1} \end{bmatrix} \tag{2.33}$$

$$\mathbf{A} = \begin{bmatrix} K_1 & K_2 & K_3 & 0 & 0 & \cdots & 0 & 0 \\ (b+d) & (a-2d) & (b+d) & -c & 0 & \cdots & 0 & 0 \\ -c & (b+d) & (a-2d) & (b+d) & -c & \cdots & 0 & 0 \\ \vdots & \ddots & \ddots & \ddots & \ddots & \ddots & \ddots & \vdots \\ 0 & 0 & \cdots & -c & b+d & a-2d & b+d & -c \\ 0 & 0 & \cdots & 0 & K_7 & K_8 & K_9 & K_{10} \\ 0 & 0 & \cdots & 0 & 0 & K_{14} & K_{15} & K_{16} \end{bmatrix} \tag{2.34}$$

$$\mathbf{B} = \begin{bmatrix} K_4 & K_5 & 0 & 0 & 0 & \cdots & 0 & 0 \\ -d & 2d-1 & -d & 0 & 0 & \cdots & 0 & 0 \\ 0 & -d & 2d-1 & -d & 0 & \cdots & 0 & 0 \\ \vdots & \ddots & \ddots & \ddots & \ddots & \ddots & \ddots & \vdots \\ 0 & 0 & \cdots & 0 & -d & 2d-1 & -d & 0 \\ 0 & 0 & \cdots & 0 & 0 & K_{11} & K_{12} & K_{13} \\ 0 & 0 & \cdots & 0 & 0 & 0 & 0 & K_{17} \end{bmatrix} \tag{2.35}$$

$$C = \tau(j),\ F = [K_6 \quad 0 \quad \cdots \quad 0]^T \tag{2.36}$$

2.5.5 State-Space Formulation

A state-space formulation of the dynamic equation of the arm can be constructed by referring to the matrix formulation. Using the notation for simulation of discrete-time linear systems, the dynamic equations of the flexible arm can be written as

$$\begin{aligned} x(n+1) &= \mathbf{P}x(n) + \mathbf{Q}u \\ y(n) &= \mathbf{R}x(n) + \mathbf{S}u \end{aligned} \tag{2.37}$$

where $P = \begin{bmatrix} A & vline & B \\ I_{\mathrm{N\times N}} & & 0_{\mathrm{N\times N}} \end{bmatrix}$, $Q = \begin{bmatrix} C \\ 0_{\mathrm{N\times 1}} \end{bmatrix}$, $R = [I_{\mathrm{N}} \quad 0_{\mathrm{N}}]$, $S = [0_{\mathrm{2N}}]$

$u = [\tau \quad 0 \quad \cdots \quad 0]^{\mathrm{T}}$, $y(n) = [x(1,n) \cdots x(N,n), x(1,n-1) \cdots x(N,n-1)]$

N represents the number of sections.

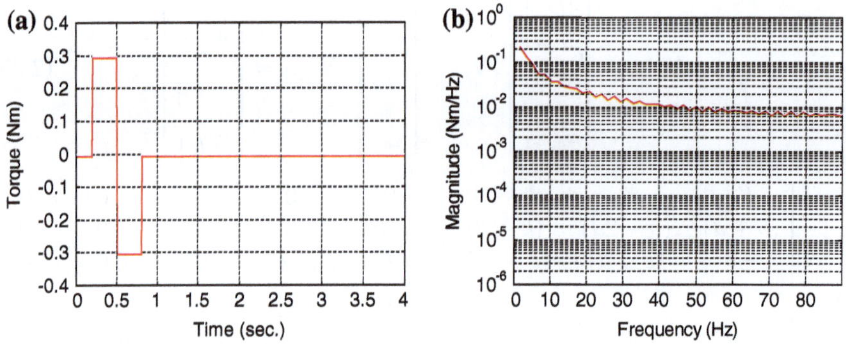

Fig. 2.7 The bang-bang torque input. Time-domain (**a**); Spectral-density (**b**)

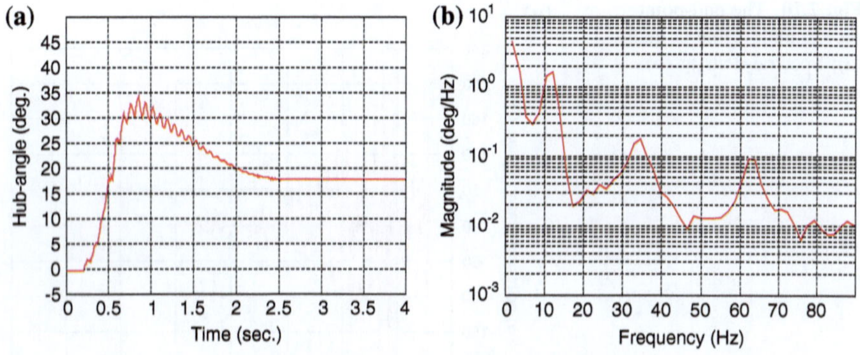

Fig. 2.8 The hub-angle. Time-domain (**a**); Spectral-density (**b**)

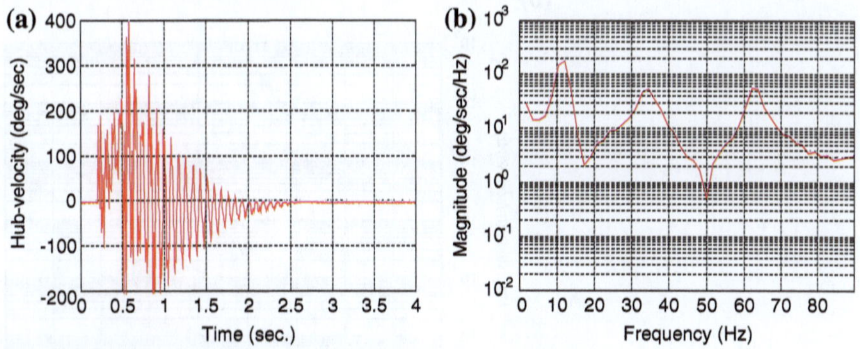

Fig. 2.9 The hub-velocity. Time-domain (**a**); Spectral-density (**b**)

2.6 Some Simulation Results

To identify an unknown process some knowledge of the process and signals is required. The simplest way is to apply a signal and record the system's response. This data set can best describe the characteristic behaviour of the flexible arm. There are various types of signals, which can be used as inputs to the system. Two types of signals, namely a bang-bang and composite PRBS are widely used for this purpose. In this experiment a bang-bang signal is used. It has been reported that the vibration of the flexible arm is dominated by the first few (commonly two) resonance modes. Anti-aliasing filters with cut-off frequency of 100 Hz are used for the four outputs namely, the hub-angle output, the hub-velocity output, the end-point acceleration output and the motor-current output.

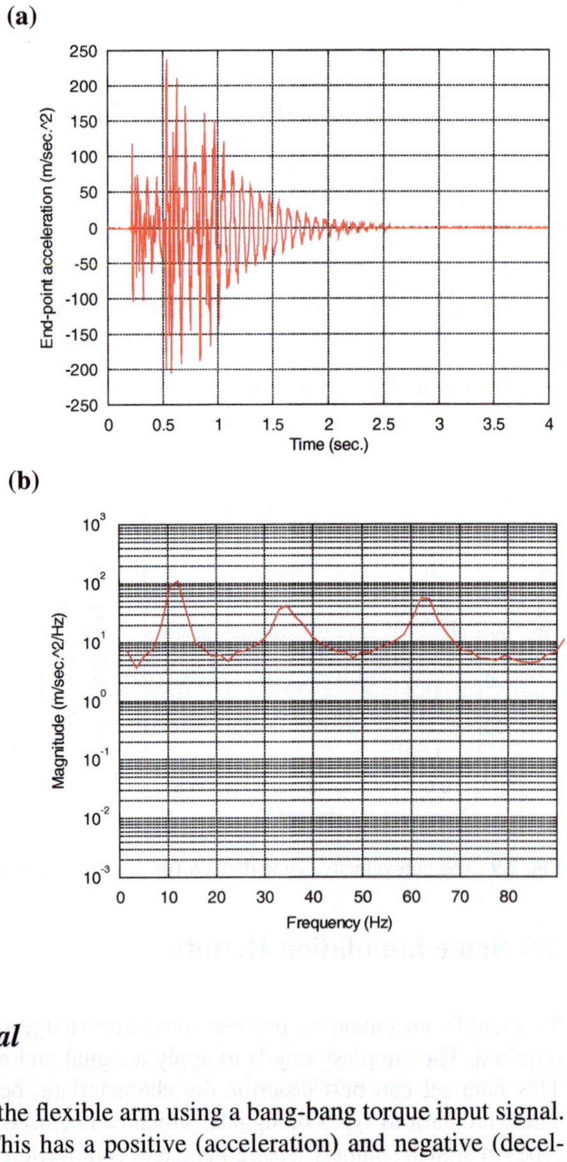

Fig. 2.10 The end-point acceleration. Time-domain (**a**); Spectral-density (**b**)

2.6.1 Bang-Bang Signal

The first attempt is to excite the flexible arm using a bang-bang torque input signal. This is shown in Fig. 2.7. This has a positive (acceleration) and negative (deceleration) period allowing the arm to, initially, accelerate and then decelerate and eventually stop at a target location. An amplitude of ±0.3 Nm and duration of 0.6 s is chosen for the bang-bang signal in this experiment. The system response as hub angle, hub velocity, end-point acceleration and motor current with the corresponding spectral densities, are observed for 4 s.

Figure 2.8 shows the trajectory of the hub displacement due to the bang-bang torque input. Theoretically, the energy of the corresponding bang-bang torque would equally distribute for acceleration and deceleration phases of the system.

Fig. 2.11 The motor current. Time-domain (**a**); Spectral-density (**b**)

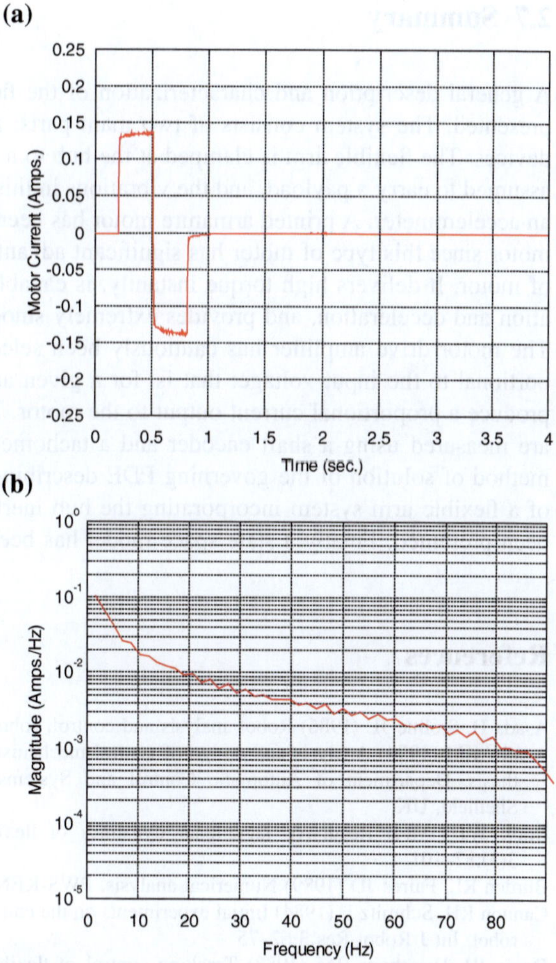

This means that there would be no further rigid body movement after the braking phase of the bang-bang torque has ended, although vibrations or oscillations might take place during this period. Due to problems, which commonly occur under real conditions, for example, due to shaft motor frictions, which lead to the dead zone problem, the response of the hub to the bang-bang torque gives a different behaviour. The displacement has reached 30–35° but then moved back to 18°.

Figures 2.9, 2.10 and 2.11 show the hub velocity, the end-point acceleration and the motor current output with corresponding spectral densities.

2.7 Summary

A general description and characterization of the flexible arm system have been presented. The system consists of two main parts: a flexible arm and measuring devices. The flexible arm is clamped at the hub to a drive motor. The end-point is assumed to carry a payload, and the vibrations in this position, are measured using an accelerometer. A printed armature motor has been carefully chosen as the drive motor since this type of motor has significant advantages compared to other types of motor. It delivers high torque instantly, is capable of producing rapid acceleration and deceleration, and provides extremely smooth torque with no 'cogging'. The motor drive amplifier has cautiously been selected to deliver a current proportional to the input voltage: that is, for a given amount of input voltage it can produce a proportional current output to the motor. The hub position and velocity are measured using a shaft encoder and a tachometer respectively. A numerical method of solution of the governing PDE describing the characteristic behaviour of a flexible arm system incorporating the hub inertia, payload and damping has been presented. Finally a state space model has been presented.

References

Asada H, Stoline JE (1986) Robot analysis and control. John Wiley, New York

Azad AKM (1994) Analysis and design of control mechanisms for flexible arm systems, Ph.D. thesis, Department of Automatic Control and Systems Engineering, The University of Sheffield, UK

Book WJ (1984) Recursive lagrangian dynamics of flexible arm arms. Int J Robot Res 3(3):87–101

Burden RL, Faires JD (1989) Numerical analysis. PWS-KENT Publishing Company, Boston

Cannon RH, Schmitz E (1984) Initial experiments on the end-point control of a flexible one-link robot. Int J Robot Res 3:62–75

Davis JH, Hirschorn RM (1988) Tracking control of flexible robot link. IEEE Trans Autom Control 33:238–248

Fu KS, Gonzalez RC, Lee CSG (1987) Robotics: control, secsing, vision and intelligence. McGraw-Hill, New York

Hastings GG, Book WJ (1987) A linear dynamic model for flexible robotics arm. IEEE Control Syst Mag 7:61–64

Hastings GG, Ravishankar BN (1988) An experimental system for investigation of flexible link experiencing general motion. In: Proceedings of the conference on decision and control, pp 1003–1008

Khosla PK, Kanade T (1988) Experimental evaluation of non-linear feedback and feed forward control schemes for arms. Int J Robot Res 7(1):790–798

Kuo BC, Tal J (ed) (1978) DC motors and control systems. SRL Publishing Company, Champaign, Illinois

Lapidus L (1982) Numerical solution of partial differential equations in science and engineering. John Wiley, New York

Meirovitch L (1967) Analytical methods in vibrations. Macmillan, New York

Schilling RJ (1990) Fundamentals of robotics analysis and control. Prentice Hall, Englewood Cliffs

Seraji H, Moya MM (1987) Position control for non-linear multiple link robots, NASA technical brief 11(3):119

Tokhi MO, Azad AKM (1995) Real-time finite difference simulation of a single-link flexible arm system incorporating hub inertia and payload. Proc IMechE-I J Syst Control Eng 209(I1):21–33

Tokhi MO, Azad AKM (1996a) Modeling of a single-link flexible arm system: theoretical and practical investigations. Robotica 14:91–102

Tokhi MO, Azad AKM (1996b) Control of flexible arm systems. Proc Inst Mech Eng 210:113–130

Tokhi MO, Azad AKM (1997) Design and development of an experimental flexible arm system. Robotica 15(3):283–292

Tse FS, Morse IE, Hinkle TR (1980) Mechanical vibrations theory and applications. Allyn and Bacon Inc., Boston

Chapter 3
Control Systems

3.1 Introduction

Control of mechanical systems is among one of the most active fields of research due to the diverse applications of mechanical systems in automotive, aerospace, manufacturing and robotics and related industries. Though, the study of mechanical systems dates back to Euler and Lagrange in the 1700s, it was not until 1850s that mechanical control systems came to the picture in the regulation of steam engines. During the past century, a series of scientific, industrial, and military applications motivated rigorous analysis and control design for mechanical systems. On the other hand, theoretically challenging nature of analysis of the behaviour of non-linear dynamical systems attracted many mathematicians to study control systems. As a result, the efforts of engineers and scientists together led to creation of Linear Control, Optimal Control, Adaptive Control, and Nonlinear Control theories. In the 1950s, control theory began new era of development. Powerful techniques were developed that allowed treating multivariable, time-varying systems, as well as many nonlinear problems. In the 1960s, Richard Bellman and Rudolf Kalman in the United States and L. Pontryagin in the former Soviet Union contributed in tremendously to form the basis for modern and optimal control theories, which continues to this day.

Over the past six decades there have been remarkable developments in linear control theory and has been extensively used in applications. Linear control design methods have been hugely successful and hence it must be true that many systems can be well approximated by linear models. On the other hand, there are well known examples of nonlinear practical systems. A common engineering approach to these kinds of problem is to base the design on a set of linearized models valid at a set of representative operating conditions. Indeed, over the last few years the more general topic of nonlinear control has attracted substantial research interest.

Nonlinear control systems are those control systems where nonlinearity plays a vital role, either in the controlled process (or plant) or in the controller itself. Nonlinear plants arise naturally in numerous engineering and natural systems, including mechanical and biological systems, aerospace and automotive control,

N. Siddique, *Intelligent Control*, Studies in Computational Intelligence 517,
DOI: 10.1007/978-3-319-02135-5_3, © Springer International Publishing Switzerland 2014

industrial process control, and many others. Nonlinear control theory is concerned with the analysis and design of nonlinear control systems. It is closely related to nonlinear systems theory in general.

State-space approach is the most widely used method for describing (modelling) nonlinear dynamical system emphasising the behaviour of the state variables of the system. An alternative approach to mathematical modelling of dynamical system is input–output approach. An input–output model relates the output of the system directly to the input, without any knowledge about the internal structure that is represented by the state equations. A control system deals with problem of obtaining desired behaviour of a dynamical system. A dynamical system is defined as an aggregation of time varying quantities that identify a goal of interest to meet. A block diagram of a dynamical system is shown in Fig. 3.1.

In Fig. 3.1, $u(t)$ is the control input to the system, $x(t)$ is state of the system and $y(t)$ is the output that is visible from the outside of the system. The input u belongs to a space that map the time interval $[0, \infty)$ into the Euclidean space R^m, i.e. $u : [0, \infty) \rightarrow R^m$. The dynamic system in Fig. 3.1 is open-loop as the control input does not depend explicitly on the output. In general, there are four steps to control a system:

- Develop the mathematical description of the system
- Design the state reconstruction in order to reconstruct the variables needed to control the system
- Design the control
- Close the loop on the real system

In closed-loop system, the control input depends on the output rather than time. The term closed-loop refers to the fact that the output is used to feed back the measurement information into the system as part of the control strategy or algorithm. A closed-loop system is shown in Fig. 3.2.

In Fig. 3.2, typically G represents the system or plant, H represents a controller, y is the control variable and u is the control input to the plant. G is some mapping or operator that specifies y in terms of u. Closed-loop control is further classified according to exact feedback signal. If the closed-loop control input u depends explicitly on the output y, it is referred to as output feedback control. The closed-loop control is called state variable feedback, if all the states are measured and feedback.

Open-loop and closed-loop controls are important concepts in automatic control systems. Open-loop controllers exist because they are comparatively cheap and easy to implement as they do not need to incorporate feedback information into control mechanism. For enhanced performance, closed-loop feedback controllers are employed.

Fig. 3.1 Dynamic system

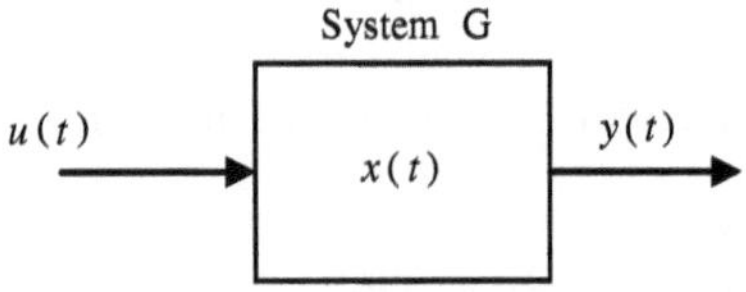

Fig. 3.2 Closed-loop system

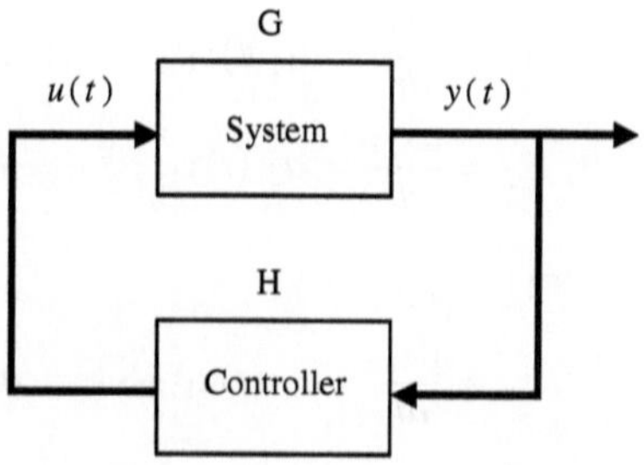

3.2 Control Systems

The theory of nonlinear dynamical systems or nonlinear control systems (if control inputs are involved) has been advanced since the nineteenth century. Nonlinear control theories are applied to variety of systems ranging from pure control engineering systems to social and biological systems. A continuous-time nonlinear control system can be described by a set of differential equations of the form

$$\begin{aligned} \dot{x}_1 &= \tfrac{dx_1}{dt} = f_1(x_1, x_2, \ldots, x_n, u_1, u_2, \ldots, u_m, t) \\ \dot{x}_2 &= \tfrac{dx_{21}}{dt} = f_2(x_1, x_2, \ldots, x_n, u_1, u_2, \ldots, u_m, t) \\ &\vdots \\ \dot{x}_n &= \tfrac{dx_n}{dt} = f_n(x_1, x_2, \ldots, x_n, u_1, u_2, \ldots, u_m, t) \end{aligned} \tag{3.1}$$

where $x = x(t)$ is the state of the system, $x \in R^n$, u is the control input, $u \in R^m$, often $m \leq n$, $t \in [t_0, \infty)$ and $f(.)$ is a Lipschitz or continuously differentiable nonlinear function. These equations expressed the time-derivative of each of the state variables as general functions of all the state variables, inputs, and possibly time. The equations in (3.1) can be written in vector notation form as

$$\dot{x} = \frac{dx}{dt} = f(x, u, t) \tag{3.2}$$

where $x = [x_1, x_2, \ldots, x_n]^T$ and $u = [u_1, u_2, \ldots, u_m]^T$. When t does not appear explicitly in the Eq. (3.2), the system is said to be time-invariant.

The factual thing is that the real world behaves in a nonlinear way—at least it is true when it is considered over wide operating ranges. That is to say that the state equations in (3.2) are nonlinear for most real systems. The differential equation in (3.2) is only a representation of an approximate model of the real world as because an accurate model is not known or too complicated to be described by a set of differential equations. In the state-space model of a linear process, the Eq. (3.1) can take the special form:

$$\begin{aligned}
\dot{x}_1 &= \frac{dx_1}{dt} = a_{11}(t)x_1 + \cdots + a_{1n}(t)x_n + b_{11}(t)u_1 + \cdots + b_{1m}(t)u_m \\
\dot{x}_2 &= \frac{dx_{21}}{dt} = a_{21}(t)x_1 + \cdots + a_{2n}(t)x_n + b_{21}(t)u_1 + \cdots + b_{2m}(t)u_m \\
&\vdots \\
\dot{x}_n &= \frac{dx_n}{dt} = a_{n1}(t)x_1 + \cdots + a_{nn}(t)x_n + b_{n1}(t)u_1 + \cdots + b_{nm}(t)u_m
\end{aligned} \tag{3.3}$$

The Eq. (3.3) can be written in vector notation form as

$$\dot{x} = \frac{dx}{dt} = \mathbf{A}(t)x + \mathbf{B}(t)u \tag{3.4}$$

Where the matrices $\mathbf{A}(t)$ and $\mathbf{B}(t)$ are given by

$$\mathbf{A}(t) = \begin{bmatrix} a_{11}(t) & \cdots & a_{1n}(t) \\ a_{21}(t) & \cdots & a_{2n}(t) \\ \vdots & \ddots & \vdots \\ a_{n1}(t) & \cdots & a_{nn}(t) \end{bmatrix} \quad \mathbf{B}(t) = \begin{bmatrix} b_{11}(t) & \cdots & b_{1m}(t) \\ b_{21}(t) & \cdots & b_{2m}(t) \\ \vdots & \ddots & \vdots \\ b_{n1}(t) & \cdots & b_{nm}(t) \end{bmatrix} \tag{3.5}$$

the matrix $\mathbf{A}(t)$ is a square matrix. In most processes it is often $m \leq n$ and the matrix $\mathbf{B}(t)$ should not necessarily be a square matrix. Though the concept of the state of a system is fundamental, there are situations where the control designer may not be interested in the state rather interested in the observation given by

$$y = g(x, u, t) \tag{3.6}$$

where $y = y(t)$ is the output of the system, $y \in R^l$, $1 \leq l \leq n$ and $g(.)$ is a continuous nonlinear function. In a linear system, the output vector $y = [y_1, \ldots, y_l]^T$ is assumed to be a linear combination of the state and the input given by

$$y = \mathbf{C}(t)x + \mathbf{D}(t)u \tag{3.7}$$

Where the matrices $\mathbf{C}(t)$ and $\mathbf{D}(t)$ are given by

$$\mathbf{C}(t) = \begin{bmatrix} c_{11}(t) & \cdots & c_{1n}(t) \\ c_{21}(t) & \cdots & c_{2n}(t) \\ \vdots & \ddots & \vdots \\ c_{l1}(t) & \cdots & c_{l,n}(t) \end{bmatrix} \quad \mathbf{D}(t) = \begin{bmatrix} d_{11}(t) & \cdots & d_{1m}(t) \\ d_{21}(t) & \cdots & d_{2m}(t) \\ \vdots & \ddots & \vdots \\ d_{l1}(t) & \cdots & d_{lm}(t) \end{bmatrix} \tag{3.8}$$

For $n, l > 1$, the system (3.4) and (3.7) is called a multi–input–multi–output (MIMO) and for $n, l = 1$, it is called a single–input–single–output (SISO) system. It is to be noted that a special case of system (3.1) and (3.6), with or without control, is said to be autonomous if the variable t does not appear independently from the state vector in the system function $f(.)$. For a time-invariant system, the dynamic equations in (3.4) and (3.7) are given by

$$\dot{x} = \mathbf{A}x + \mathbf{B}u \tag{3.9}$$

$$y = \mathbf{C}x + \mathbf{D}u \tag{3.10}$$

where **A**, **B**, **C** and **D** are constant matrices. The outputs of a system are those that can be observed or measured by means of sensors. The presence of the matrix **D** in Eq. (3.10) means that there is direct relation between the input and the output. There is no general reason for **D** to be absent in system description but it is omitted in majority of applications to reduce the complexity of much of the theory.

The control system described by the Eqs. (3.9)–(3.10) has a number of properties that are advantageous for applications and analysis of control systems. The following notions from system theory are useful:

(1) An equilibrium of the system (3.1) at the origin of R^n, if exists, is a solution x^* of the algebraic equation $f(x^*) = 0$ and $\dot{x}^* = 0$, which means that an equilibrium of a system must be a constant state.
(2) An equilibrium point x^* of the system (3.1) at the origin of R^n is weakly stable if all solutions x^* of the algebraic equation $f(x^*) = 0$ and $\dot{x}^* = 0$ start near x^* and stay near it, i.e. the system has a unique and constant equilibrium.
(3) It is asymptotically stable if, in addition, $x(t)$ converges to $\lim_{t\to\infty} x(t) = x^*$ whenever started near enough to it. If this convergence occurs for any initial state then x^* is globally asymptotically stable. Note that a nonlinear system may have several equilibrium points, each with different stability properties.
(4) The controllability is a closely-related concept that applies to the state equation $\dot{x} = f(x, u)$ concern the possibility of reaching a given state from any other state (controllability) by choosing appropriate controls. The controllability is directly given by the rank of $\left(\mathbf{B}, \mathbf{AB}, \mathbf{A^2B}, \ldots, \mathbf{A^{n-1}B}\right)$.
(5) Observability concerns the ability to distinguish between two (initial) states based on proper choice of input and observation of the system output. This concept, roughly, indicates whether a feedback controller that uses only the output y can fully control the state dynamics. The observability is directly given by the rank of $\left(\mathbf{C}, \mathbf{CA}, \mathbf{CA^2}, \ldots, \mathbf{CA^{n-1}}\right)$.
(6) The stability can directly be given by the poles of the transfer function $G(s)$ or by the eigenvalues of the system matrix **A**. The transfer function of the system is given by

$$G(s) = \mathbf{C}[s\mathbf{I} - \mathbf{A}]^{-1}\mathbf{B} + \mathbf{D} \tag{3.11}$$

Nonlinear system has a number of properties that make it difficult for control design such as equilibrium of nonlinear system can be unique, multiple, infinite or even not exist. The controllability and observability are very hard to prove, the stability may be hard to prove, and frequency analysis is almost impossible. Another aspect is that the local properties are not equal to the global properties and the time behaviour is dependent of the initial condition. There is no systematic approach for building control law. Therefore, in general, the control of nonlinear system is not an easy task. Numerous methods and approaches exist for the analysis and design of nonlinear control systems. A brief overview of some prominent ones is given in the above discussion. Interested readers are directed to the textbooks (Khalil 2002; Najmeijer and van der Schaft 2006; Sastry 2004; Vidyasagar 2002).

3.3 Control of Flexible Arm

Control of flexible arms has been of significant research interest in recent years. Much of the research work has been done on single-link flexible arms, which may have limited industrial application. There are applications in industry, which require lightweight, small cross-section, and arms. A lightweight and small cross-section arm has the advantage that it reduces the large inertial forces during motion. However, it exhibits a significant degree of vibration in the structure at high speeds, which require good controls for accurate end-point positioning. To allow this, an accurate mathematical model of a flexible arm system needs to be developed.

To guarantee that the arm links have enough rigidity so that structural vibration is avoided, each link of the arm is made to be thick and strong. This adds considerable mass to the arm, leading to higher material cost and increased energy consumption in operating the arm. As compared with traditional robot arms, constructed with rigid links, flexible robot arms are not only able to move larger payloads without increasing the mass of the linkages, but also have significant advantages such as: they require smaller actuators, have less link weights, consume less power and are more manoeuvrable. Flexible arms have not been widely used in industry due to their control requirements and that the performance is severely deteriorated by structural deformation, especially in the flexible links where the deformation is oscillatory. Thus, the flexibility of the links must be considered. Some of the approximate methods for vibration analysis of flexible systems (Meirovitch 1967) have been applied to the modelling of flexible arms. Book (1984) has addressed 4×4 coordinate transformation matrices to represent the elastic deformation, and has proposed a recursive Lagrange assumed mode method to construct the dynamic equations. Judd and Falkenburg (1985) have constructed a lumped mass and spring model and have derived the dynamic equations of a two-link, planer, flexible arm using Euler–Lagrange formulation. King et al. (1987) has developed a nonlinear inverse dynamic equation in recursive form for arms with an arbitrary number of flexible links. They have chosen the Newton–Euler assumed mode method using angular velocities instead of transformation matrices. Li and Sankar (1993) have proposed a systematic method for modelling and dynamic computation of flexible arms using the Lagrangian assumed mode method. The link deflection in this case is described by a truncated modal expansion.

The term 'control', in general, can characterise an open-loop strategy based on a more or a less accurate model of the system under investigation or a feedback law making use of the control deviation. In some cases, it may also characterise a combination of these two categories; i.e., an open-loop strategy for the gross motion with an underlying feedback accounting for small deviations.

Two main approaches can be distinguished when considering the control of flexible arm systems. In the first approach, a rigid body structure is assumed and a mathematical model is developed through computation of the necessary geometric, kinematic or kinetic quantities. To obtain a satisfactory model, an investigation is required to reveal the accuracy of the identified parameters for which necessary measurements to yield information on the deflections have to be carried out. The arm deflection mainly depends on the modes of vibration, length of arm and shear deformation with end-point load. Figure 3.3 shows the deflection of the link considering four modes of vibration and the shear deformation with an end-point load of 0.187 kg.

The second approach accounts, in addition to the factors in the first approach, for deviations caused by the elastic properties of the arm. Thus it requires additional measurements, for example by strain gauges, optical sensors, accelerometers, etc. These measurements are to compensate for deviations of the identified parameters caused by elasticities, and thus are used to improve the control performance.

Due to elastic properties of the system, the development of a mathematical description and subsequent control of a flexible arm is a complicated task. A considerable amount of basic research has been carried out on the modelling and control of flexible arms for the last 30 years. The control problem, to achieve high performance, is to acquire the ability to dampen the oscillations of the structure. This is made difficult by the presence of a large (infinite) number of modes of vibration in the structure, which are in general lightly damped.

One of the most promising techniques for flexible arm control used to date is input command shaping, where the system inputs, e.g. motor voltage, are shaped in such a manner that minimal energy is injected into the flexible modes of the system (Poerwanto 1998; Azad et al., 2008; Singhose and Seering, 2008). The difficulty of such command shaping techniques is that they are open-loop strategies. Open-loop control strategy requires relatively precise knowledge of the physical and vibrational property of the flexible arm system. Also the approach does not account for changes in the system once the control input is developed (Azad 1994; Moudgal et al. 1995). An alternative approach is the closed-loop control, which differs from open-loop control in that it uses measurements of the system's state and changes the actuator input accordingly to reduce the system oscillation. Many open loop, closed-loop and non-linear adaptive control strategies have been proposed for flexible arm systems (Jain and Khorrami 1998; Tokhi and Azad 1996a, b; Tokhi and Azad 1995; Yang et al. 1997). It is well known that the primary difficulty of such control approaches lies in the fact that they all require precise mathematical model of the system dynamics. Although the requirement that a very flexible arm must undergo only very small vibration is too stringent for most practical applications,

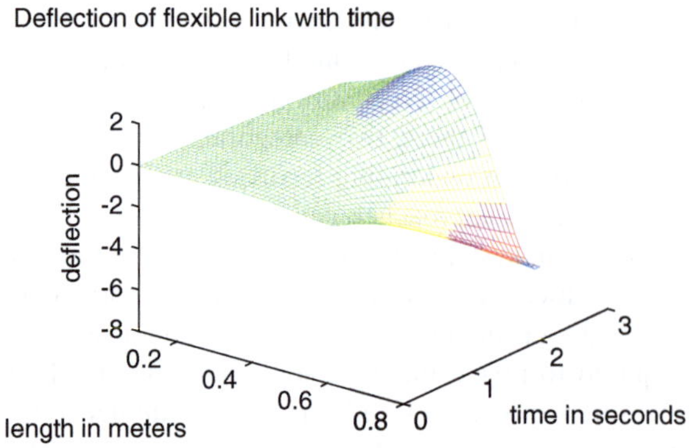

Fig. 3.3 Arm deflection as a function of arm length

the problem of non-linearity has been avoided by linearization to approximate mathematical models with the assumption of small deflections.

As mentioned previously, a considerable amount of research on flexible arm systems is limited to single-link arms, due to the complexity and non-linearity of multi-link arm systems resulting from their greater degrees of freedom and the increased interactions between gross and deformed motions. Thus, with the current state of research work, the complex behaviour of the system makes it almost impossible to control a flexible arm successfully with higher degrees of freedom and larger angles (Poerwanto 1998; Siddique 2002). Vibration control for flexible arm systems are generally addressed in two ways:

- Passive control
- Active control

Passive control utilises the absorption property of matter and thus is realised by a fixed change in the physical parameters of the structure, for example adding viscoelastic materials to increase the damping properties of the flexible arm. However, it has been reported that the control of vibration of a flexible arm by passive means is not sufficient by itself to eliminate structural deflection (Book et al. 1986).

Active control utilises the principle of wave interference. It is realised by artificially generating anti-source(s) (actuator(s)) to interfere with the disturbances and thus result in reduction in the level of vibrations. Active control of flexible arm systems can in general be divided into two categories:

- Open-loop control and
- Closed-loop control

Open-loop control involves altering the shape of actuator commands by considering the physical and vibrational property of the flexible arm system. The approach does not account for changes in the system once the control input is developed. Closed-loop control differs from open-loop control in that it uses measurements of the system's state and changes the actuator input accordingly to reduce the system oscillation.

3.4 Open-Loop Control

Recently, open-loop control methods have been considered in control where the control input is developed by considering the physical and vibrational property of the flexible arm system. Although, the mathematical theory of open-loop control is well established (Athans and Ealb 1966; Cesari 1983; Citron 1969; Dellman et al. 1956; Lee 1960; Sage and White 1977; Singh et al. 1989), few successful applications in the control of distributed parameter flexible arm systems have been reported.

The goal is to develop methods to reduce motion induced vibration in flexible arm systems during fast movements. The assumption is that the motion itself is the main source of system vibration. Thus, torque profiles which do not contain energy at system natural frequencies do not excite structural vibration and hence require no additional settling time. The procedure for determining shaped inputs that generate fast motions with minimum residual vibration has been addressed by various researchers (Azad et al. 2008; Bayo 1988; Meckl and Seering 1985; Singhose and Seering 2008). Also some simulation results using bang–bang control have been presented in Chap. 2.

3.5 Closed-Loop Control

Effective control of a system always depends on accurate real-time measurement of control variables and the corresponding control effort. The common approach is to measure the hub angle θ, hub velocity $\dot{\theta}$ and end-point acceleration α. The control strategy that uses the hub angle θ, hub velocity $\dot{\theta}$ as feedback is called collocated control and the control strategy that uses end-point acceleration α as feedback is called non-collocated control. The implementation of the collocated and non-collocated controllers imposes a number of problems associated with the reasonable accurate measurement of the variables. An appreciable amount of work carried out in the control of flexible arm systems involves the utilisation of strain gauges, mainly to measure mode shapes at the end-point. Hastings (Hastings and Book 1985; Hastings and Ravishankar 1988) has looked into the real-time control of flexible arm by using two sets of strain gauges and a collocated potentiometer and tachometer set. It appears that the strain gauge measurement is very simple and relatively inexpensive to use. However, the technique may place more stringent requirements on the dynamic modelling and control tasks. Strain gauges have the disadvantage of not giving a direct measurement of manipulator displacement, as they can only provide local information. Thus, displacement measurement by using strain gauges requires more complex and possibly time consuming computations which can lead to inaccuracies (Hastings and Ravishankar 1988). To solve the problem of displacement measurement, as encountered in using strain gauges only, attempts have been made to develop schemes that incorporate end-point measurements as well (Cannon and Schmitz 1984; Kotnik et al. 1988; Schmitz 1985). Schmitz (1985) has introduced an end-point measurement scheme in addition to the strain gauge and hub angle sensor and used these measurements to design a controller. A control law partitioning scheme which uses end-point sensing device has been reported by Rattan et al. (1990). The scheme uses end-point position signal in an outer loop controller to control the flexible modes, whereas the inner loop controls the rigid body motion independent of the dynamics of the manipulator. Performance of the scheme has been demonstrated in both simulation and experimental trials incorporating the first two flexible modes. A combined feedforward and feedback method in which the end-point position is sensed by an accelerometer and fed back to the motor controller, operating as a velocity servo, has been proposed by Wells and Schueller (1990) in the control a flexible arm.

In the investigations carried out on the control of flexible manipulator systems, as discussed in the preceding paragraphs, the only non-collocated sensor/actuator pairs that have successfully been employed include the motor torque with either the manipulator strain or global/local end-point position. However, practical realisation of both methods has associated short-term and long-term drawbacks. It has been shown by Stadenny and Belanger (1986) that, if a state space description of the closed-loop dynamics is available, it is possible to use acceleration feedback to stabilise a rigid arm. Kotnik and co-workers have carried out a comparative study on the control of a flexible arm using acceleration feedback to design the compensator and the end-point position feedback using a design based on a full-state feedback observer (Kotnik et al. 1988). It is shown that the controller using end-point position feedback exhibits a relatively slow and rough response in comparison with the acceleration feedback controller; the difference becoming more noticeable with increasing slewing angle. Moreover, acceleration feedback produces relatively higher overshoot. The robustness of the controller to system uncertainties has not been discussed.

From the discussion in the preceding section, it is apparent that the practical realisations of both collocated and non-collocated methods have associated short term and long term drawbacks. To tackle the various problems associated with controller design approaches for flexible arm systems, a control strategy that uses the collocated (hub angle and hub velocity) and non-collocated (end-point acceleration) feedback is suggested at this stage.

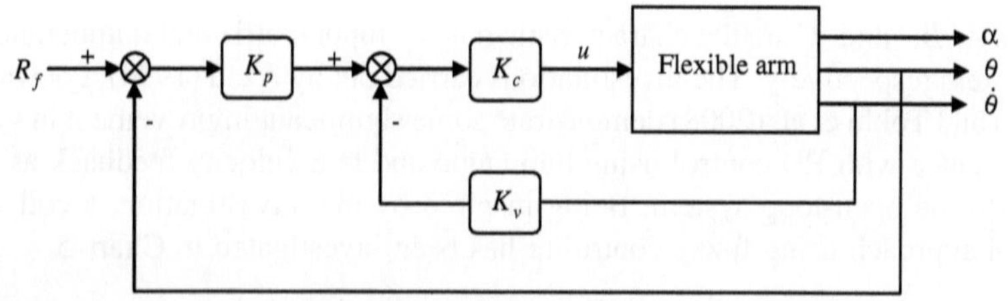

Fig. 3.4 Joint based collocated controller for flexible arm

3.5.1 Joint Based Collocated Controller

A common strategy in the control of manipulator systems involves the utilisation of proportional and derivative (PD) feedback of collocated sensor signals. Such a strategy is can be adopted for investigation here. The controller proposed provides a base line controller that will later can be used to guess the performance of other more advanced control strategies.

A block diagram of the PD controller is shown in Fig. 3.4, where K_p and K_v are the proportional and derivative gains, θ represents hub angle, $\dot{\theta}$ represents hub velocity, α represents end-point acceleration, R_f is the reference hub angle and K_c is the gain of the motor amplifier. Here the motor/amplifier set is considered as a linear gain K_c, as the set is found to function linearly in the frequency range of interest. To design the PD controller, a linear state-space model of the flexible arm has to be obtained by linearizing the system equations of motion of the system. The first two flexible modes of the manipulator were assumed to be dominantly significant. The control signal $u(s)$ in Fig. 3.4 can thus be written as

$$u(s) = K_c\left[K_p\left(R_f(s) - \theta(s)\right) - K_v\dot{\theta}\right] \tag{3.12}$$

where, s is the Laplace variable. The closed-loop transfer function is, therefore, obtained as

$$\frac{\theta(s)}{R_f(s)} = \frac{K_pH(s)K_c}{1 + K_cK_v\left(s + K_p/K_v\right)H(s)} \tag{3.13}$$

where, $H(s)$ is the open-loop transfer function from the input torque to hub angle, given by

$$H(s) = C(sI - A)^{-1}B \tag{3.14}$$

where, A, B, and C are the characteristic matrix, input matrix and output matrix of the system respectively. The investigations carried out by Azad (1994), Poewaranto (1998) and Tokhi et al. (2008) demonstrate some significant improvement in system performance with PD control using hub angle and hub velocity feedback as compared to the open-loop system. Being inspired by this investigation, a collocated control approach using fuzzy controller has been investigated in Chap. 5.

3.5.2 Hybrid Collocated and Non-Collocated Controller

A block diagram of the control structure, incorporating a combined collocated and non-collocated controller, is shown in Fig. 3.5. The controller design utilises end-point acceleration feedback through a PID control scheme. Moreover, the hub angle and hub velocity feedback are also used in a PD configuration for control of the rigid body motion of the manipulator. The control structure utilised thus comprises of two feedback loops: one using the filtered end-point acceleration as input to one control law, and the other using the filtered hub angle and hub velocity as input to a separate control law. These two loops are then summed to give a command motor input voltage, which produces a torque.

Consider first the rigid body control loop, in which the hub angle θ and hub velocity $\dot{\theta}$ are the output variables. The open-loop transfer function can be obtained from Eq. (3.14). To design the controller in this loop, a low-pass filter is required for both θ and $\dot{\theta}$ so that the flexible modes are attenuated before reaching the controller input.

The flexible motion of the flexible arm is controlled using the end-point acceleration feedback through a PID controller. The transfer function of the flexible manipulator with end-point acceleration as output is obtained from Eq. (3.14). The end-point acceleration is fed back through a low-pass filter. The values of proportional, derivative and integral gains are adjusted using the

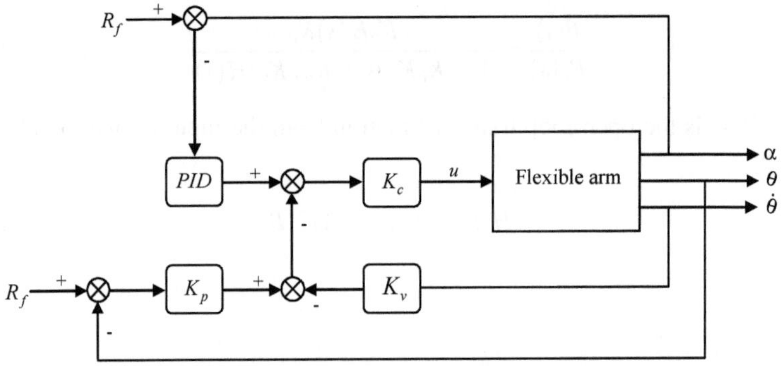

Fig. 3.5 Hybrid collocated and non-collocated controller for flexible arm

Ziegler-Nichols procedure (Warwick 1989). The investigations carried out by Azad (1994) and Poewaranto (1998) demonstrate the significant improvement in system performance with PD control.

A problem associated with the hybrid collocated and non-collocated controller designed for flexible arm system, however, is that the control effort at the manipulator input produces a spike at the beginning of the move. This may cause damage to the actuator and/or to the flexible arm system itself.

3.6 Alternative Control Approaches

To reduce system complexity, several approximations to the mathematical model of the system can be made. However, in practice, it is not possible to construct a mathematical model using traditional methods, which characterises the system completely and accurately in terms of its non-linear behaviour and other physical phenomena. These limitations of conventional model-based control mechanisms for flexible robot arm systems have stimulated the development of intelligent control mechanisms incorporating adaptive controls, neural networks based control and fuzzy logic based control. Thus, an investigation into the development of an intelligent control mechanism using fuzzy logic and neural networks is intended in this research monograph.

Many other techniques from control engineering are applicable to nonlinear systems, some of which may be considered as separate fields or non-traditional control engineering. Among these are:

- Optimal Control: The control objective is to minimize a pre-determined cost function. The basic solution tools are dynamic programming and variational methods (Calculus of Variations and Pontryagin's maximum principle). The available solutions for nonlinear problems are mostly numeric.
- Model Predictive Control: An approximation approach to optimal control, where the control objective is optimized on-line for a finite time horizon. Due to computational feasibility this method has recently found wide applicability, mainly in industrial process control.
- Adaptive Control: A general approach to handle uncertainty and possible time variation of the controlled system model. The controller parameters are tuned on-line as part of the controller operation using various techniques of estimation and learning algorithms. Adaptive control systems have evolved as an attempt to avoid degradation of dynamic performance of a control system in the presence of unmodelled dynamics or a change in system parameters. While a feedback control system is oriented towards the elimination of the effect of state perturbations, the adaptive control system is oriented towards the elimination of the effect of structural perturbations upon the performance of the control system. These structural perturbations are essentially caused by the variations of the dynamic parameters of the controlled plant.

- Learning Control: Learning control techniques were popular in the 1960s when cybernetics appeared. The self-learning and adaptive methods were developed for problems with stochastic properties in control systems
- Intelligent Control: Intelligent control has gained wide recognition since it first emerged in 1960s. Intelligent control systems should posses the capability of learning, reasoning, adaptability and self-organisation within a wide range of variation and uncertainty in the environment and tasks.

3.6.1 Intelligent Control Approaches

The basic difference between the intelligent control and traditional control (few of them discussed above in brief) is that the control strategy in intelligent control does not demand any precise model of the system. Therefore, intelligent control approaches are getting more attention from research community in applying them to systems which are complex, nonlinear and uncertain. A brief introduction to intelligent control has been made in Chap. 1. Among them are neural networks and fuzzy logic based controllers are widely accepted and applied controllers.

Neural network controllers are a particular class of adaptive control systems, where the controller is in the form of an artificial neural network. Neural network controllers are characterised by adaptation, learning and self-organisation and interpolation, robustness and plasticity, fault and noise tolerance. Narendra and Parthasarathy (1990) demonstrated how neural networks can be used effectively for the identification and control for nonlinear dynamical systems of the form (3.9)–(3.10). Neural networks implementation has been reported by other researchers as well (Sarangapani 2006).

The main features of fuzzy control in complex dynamical systems include the transparency and local representation of operator knowledge, the use of qualitative reasoning imitating the human operator, the possibility of control without an analytical model of the plant, robustness against noise and parameter variance of the plant, generalisation and interpolations. Fuzzy logic controller implements a (often heuristic) set of logical (or discrete) rules for formulating the control signal based on the observed outputs. Fuzzification, inferencing and defuzzification procedures are used to obtain a smooth control law from discrete rules. Based on the inferencing mechanism, different types of controller can be constructed, namely, Mamdani-type, Takagi–Sugeno-tyoe and Tsukamoto-type.

These features of neural network and fuzzy controllers could be exploited in the control of flexible arm systems. These control techniques have been investigated in the next few chapters.

3.7 Summary

In this chapter, a brief theoretical introduction to nonlinear control is presented, which will be useful in understanding the controller design and development and further discussion on stability issues presented in Chap. 9. The goal is to develop methods to reduce motion induced vibration in flexible arm systems during fast movements. Therefore, different control schemes such as open-loop and closed-loop control strategies are studied. Application of open-loop control strategy is limited. The problems of closed-loop control strategies are associated with measurement of the control variables and based on the measurement using suitable sensor mechanism, collocated and non-collocated control approaches have been suggested depending on the accuracy of the available models. This induces further condition on the controller development to find suitable tradeoffs. One possible way is to apply intelligent control methodologies, which will be addressed in the subsequent chapters.

References

Athans M, Elab PL (1966) Optimal control: a introduction to the theory and its applications. McGraw-Hill, New York, pp 504–561

Azad AKM (1994) Analysis and design of control mechanisms for flexible manipulator systems. Ph.D. thesis, Department of Automatic Control and Systems Engineering, The University of Sheffield, UK

Azad AKM, Shaheed MH, Mohammed Z, Tokhi MO, Poerwanto H (2008) Open-loop Control of Flexible Manipulators using Command-generation techniques, Book Chapter 8: Flexible Robot Manipulators - Modelling, Simulation and Control, Edt. MO Tokhi and AKM Azad, IET, London, UK

Bayo E (1988) Computed torque for the position control of open-loop flexible robots. In: Proceeding of IEEE international conference on robotics and automation, Philadelphia, 25–29 April, pp 316–321

Book WJ (1984) Recursive lagrangian dynamics of flexible manipulator arms. Int J Robot Res 3(3):87–101

Book WJ, Alberts TE, Hastings GG (1986) Design strategies for high-speed lightweight robots. Computers in mechanical engineering, 5(2):26–33

Cannon RH, Schmitz E (1984) Initial experiments on the control of a flexible one-link robot. Int J Robot Res 3(3):62–75

Cesari L (1983) Optimisation theory and applications: problems with ordinary differential equations. Springer, New York

Citron SJ (1969) Elements of optimal control. Holt, Rinehart and Winston, New York

Dellman R, Glicksber I, Gross O (1956) On the bang–bang control problem. Q Appl Mech 14(1):11–18

Fu KS, Gonzalez RC, Lee CSG (1987) Robotics: control, sensing, vision, and intelligence. McGraw-Hill, Singapore

Hastings GG, Book WJ (1985) Experiments in optimal control of a flexible arm. In: Proceedings of American control conference, vol 2, pp 728–729

Hastings GG, Ravisanker BN (1988) An experimental system for investigation of flexible link experiencing general motion. In: Proceedings of the conference on decision and control, pp 1003–1008
Hollerbach JM (1980) A recursive lagrangian formulation of manipulator dynamics and a comparative study of dynamics formulation complexity. IEEE Trans Syst Man Cybern 10(11):730–736
Isidori A (1995) Nonlinear control systems, 3rd edn. Springer, London
Jain S, Khorrami F (1998) Robust adaptive control of flexible joint manipulators. Automatica 34(5):609–615
Judd RP, Falkenburg DR (1985) Dynamics of nonrigid robot linkages. IEEE Trans Automatic Control 30(5):499–502
Khalil HK (2002) Nonlinear systems, 3rd edn. Prentice Hall, Upper Saddle River
King JO, Gourishankar VG, Rink RE (1987) Lagrangian dynamics of flexible manipulators using angular velocities instead of transformation matrices. IEEE Trans Syst Man Cybern 11:1059–1068
Kotnok PT, Yorkovich S, Ozguner U (1988) Acceleration feedback for control of a flexible manipulator arm. J Robot Syst 5(3):181–196
Lee EB (1960) Mathematical aspects of the synthesis of linear minimum response-time controller, IRE Trans Autom Control AC-5(4):283–290
Lee JW, Huggins JD, Book WJ (1988) Experimental verification of a large flexible manipulator. In: Proceedings of American control conference, vol 2, Atlanta, GA, pp 1021–1028
Levine WS (1996) The control handbook. CRC Press, Florida
Li C-Jin, Sankar TS (1993) Systematic methods for efficient modeling and dynamics computation of flexible robot manipulators. IEEE Trans Syst Man Cybern 23(1):77–95
Meirovitch L (1967) Analytical methods in vibrations. The Macmillian Company, New York
Meckl PH, Seering WP (1985) Active damping in a three-axis robotic manipulator. J Vibr Acoust Stress Reliab Des 107(1):38–46
Morris AS, Madani A (1995) Inclusion of shear deformation term to improve accuracy in flexible-link robot modeling. Department of Automatic Control and Systems Engineer, University of Sheffield, Research Report 583
Moudgal VG, Kwong WA, Passino KM, Yurkovich S (1995) Fuzzy learning control of a flexible-link robot. IEEE Trans Fuzzy Syst 3(2):199–210
Najmeijer H, van der Schaft AJ (2006) Nonlinear dynamical control systems, 3rd edn. Springer, New York
Narendra KS, Parthasarathy K (1990) Identification and control of dynamical systems using neural networks. IEEE Trans Neural Networks 1(1):4–27
Poerwanto H (1998) Dynamic simulation and control of flexible manipulator systems. Ph. D Thesis, Department of Automatic Control and Systems Engineering, Sheffield University, England
Rattan KS, Feliu V, Brown HB (1990) Tip position control of flexible arms. In: IEEE International conference on robotics and automation, vol 3, pp 1803–1808
Saga AZ, White CC (1977) Optimum systems control. Prentice-Hall, Englewood Cliffs, pp 103–107
Sarangapani J (2006) Neural network control of nonlinear discrete-time systems. CRS Press, Taylor & Francis, Boca Ratan, London
Sastry S (2004) Nonlinear systems: analysis, stability and control. Springer, New York
Schmitz E (1985) Experiments on the end-point position control of a very flexible one-link manipulator. Ph.D. Thesis, Stanford University, USA
Shimkin N (2009) Nonlinear control systems. In: Binder MD, Hirokawa N, Windhorst U (eds) Encyclopedia of neuroscience. Springer-Verlag GmbH Berlin, Heidelberg
Siddique, NH (2002) Intelligent Control of Flexible-link Manipulator Systems, PhD Thesis, Department of Automatic Control and Systems Engineering, The University of Sheffield. England, UK

Singh G, Kabamba PT, McClamroch NH (1989) Planner time-optimal rest-to-rest slewing manoeuvre of flexible spacecraft. J Guidance Control Dyn 12(1):71–81

Singhose, WE, Seering WP (2008) Control of Flexible Manipulators with Input Shaping Techniques, Book Chapter 9: Flexible Robot Manipulators - Modelling, Simulation and Control, Edt. MO Tokhi and AKM Azad, IET. London, UK

Slotine J–J, Li W (1991) Applied nonlinear control. Prentice-Hall, New Jersey

Stadenny J, Belanger P (1986) Robot manipulator control by acceleration feedback: stability, design and performance issues. In: Proceedings of the IEEE conference on decision and control, Athens, Greece, pp 80–85

Theodore RJ, Ghosal A (1997) Modeling of flexible-link manipulators with prismatic joints. IEEE Trans Syst Man Cybern Part B 27(2):296–305

Tokhi MO, Azad AKM (1995) Real-time finite difference simulation of a single-link flexible manipulator system incorporating hub inertia and payload. Proc IMechE-I: J Syst Control Eng 209(2):21–33

Tokhi MO, Azad AKM (1996a) Modeling of a single-link flexible manipulator system: theoretical and practical investigations. Robotica 14:91–102

Tokhi MO, Azad AKM (1996b) Control of flexible manipulator systems. Proc Inst Mech Eng 210:113–130

Tokhi MO, Azad AKM, Shaheed MH, Poerwanto H (2008) Collocated and Non-collocated Control of Flexible Manipulators, Book Chapter 12: Flexible Robot Manipulators - Modelling, Simulation and Control, Edt. MO Tokhi and AKM Azad, IET. London, UK

Vidyasagar M (2002) Nonlinear systems analysis, 2nd edn. SIAM Classics in Applied Mathematics, Philadelphia

Warwick K (1989) Control systems: an introduction. Prentice-Hall, UK

Wells RL, Schueller JK (1990) Feedforward and feedback control of a flexible robotic arm. IEEE Control Syst Maga 10(1):9–15

Yang JH, Lian FL, Fu LC (1997) Non-linear adaptive control for flexible-link manipulators. Trans Robot Autom 13(1):140–148

Chapter 4
Mathematics of Fuzzy Control

4.1 Fuzzy Logic

Logic is the study of methods of reasoning, where reasoning means obtaining new propositions from existing propositions. In classical logic, a proposition, p, is either true or false, that is, the truth value of a proposition is either 1 or 0. The classical two-valued logic has been dominating the scientific world for over a century. But there are many real world problems where the traditional two-valued logic didn't work out well or failed to be applicable due to the fact of absolute truth values. Fuzzy logic is a transition from absolute truth to partial truth that generalizes classical two-value logic by allowing the partial truth values of a proposition represented by a number in the interval of $[0, 1]$. This generalization allows performing approximate reasoning, i.e. deducing imprecise conclusions, i.e. fuzzy propositions, from a collection of imprecise premises. Fuzzy logic is not fuzzy rather it is a precise logic of imprecision and approximate reasoning (Zadeh 1975a, b, 1979, 2008). Zadeh argues that fuzzy logic may be viewed as an attempt at formalization or mechanization of two human capabilities. Firstly, human are the capable to converse, reason and make rational decisions on imprecision, uncertainty, incompleteness of information, conflicting information, partiality of truth and partiality of possibility in an environment of imperfect information. Secondly, they are capable to perform a wide variety of physical and mental tasks without any measurements and any computations (Zadeh 1999, 2000, 2001, 2002). The core constituent of fuzzy logic is the fuzzy set that describes the truth values of a proposition in form of a function.

4.2 Fuzzy Sets

A fuzzy set A in a universe of discourse U is characterised by a membership function (MF) $\mu_A(x)$ which takes the values within the interval $[0, 1]$ (Zadeh 1965). A fuzzy set A in U may be represented as a set of ordered pairs of a generic element x and its membership value, that is,

N. Siddique, *Intelligent Control*, Studies in Computational Intelligence 517,
DOI: 10.1007/978-3-319-02135-5_4,

$$A = \{(x, \mu_A(x)) | x \in U\} \tag{4.1}$$

when U is continuous, A is generally written as

$$A = \int_U \frac{\mu_A(x)}{x} \tag{4.2}$$

Here the integral sign does not represent integration. It denotes collection of all points $x \in U$ with associated MF $\mu_A(x)$. For discrete U, A is written as

$$A = \sum_U \frac{\mu_A(x)}{x} \tag{4.3}$$

Here the summation sign does not represent arithmetic addition. It denotes collection of all points $x \in U$ with associated MF $\mu_A(x)$. A fuzzy set has a one-to-one correspondence with its membership function. That is, there must be a unique membership function associated with a fuzzy set and conversely an MF must represent a fuzzy set.

4.3 Membership Functions

A fuzzy set is characterised by an MF. Among the different constituents of a fuzzy system, the MFs are the central as they need to be representative of the input–output space of the system. Therefore, the MFs directly affect the modelling accuracy and the system performance. In fuzzy system, the input–output behaviour of the system is represented by the rule-base. Consequently, the interpretability of the fuzzy rule-base primarily relies upon the MFs. Therefore, determining appropriate MFs is a decisive task in the realization of a well-behaved fuzzy system. By appropriate MFs it is meant particularly the shape of MFs within the universe of discourse. Eventually, the shape of the MFs plays a critical role in fuzzy system and fuzzy control. Therefore, defining appropriate MFs is important towards the development of any application involving fuzzy logic. MFs can have any form of regular or irregular shapes as long as they are convenient to be described mathematically. A convenient and concise way to construct an MF is to parameterise it and then express the MF mathematically in terms of parameters. MFs that are highly irregular shaped cannot be parameterised so easily, even if it can, computation would be excessive. A common practice is that the designer adopts regular shaped of known parameterised MFs such as triangular, trapezoidal, sigmoidal, Gaussian and bell-shaped MFs. Some common and widely used MFs are discussed in the following.

4.3.1 Piecewise Linear MF

Piecewise linear functions are the simplest form of MFs. Triangular and trapezoidal MFs are the most widely used among them. These MFs can be either symmetric or asymmetric shaped.

Triangular MF

Triangular MFs are very common in fuzzy system computation because of their simplicity and ease of computation. A triangular MF is specified by three parameters $\{a, b, c\}$ and defined by

$$\mu(x) = \max\left(\min\left(\frac{x-a}{b-a}, \frac{c-x}{c-b}\right), 0\right) \tag{4.4}$$

The parameters $\{a, b, c\}$ with $a < b < c$ determine the x coordinates of the three corners of the underlying triangular MF. The parameters a and c locate the 'feet' of the triangle and the parameter b locates the peak. A symmetric triangular MF is shown in Fig. 4.1a.

Trapezoidal MF

A trapezoidal MF has the shape of a truncated triangle. A trapezoidal MF is specified by four parameters $\{a, b, c, d\}$ and defined by

$$\mu(x) = \max\left(\min\left(\frac{x-a}{b-a},\ 1,\ \frac{d-x}{d-c}\right), 0\right) \tag{4.5}$$

The parameters $\{a, b, c, d\}$ with $a < b \leq c < d$ determine the x coordinates of the four corners of the underlying trapezoidal MF. The parameters a and d locate the 'feet' and b and c locate the 'shoulder' of the trapezoid. A trapezoidal MF can have either narrow 'shoulder' or wide 'shoulder'. A symmetric trapezoidal MF is shown in Fig. 4.1b.

Due to simple formulae and computational efficiency, linear MFs, i.e. both triangular and trapezoidal MFs, are extensively used, especially in real-time applications. Although trapezoidal type MF has often been used in fuzzy control literature, triangular MFs are most commonly used almost intuitively for all the variables.

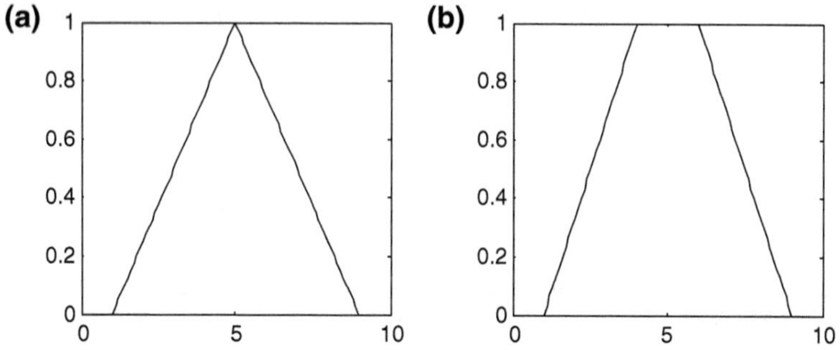

Fig. 4.1 Linear MFs Triangular MF = [1 5 9] (**a**); Trapezoidal MF = [1 4 6 9] (**b**)

4.3.2 Nonlinear Smooth MF

Piecewise linear MFs may not always be suitable for all applications. Therefore, nonlinear smooth functions are also in widespread use in fuzzy systems and control. Gaussian, bell-shaped, z-shaped, s-shaped and sigmoidal functions are very common among them.

Gaussian MF

Gaussian MFs are also popular among fuzzy systems researchers due to the fact that it can be specified by only two parameters $\{m, \sigma\}$ and defined by

$$\mu(x) = \exp\left[-\frac{1}{2}\left(\frac{x-m}{\sigma}\right)^2\right] \tag{4.6}$$

The parameters m and σ represent the centre and width of the Gaussian MF respectively. Gaussian MF is smooth, symmetric and non-zero at all points as illustrated in Fig. 4.2a.

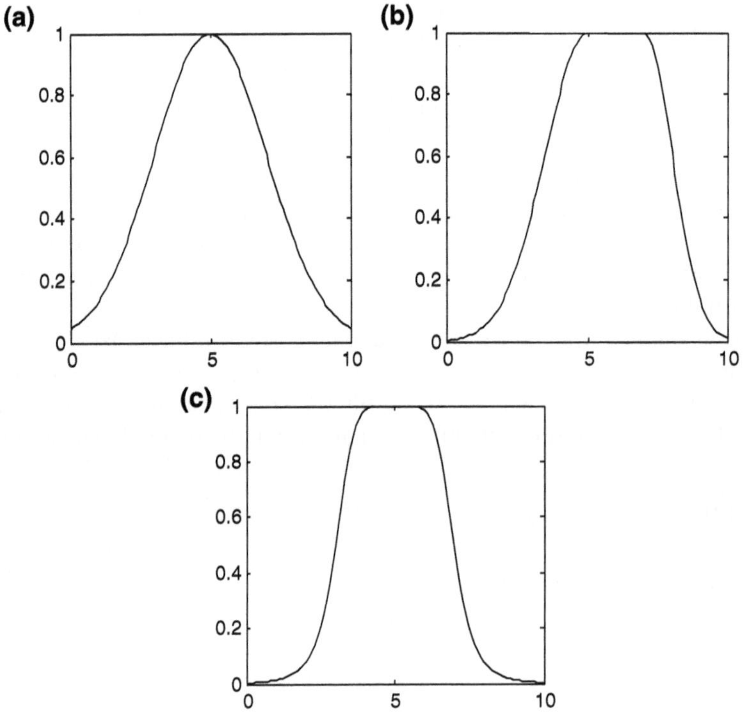

Fig. 4.2 Nonlinear MFs Gaussian MF = [2 5] (**a**); Two-sided Gaussian MF = [1.5 5 1 7] (**b**); Bell-shaped MF = [2 3 5] (**c**)

Two-Sided Gaussian MF

In two-sided Gaussian MF, there are two Gaussian functions. The first function $f_1(x)$, described by the parameters (σ_1, m_1) determines the shape of the left-side curve. The second function $f_2(x)$ described by the parameters (σ_2, m_2), determines the shape of the right-side curve.

$$f_1(x) = \begin{cases} \exp\left[-\frac{1}{2}\left(\frac{x-m_1}{\sigma_1}\right)^2\right], & x \le m_1 \\ 1 & \text{otherwise} \end{cases} \tag{4.7}$$

$$f_2(x) = \begin{cases} 1 & x \le m_2 \\ \exp\left[-\frac{1}{2}\left(\frac{x-m_2}{\sigma_2}\right)^2\right], & \text{otherwise} \end{cases} \tag{4.8}$$

Then the two-sided Gaussian membership function is defined by

$$\mu(x) = f_1(x) * f_2(x) \tag{4.9}$$

The argument x must be a real number or a non-empty vector of strictly increasing real numbers, and $\{\sigma_1, m_1\}$ and $\{\sigma_2, m_2\}$ must be real numbers. Two-sided Gaussian membership function always returns a continuously differentiable curve with values within the range of [0, 1]. Usually when $m_1 < m_2$, the two-sided Gaussian function is a normal membership function and has a maximum value of 1, with the rising curve identical to that of $f_1(x)$ and a falling curve identical to that of $f_2(x)$ defined above. If $m_1 \ge m_2$, the two-sided Gaussian membership function is a subnormal membership function and has a maximum value less than 1. The two-sided Gaussian MF is asymmetric and it is illustrated in Fig. 4.2b.

Bell-Shaped MF

A bell shaped MF is specified by three parameters $\{m, \sigma, a\}$ shown in Fig. 4.2c and defined by

$$\mu(x) = \frac{1}{1 + \left|\frac{x-m}{\sigma}\right|^{2a}} \tag{4.10}$$

The parameters m and σ represent the centre and width of the bell shaped MF respectively. Parameter a, usually positive, controls the slope of the MF at crossover point. Bell-shaped MF is symmetric and has the features of being smooth and non-zero at all points.

4.3.3 Sigmoidal MF

A sigmoidal MF, which is either open left or right, are useful in many applications. A sigmoidal MF is specified by two parameters $\{a, c\}$ shown in Fig. 4.3a, b and defined as

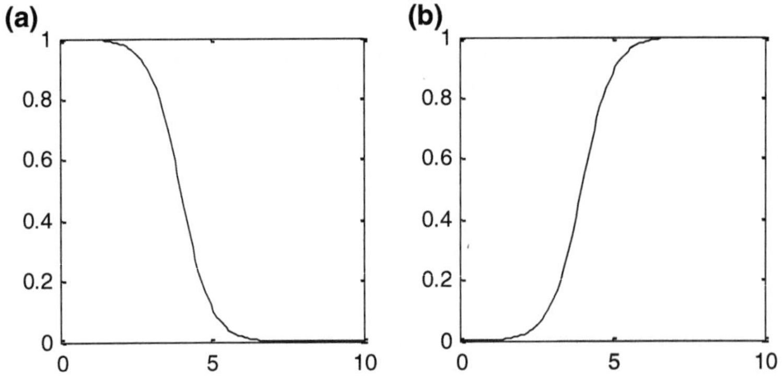

Fig. 4.3 Sigmoidal MF *left open* MF = [−2 4] (**a**); *right open* MF = [2 4] (**b**)

$$\mu(x) = \frac{1}{1 + e^{-a(x-c)}} \tag{4.11}$$

The parameter c is the centre of the sigmoidal function. The sign of the parameter a determines the spread of the sigmoidal membership function, i.e., inherently open to the right or to the left. Thus the parameter a is appropriate for representing concepts of linguistic hedges such as 'very large' or 'more or less small'. If a is positive number, the MF will open to the right. If a is negative number, the MF will open to the left. This simple property of the sigmoidal MF helps to represent the fuzzy concepts such as 'very large positive' or 'very large negative' in linguistic terms. Further parameterised sigmoidal MFs and their usage can be found in Jang et al. (1997).

Closed symmetrical or asymmetrical sigmoidal MFs (i.e. not open to the right or left) can be constructed by using either the difference or product of the two sigmoidal MFs described above. The MF formed by the difference between the two sigmoidal MFs is defined as difference-sigmoid, and the MF formed by the product of these is defined as product-sigmoid. Both the MFs in this family are smooth and non-zero at all points.

Difference Sigmoidal MF

The difference sigmoidal function depends on four parameters, a1, c1, a2, and c2, and is the difference between two sigmoidal functions defined by

$$f_1(x, a_1, c_1) - f_2(x, a_2, c_2) = \frac{1}{1 + e^{-a_1(x-c_1)}} - \frac{1}{1 + e^{-a_2(x-c_2)}} \tag{4.12}$$

The difference sigmoidal function is shown in Fig. 4.4a. The parameters of difference sigmoidal are listed as $[a_1, c_1, a_2, c_2]$.

Product Sigmoidal MF

The product sigmoidal function is simply the product of two sigmoidal curves defined by

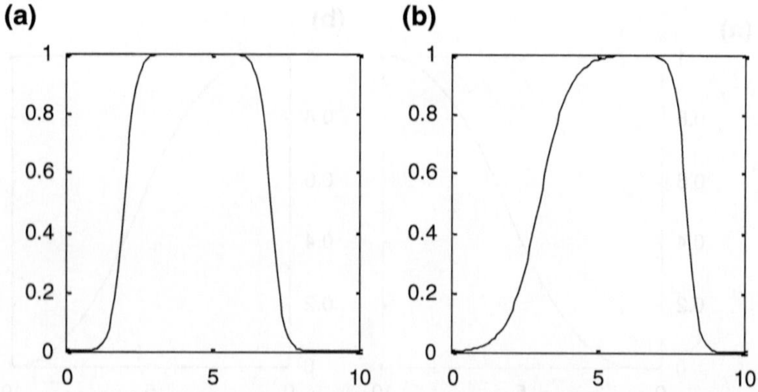

Fig. 4.4 Sigmoidal MF Difference sigmoidal MF = [5 2 5 7] (**a**); Product sigmoidal MF = [2 3 −5 8] (**b**)

$$f_1(x, a_1, c_1) * f_2(x, a_2, c_2) = \frac{1}{1 + e^{-a_1(x-c_1)}} * \frac{1}{1 + e^{-a_2(x-c_2)}} \tag{4.13}$$

The product sigmoidal function is shown in Fig. 4.4b. The parameters of product sigmoidal are listed as $[a_1, c_1, a_2, c_2]$.

In neuro-fuzzy control, particularly where neural network techniques are used to tune and implement a fuzzy controller, generally sigmoidal type MFs have been found very useful.

4.3.4 Polynomial or Spline-Based Functions

Three polynomial or spline-based MFs are also in wide use. They are defined as polynomial-Z, polynomial-S and polynomial-π. They are named according to their shapes. The Z- and S-shaped MFs are always asymmetric but the function π-shaped may be symmetric or asymmetric.

Z-Shaped and S-Shaped MF

There are also two spline-based functions z-shaped and s-shaped MFs. They are so named because of their Z-shape and S-shape (also called Zadeh's S-function) (Driankov et al. 1993). The parameters a and b locate the extremes of the sloped portion of the curve. Z-shaped MF is defined by

$$\mu(x) = \begin{cases} 1, & x \le a \\ 1 - 2\left(\frac{x-a}{b-a}\right)^2, & a \le x \le \frac{a+b}{2} \\ 2\left(b - \frac{x}{b-a}\right)^2, & \frac{a+b}{2} \le x \le b \\ 0, & b \le x \end{cases} \tag{4.14}$$

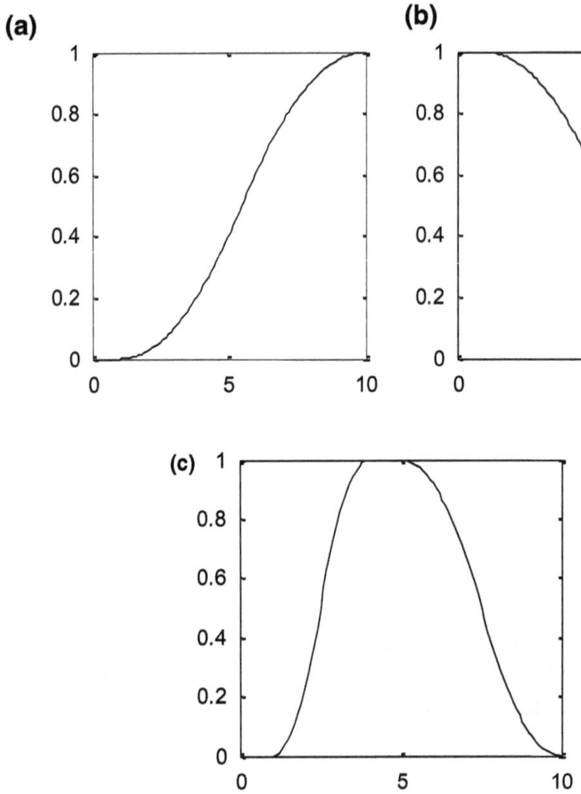

Fig. 4.5 Spline based functions S-shaped MF = [1 10] (**a**); Z-shaped MF = [1 10] (**b**); π-shaped MF = [1 4 5 10] (**c**)

Zadeh's S-shaped function is defined by

$$\mu(x) = \begin{cases} 0, & x \le a \\ 2\left(\frac{x-a}{b-a}\right)^2, & a < x \le \frac{a+b}{2} \\ 1 - 2\left(\frac{x-b}{b-a}\right)^2, & \frac{a+b}{2} < x \le b \\ 1, & x > b \end{cases} \tag{4.15}$$

The S-shaped and Z-shaped functions are shown in Fig. 4.4a, b. The parameters of S-shaped and Z-shaped functions are listed as $[a, b]$.

Π-Shaped MF

The π-shaped membership function is a spline-based curve. It is named π-shape because of its shape. The parameters a and d locate the 'feet' of the curve, while b and c locate its 'shoulders'. The membership function is a product of S-shaped and Z-shaped membership function defined by

$$\mu(x) = \begin{cases} 0, & x \le a \\ 2\left(\frac{x-a}{b-a}\right)^2, & a \le x \le \frac{a+b}{2} \\ 1 - 2\left(\frac{x-b}{b-a}\right)^2, & \frac{a+b}{2} \le x \le b \\ 1 & b \le x \le c \\ 1 - 2\left(\frac{x-c}{d-c}\right)^2, & c \le x \le \frac{c+d}{2} \\ 2\left(\frac{x-d}{d-c}\right)^2, & \frac{c+d}{2} \le x \le d \\ 0, & x \ge d \end{cases} \tag{4.16}$$

π-shaped MF is shown in Fig. 4.5c. The parameters of π-shaped functions are listed as $[a, d, b, c]$.

4.3.5 Irregular Shaped MF

In general, the type of the MFs for an application is chosen by an expert, which very often constraining their modelling capability. Clearly the predefined shape of the MF may not be adequate or sufficiently flexible to represent various types of data distribution. In most of the applications, designers choose symmetric MFs. Symmetric MFs do not always suit applications as they may not represent the actual data distribution. Some designers, therefore, favour asymmetric MFs as they are flexible and may represent data distribution better (Russo 1998; Murata et al. 1999). For example, an asymmetric sigmoidal MF is shown Fig. 4.4b, an asymmetric π-shaped MF is shown in Fig. 4.5c, and asymmetric triangular and traperzoidal MFs are shown in Fig. 4.6a, b.

In certain applications, an aggregation of several basic-shaped MFs may be useful to capture the underlying data distribution exhibited by the numerical observations Huang et al. (2006). Because the representation capability of the common shaped MFs is limited by the basic shape constraint, more generic forms of MFs should be explored further. Klir and Yuan (1995) provided some examples

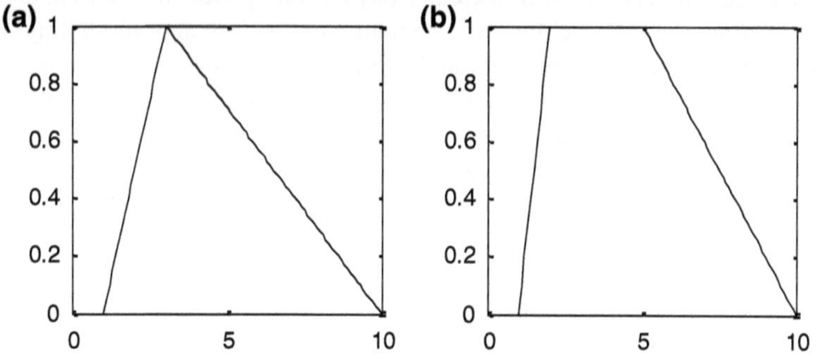

Fig. 4.6 Asymmetric MFs asymmetric triangular MF = [1 3 10] (**a**); asymmetric trapezoidal MF = [1 2 5 10] (**b**)

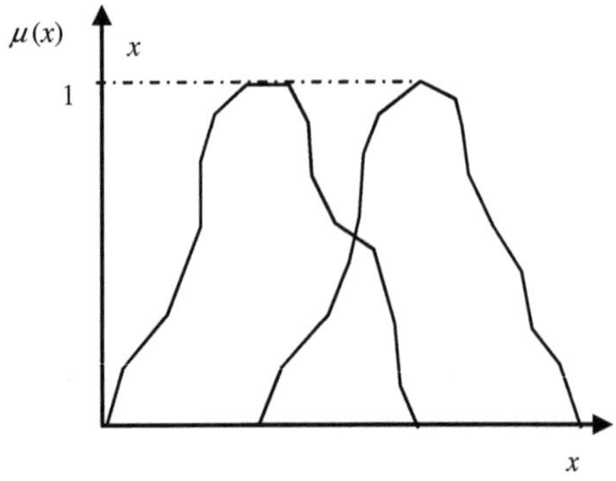

Fig. 4.7 MFs of irregular shape

of piecewise linear MFs determined by several sampling points. Given enough sampling points, MFs can be defined that represent approximately close enough to the common type MFs. Examples of irregular shaped MFs are shown in Fig. 4.7. Piecewise linear MFs are also more flexible as they can represent shapes that are not possible for common type MFs (Huang et al. 2006).

Another approach for enhancing the modelling ability of MFs is that of Type-2 membership functions, first proposed by Zadeh (1975a, b) and further developed by Mendel (2001). However, Type-2 MFs are difficult to apply for a general class of fuzzy systems because of their inherent complex nature and they are also not well defined mathematically as of yet. Although both piecewise linear and Type-2 MFs possess great representation capability, they can hardly be found in existing applications in the literature on fuzzy systems and control. The reason is that it is near impossible to manually specify such MFs and very difficult to derive them from training data (Huang et al. 2006). Type-2 fuzzy system is beyond scope of this book and the Type-2 MFs will not be discussed any further. Interested readers are directed to Castilo and Melin (2008); Mendel (2001); Pedrycz and Gomide (1998).

So far in the literature, there has been very little study on the systematic analysis, evaluation, comparison of different types of MFs for determining the type of MFs to be used in fuzzy control and other applications. The general consensus is that it is rather application dependent. In the early days of fuzzy system research, designers used to define the MFs either by manually extracting and modelling human expertise or through a tedious trial-and-error process. However, in most complicated real-world problems, the number of inputs is large and the search space is complex, which makes manual derivation of MFs very difficult or just impossible for experts. Therefore, the researchers have been applying different techniques that can automatically derive appropriate shape of the MFs from data. The shape of the MFs mainly depends on the estimated parameters, which greatly influences the performance of a fuzzy system. Heuristic selection of parameters of MFs are widely used and practised in fuzzy control applications. Beside that many other techniques have been proposed which reflect the actual data distribution by using supervised and unsupervised learning algorithms. Learning MFs are not only important but also feasible where some input/output data are available. There are different

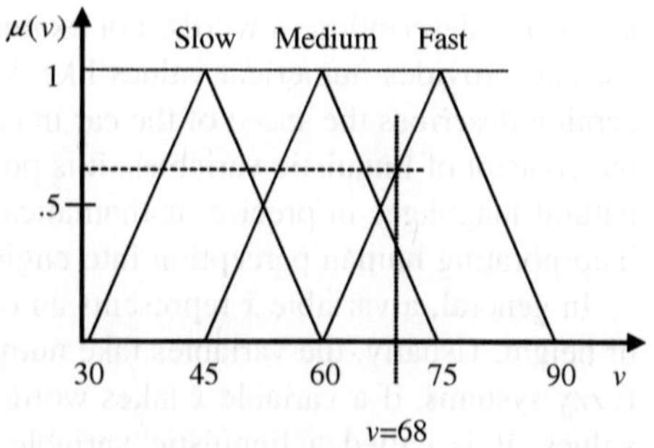

Fig. 4.8 Relation between numerical and linguistic variables in an overlapping MFs

approaches to the determination of the parameters of the MFs, such as heuristic selection, clustering approach, C-means clustering approach (Bezdek 1981; Bezdek et al. 1999), adaptive vector quantization (Kohonen 1989), Modified Linear Vector Quantization (MLVQ), Fuzzy Kohonen Partitioning (FKP), Pscudo Fuzzy Kohonen Partitioning (PFKP) (Ang et al. 2003), Generalized Linear Vector Quantization (GLVQ) (Pal et al. 1995) and self-organising map (Kohonen 1989). To further enhance the system performance, generated MFs can be further tuned by any other techniques such as gradient descent using neural networks for instance.

Since Zadeh introduced fuzzy sets, the main difficulties have been with the meaning and measurement of MFs as well as their extraction, modification and adaptation to dynamically changing conditions. There is no general rule for choice of membership functions and mainly depends on the problem domain. In fuzzy control applications, use of narrower MFs results in faster response but causes larger oscillations, overshoot and settling time. Gaussian and bell-shaped membership functions involve calculation of exponential terms and require substantial processing time. Trapezoidal MFs have four parameters that result in increase of optimisation parameters and lengthen the procedure. Triangular MFs are simple with three parameters. Therefore, they are the best choice and used for simplicity.

In general, if cross-point of two adjacent MFs is 0.5 as shown in Fig. 4.8, then a fuzzy controller provides faster rise-time, significantly less overshoot and less undershoot. In such cases, the shape of MFs does not play a dominant role in the performance of a fuzzy system. It is found that trapezoidal shape of MFs causes slower rise-time (Driankov et al. 1993). Though these results are empirical in nature but in general an overlap of two adjacent MFs at cross-point of 0.5 provides best results, which has been reported by many researchers (Boverie et al. 1991; Yager and Filev 1994; Margaliot and Langholz 2000; Kovacic and Bogdan 2006).

4.4 Linguistic Variables

The notion of linguistic variables is the most fundamental elements in human perception and its representation. Sensors usually measure a variable and provide a numerical value to users. Human perception evaluates a variable in linguistic terms, i.e. description in words. For example, a radar gun measures the speed of a car and provides numerical values like 33, 62 mph, etc. whereas the human perception describes the speed of the car in natural language like slow, fast etc. Using the concept of linguistic variables, it is possible to formulate vague descriptions in natural languages in precise mathematical terms. Thus, linguistic variables allow incorporating human perception into engineering systems in a systematic manner.

In general, a variable x represents an entity such as temperature, speed, weight or height. Usually, the variables take numerical values of real or complex type. In fuzzy systems, if a variable x takes words or sentences in natural languages as its values, it is called a linguistic variable. The words or linguistic variables are characterised by fuzzy sets defined in the universe of discourse in which the variable is defined. This is a simple and intuitive definition for linguistic variables. In other words, a linguistic variable is used as labels of fuzzy subsets (Zadeh 1972, 1975a, b, 1976). Such linguistic variables serve as a means of approximate characterisation of systems which cannot be described precisely by numerical values or other traditional quantitative terms. For example, speed is represented by the variable $v \in [v_{\min} = 30, v_{\max} = 90]$. If v is a linguistic variable, its values can be slow, medium, fast, not slow, very fast, not very slow etc. In this case, fast is a linguistic value of speed and is imprecise compared with exact numeric value such as 'speed is 68 mph'. The relation between a numerical variable $v = 68$ and linguistic variables Slow, Medium and Fast is illustrated graphically in Fig. 4.8.

The above definition gives a simple and intuitive definition for linguistic variables. In the literature of fuzzy theory, a more formal definition of linguistic variables was usually employed (Zadeh 1973, 1975a, b). In general, a linguistic variable is characterised by a quintuple $\{X, T, U, G, M\}$ where X is the name of the variable e.g. Speed, T denotes the term set of X, i.e. the set of names of linguistic labels of X over a universe of discourse U, e.g., slow, medium, fast etc., G is the syntactic rule or grammar for generating names, and M is the semantic rule for associating with each X its meaning, $M(X) \subseteq U$ (Zadeh 1975a). From the above definitions, the linguistic variables can be seen as extensions of numerical variables in the sense that they are allowed to take fuzzy sets as their values. The fuzzy sets for three linguistic values are illustrated in Fig. 4.8.

4.5 Features of Linguistic Variables

Theoretically, the term set $T(X)$ is infinite but in practical applications $T(X)$ is defined with a small number of terms, so that each elements of $T(X)$ defines a mapping between each element and the function $M(X)$, which associates a meaning with each term in the term set. Let the term set of the linguistic variable Speed be $\{slow,\ medium,\ fast\}$ within the universe of discourse $U = [0,\ 120]$. The term set can be expressed as

$$T(Speed) = \{slow,\ medium, fast\} \tag{4.17}$$

Fig. 4.9 MF of the linguistic variable for 'Slow or Medium'

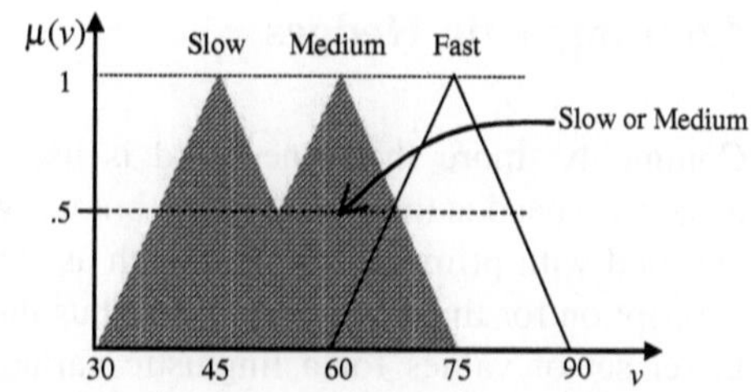

Fig. 4.10 MF of the linguistic variable for "Medium and Fast"

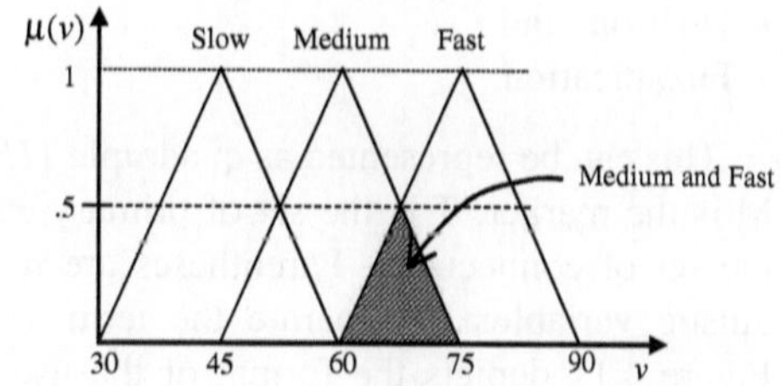

The semantic rule of linguistic variables can be expressed using context-free grammar. For example

$$T = \{slow, very\ slow, very\ very\ slow, \ldots\} \tag{4.18}$$

Using context-free grammar the above expression can be written as

$$\begin{aligned} T &\rightarrow slow \\ T &\rightarrow very\ T \end{aligned} \tag{4.19}$$

Here 'very' is called a linguistic hedge, which is used to derive new linguistic variables. Linguistic hedge will be discussed further in the next section.

A linguistic variable is thus a perception expressed in natural language using words or sentences, e.g. Slow OR Medium, Medium AND Fast. The words such as 'slow', 'medium' and 'fast' together with the connectives 'OR' and 'AND' created new expressions, which is completely a new set comprises of the term sets. The linguistic variable 'Slow OR Medium' is shown graphically in Fig. 4.9. It is the shaded area representing the union of the membership functions for 'Slow OR Medium'.

The linguistic variable 'Medium AND Fast' is shown graphically in Fig. 4.10. It is the shaded area representing the intersection of the membership functions Slow and Medium.

In the above examples, OR and AND are connectives, which play an important role in the description of linguistic variables. It can be seen from the Figs. 4.9 and 4.10 that they are used to derive new linguistic variables from the term sets. The role of connectives is discussed further in the use of linguistic hedges section presented in the next section.

4.6 Linguistic Hedges

Commonly, more than one word is used to describe a linguistic variable, for example, speed is more or less fast but not very fast. Here 'more or less' and 'very' are used with primary term sets such as 'fast' or 'slow' to produce more accurate description for linguistic variables. Thus the purpose of the hedges is to generate a larger set of values for a linguistic variable from a small collection of primary terms (Zadeh 1972). Hedges are realised on primary terms through the processes:

- Intensification or concentration,
- Dilation, and
- Fuzzification

This can be represented as quadruple $\{H, M, T, C\}$ where H is the set of hedges, M is the marker, T is the set of primary terms, e.g., slow, medium, fast etc., C is the set of connectives. Parentheses are used as markers in the definition of linguistic variables to separate the term set from the hedge, e.g. Very (Small). Figure 4.11 depicts the format of the use of the different term sets, hedges and connectives for defining linguistic variables.

An example applying the structure shown in Fig. 4.11 would be: Big but Not Very (Big). Here 'Big' is a primary term set, 'But' is a connective (which means AND in this case) and 'Very' is a hedge. 'Not' is a complement operation on term set. Parenthesis '()'is used as a marker to separate hedge and term set.

The application of the linguistic hedges 'very' and 'more or less' are demonstrated through the concentration (or intensification) and dilation process in Fig. 4.12.

The hedge 'very' is a concentration (or intensification) operation defined as

$$\text{Very (Nice)} = \text{Nice}^2 = [\mu_{Nice}(x)]^2 \tag{4.20}$$

The hedge 'more or less' is a dilation operation defined as

$$\text{More or less (Nice)} = \text{Nice}^{1/2} = [\mu_{Nice}(x)]^{1/2} \tag{4.21}$$

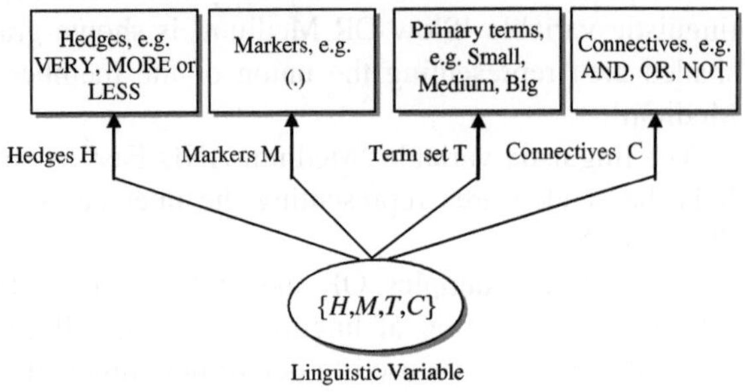

Fig. 4.11 Structure of linguistic variables with hedges

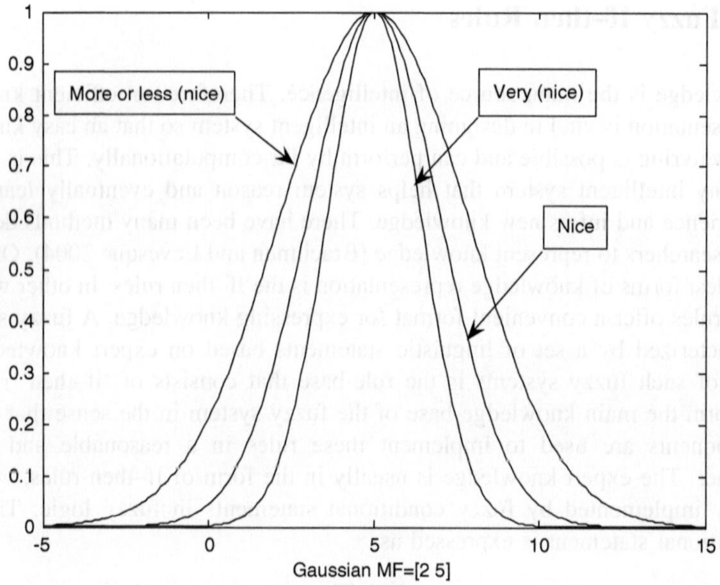

Fig. 4.12 Dilation and concentration (or intensification)

The operation of multiple hedges can result in the same primary fuzzy set. For example, the operation of the hedges 'more or less very nice' is represented by the following expression.

$$\text{More or less (Very (Nice)} = \text{More or less}\left(\text{Nice}^2\right) = \left(\text{Nice}^2\right)^{1/2} = \text{Nice} \tag{4.22}$$

It can be seen that the operation of the hedges on the primary term set 'Nice' in (4.22) resulted in the same primary fuzzy set 'Nice'. A linguistic variable can be used with more than one hedge, for example

Almost very good but generally above good
Close to 100m but not very high
Not more than about zero

The equivalent versions of the above linguistic variables with markers can be expressed as

Almost (Very (good)) AND Generally (above good)
Close (100m) AND Not (Very (High))
Not (More than (About zero))

Linguistic variables and hedges allow constructing mathematical models for expression of natural language. These models can then be used to formulate process rules, computer programs and simulate behaviour of real world process.

4.7 Fuzzy If–then Rules

Knowledge is the main source of intelligence. Therefore, an efficient knowledge representation is vital in designing an intelligent system so that an easy knowledge manoeuvring is possible and can perform by far computationally. This is essential for any intelligent system that helps system reason and eventually learns from experience and infers new knowledge. There have been many methods developed by researchers to represent knowledge (Brachman and Levesque 2004). One of the simplest forms of knowledge representation is the If–then rules. In other words If–then rules offer a convenient format for expressing knowledge. A fuzzy system is characterized by a set of linguistic statements based on expert knowledge. The core of such fuzzy systems is the rule-base that consists of 'if–then' rules and conform the main knowledge-base of the fuzzy system in the sense that all other components are used to implement these rules in a reasonable and efficient manner. The expert knowledge is usually in the form of If–then rules, which are easily implemented by fuzzy conditional statements in fuzzy logic. The fuzzy conditional statement is expressed as

$$\text{If } \langle \textit{fuzzy proposition} \rangle, \text{Then } \langle \textit{fuzzy proposition} \rangle \tag{4.23}$$

The collection of fuzzy conditional statements forms the rule-base or the rule set of a fuzzy system.

4.7.1 Fuzzy Proposition

A fuzzy proposition is a statement which can have a value within the interval of $[0, 1]$. There are two types of fuzzy propositions: atomic proposition, and compound proposition. An *atomic proposition* is a single statement whereas a *compound proposition* is a composition of more than one atomic propositions using the connectives 'and', 'or', and 'not' which represent fuzzy intersection, fuzzy union, and fuzzy complement, respectively. For example, if x represents the speed of the car, then the following is fuzzy propositions:

$$x \text{ is } A \tag{4.24}$$

$$x \text{ is } B \textit{ or } x \text{ is } \textit{not } C \tag{4.25}$$

$$x \text{ is } \textit{not } B \textit{ and } x \text{ is } \textit{not } C \tag{4.26}$$

$$(x \text{ is } B \textit{ and } x \text{ is } \textit{not } C) \textit{ or } x \text{ is } C \tag{4.27}$$

where x is a linguistic variable, and A, B, and C are linguistic values of x (that is, A, B, and C are fuzzy sets defined within the physical domain of x) and denote the fuzzy sets 'slow', 'medium', and 'fast' respectively. Finally, the fuzzy rule-base comprises of the fuzzy propositions of the form:

$$\text{If } x_1 \text{ is } A_i \text{ and } x_2 \text{ is } B_j \text{ and } x_3 \text{ is } C_k, \text{ Then y is } Z_l. \tag{4.28}$$

where A_i, B_j, C_k and Z_l are fuzzy sets of the linguistic variables x1, x2, x3, and y in the respective universe of discourses. Let M be the number of rules in the fuzzy rule-base, that is, $r = 1, 2, 3, \ldots, M$ in (4.28). The rules in the form of (4.28) are called canonical fuzzy If–Then rules.

4.7.2 Methods for Construction of Rule-Base

In general, the fuzzy 'if–then' rules are derived by human experts by applying rules of physical laws and experience. Experts can map the inputs and outputs to generate rules for fuzzy system with few inputs and outputs. The rule-base is the mapping between the input and the output spaces defined as

$$\Phi : X_1 \times X_2 \rightarrow Y \tag{4.29}$$

where X_1 and X_2 are inputs and Y is the output. The mapping Φ can be visualised pictorially as shown in Fig. 4.13. The mapping is sometimes a kind of intuition, which requires a trial and error process to refine it.

As the number of inputs and outputs grows, the rule-base grows drastically. In such cases, it is nearly impossible for a human expert to map the inputs and outputs of the system and leads to difficulty in defining the rules for the system. Therefore, a systematic approach for constructing rule-base seems demanding. The most common approach is to partition the input space. The simple way to generate a grid partitioning is to divide each input variable into a given number of intervals. The limits of intervals do not necessarily have any physical meaning and do not

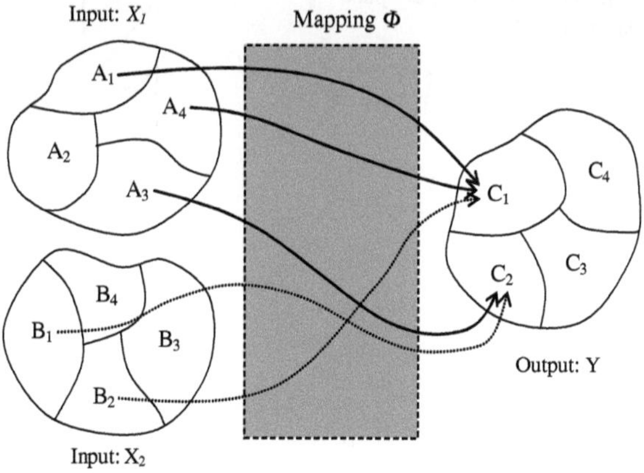

Fig. 4.13 Rule-base as a mapping

take into account the data density repartition function (Guillaume 2001; Wong and Lin 1997). The first and most intuitive approach implements all possible combinations of the given fuzzy sets as rules. For example, the error and change of error as inputs and torque as output of a fuzzy system (specifically a fuzzy controller) can be divided into suitable number of partitions, i.e. partitioned into finite number of fuzzy sets. For example:

$$\text{Error } X_1 = \{A_1, A_2, A_3, \ldots, A_N\} \subset E \tag{4.30}$$

$$\text{Change of error } X_2 = \{B_1, B_2, B_3, \ldots, B_M\} \subset \Delta E \tag{4.31}$$

$$\text{Torque } u = \{C_1, C_2, C_3, \ldots, C_L\} \subset U \tag{4.32}$$

where $X_1 \hat{=}$ error, $X_2 \hat{=}$ change of error, $u \hat{=}$ control input, A_i, $i = 1, 2, \ldots, N$, B_j, $j = 1, 2, \ldots, M$ are the input linguistic variables and C_k, $k = 1, 2, \ldots, L$ are the output linguistic variable, N, M and L are the maximum number of linguistic values (primary fuzzy sets defined by MFs). E, ΔE, and U are the universes of discourse for error, change of error and control input respectively. The rule-base of the fuzzy system is shown in Table 4.1, where C_r is defined as $C_r \in C_k = \{C_1, C_2, \ldots, C_L\}$ and each C_r is to be found from available data.

Table 4.1 Rule-base R

	Input X_2				
Input	U	B_1	B_2	...	B_M
X_1	A_1	C_r	C_r	...	C_r
	A_2	C_r	C_r	...	C_r
	⋮	⋮	⋮	⋱	⋮
	A_N	C_r	C_r	...	C_r

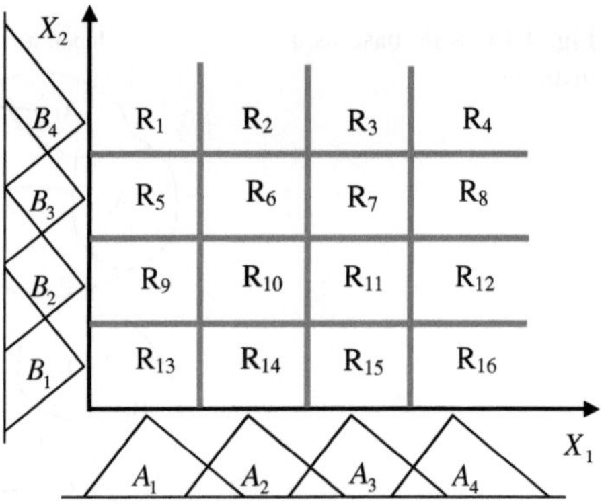

Fig. 4.14 Fuzzy input space partitioning

The nth rule for the two-input single-output system can be defined as:

$$R_n \text{ IF } (X_1 \text{ is } A_i) \text{ and } (X_2 \text{ is } B_j) \text{ THEN } u \text{ is } C_k \tag{4.33}$$

If X_1, X_2 and u are partitioned into 4 fuzzy sets, then there are 16 rules obtained from this uniform partitioning. A_i, B_j, and C_k with $i, j, k = 1, 2, \ldots, 4$ are primary fuzzy sets. Figure 4.14 shows a grid partitioning of input space for two-input single-output system, i.e. a two dimensional input space. The fuzzy rules R_n must be completed and covered by fuzzy partitioning the input space.

Initially, fuzzy rules are based on input–output data. Due to insufficient workspace coverage, some rules may never fire. The number of rules can be minimised by applying other partitioning strategies such as tree partitioning and scatter partitioning (Jang et al. 1997). However, the approach encounters problem of curse of dimensionality as the number of inputs increases.

Several methods have been proposed to generate fuzzy 'if–then' rules directly from numerical data. The generation of fuzzy if–then rules from numerical data involves (1) the fuzzy partition of a pattern space into fuzzy subspaces and (2) the determination of fuzzy if–then rule for each fuzzy partition. The classification phase follows next, where either the training data or the test data are classified using the fuzzy if–then rules generated. The performance of such a classification system depends on the choice of a fuzzy partition. If a fuzzy partition is too coarse, the performance may be low, because many patterns may be misclassified. On the other hand, if a fuzzy partition is too fine, many fuzzy if–then rules cannot be generated due to the lack of training patterns in the corresponding fuzzy subspaces. Therefore, the choice of a fuzzy partition is very important.

Unfortunately, there is no systematic approach to learning of rule-base of a fuzzy controller. Efforts have been made to automate the construction of rule-bases in various ways using clustering methods, neural networks (NNs) and genetic algorithms (GAs).

Some researchers apply product space clustering (Kosko 1992; Nie and Lee 1996; Setnes et al. 1998). Setnes et al. (1998) show that fuzzy rule-based models acquired from measurements can be both accurate and transparent by using a low number of rules. The rules are generated by product-space clustering and describe the system in terms of the characteristic local behaviour of the system in regions identified by the clustering algorithm. Nie and Lee (1996) proposed a two-step approach based on the notion of product space clustering. The rule-base is constructed in the first step by the principal algorithm. Three principal algorithms with self-organizing capability utilizing the concept of localized clustering or competitive learning are employed. Efforts have been put for the algorithms being fast, on-line, and simple with the capability of dealing with systems with nonlinear input–output relationships. The obtained rule-base may be further processed by a refining algorithm which makes the size of the rule-base smaller. Two algorithms have been developed for manipulating the obtained rule-base with novel data, one being a direct application of a fuzzy control algorithm and the other being an optimal algorithm in the sense of least square error with respect to an appropriately chosen cost function. It is also possible to apply some techniques developed in fuzzy clustering and learning vector quantization (LVQ) to rule-base construction. By replacing the simple competitive learning law in the LVQ with differential competitive learning laws, Kosko (1992) proposed an approach to building the rule-base in the product space.

Nefti (2002) purposed a merging method, which can be applied for membership functions simplification and also for rule base reductions in an unsupervised manner. Neural network-based solutions are proposed by Takagi and Hayashi (1991). Denna et al. (1999) used tabu search for the automatic definition of the fuzzy rules for a fuzzy controller and showed how the learning of fuzzy rule-base can be improved using such heuristic symbolic meta-rules.

4.7.3 Properties of Fuzzy Rules

Care must be taken while constructing fuzzy rules as there can be inconsistent or incomplete rules that will produce no useful output but will demand significant computation. Consider the rule-base in tabular form for a two-inputs and single output system shown in Fig. 4.15. The two inputs are error (e) and change of error (Δe) and single output is the control signal (u). For this simple example, four membership functions such as NB, NS, ZO and PB are used for each input and output. There are 16 rules in the rule-base table in Fig. 4.15. The rule $\{R_{16}\}$ is described as

$$R_{16}\colon \text{If}\{e \text{ is NB}\} \text{ and } \{\Delta e \text{ is PB}\}, \text{ Then } \{u \text{ is } \emptyset\} \tag{4.34}$$

That is, the rule R_{16} provides no output. The rule-base should provide an exact non-fuzzy value for the output u for each pair of crisp value of $\{e, \Delta e\}$. That is, the fuzzy rule R_i, $i = 1, 2, \ldots, 16$ must satisfy the following properties:

- Completeness
- Consistency
- Continuity

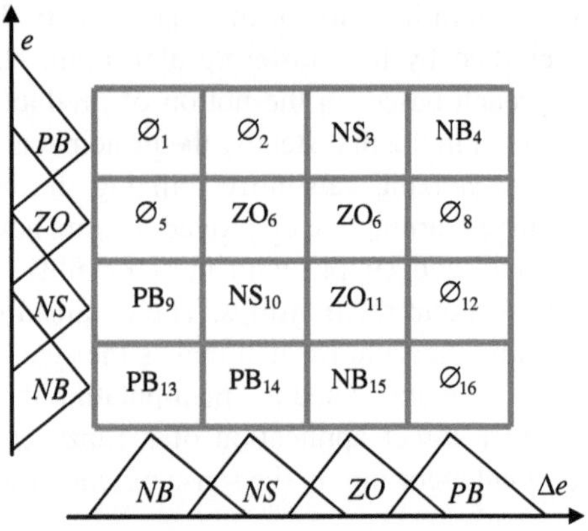

Fig. 4.15 Rule-base in tabular form

A set of fuzzy rules $\{R_i, i = 1, \ldots, 16\}$ is complete if any combination of input values results in a corresponding appropriate value for the output. It means that

$$\forall (e, \Delta e)\colon height(\Phi(e, \Delta e)) > 0 \tag{4.35}$$

where $height(.)$ of a fuzzy set is defined as $height(A) = \sup_{u \in X} \mu_A(u)$. The rule-base in Fig. 4.15 is incomplete as it does not provide any output value for the rule set $\{R_1, R_2, R_5, R_8, R_{12}, R_{16}\}$. In other words, the output for the rule set $\{R_i\}$ for $i = 1, 2, 5, 8, 12, 16$ is $\emptyset$.

Another feature of the rule-base is that all pairs of $\{e, \Delta e\}$ should provide a consistent output within the universe of discourse. A set of fuzzy rules $\{R_i, i = 1, \ldots, 16\}$ is consistent if it does not have any contradictory rule. That is, the set of fuzzy rules is inconsistent if two rules R_i and R_j with the same rule antecedent has mutually exclusive rule consequent or different consequent. For example, the two rules R_i and R_j in Eqs. (4.36) and (4.37) are conflicting as they provide different outputs for the same antecedent.

$$R_i\text{: If } e \text{ is ZO and } \Delta e \text{ is NB, Then } u \text{ is ZO} \tag{4.36}$$

$$R_j\text{: If } e \text{ is ZO and } \Delta e \text{ is NB, Then } u \text{ is NB} \tag{4.37}$$

Also the consequent fuzzy sets of the two rules are mutually exclusive, i.e. $ZO \cap NB = \emptyset$ that implies $height(ZO \cap NB) = 0$.

The consistency of rules yields another notion of continuity. That means all pairs of $\{e, \Delta e\}$ should provide a continuous output. A set of fuzzy rules $\{R_i, i = 1, \ldots, 16\}$ is continuous it does not have neighbouring rules with output fuzzy sets that have empty intersection. For example, the two rules R_{14} and R_{15} in Eqs. (4.38) and (4.39) are discontinuous because the consequent fuzzy sets are disjoint, i.e. $PB \cap NB = \emptyset$.

$$R_{14}\text{: If } e \text{ is NB and } \Delta e \text{ is NS, Then } u \text{ is PB} \tag{4.38}$$

$$R_{15}\text{: If } e \text{ is NB and } \Delta e \text{ is ZO, Then } u \text{ is NB} \tag{4.39}$$

Completeness, consistency and continuity in rule-base must be ensured while constructing it otherwise it can cause severe problem like instability or oscillatory behaviour, especially in fuzzy control systems.

4.8 Fuzzification

In fuzzy control applications, the observed data are usually crisp. Since the data manipulation in an FLC is based on fuzzy sets, fuzzification is necessary. Fuzzification performs a scale transformation of physical values of the current state variables into normalised universe of discourse. It means that it maps the normalised value of the control output variable onto physical domain. Therefore, fuzzification is defined as a mapping from an observed input space to fuzzy sets in

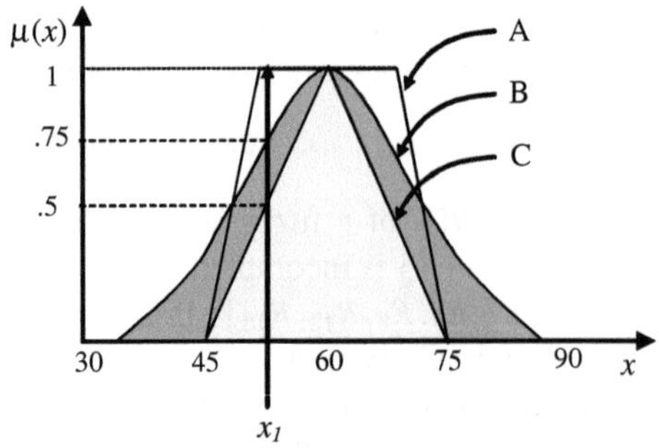

Fig. 4.16 Fuzzification of different types of MFs

certain input universe of discourse. This process consists of associating to each fuzzy set a membership function. These functions can be thought of as maps from the real numbers to the interval $I = [0, 1]$. If there are n fuzzy sets associated with a given quantity $x \in R$, such n maps $F_i : R \rightarrow I, \quad i = 1, \ldots, n$ are defined. They determine to what extent the label associated with fuzzy set Fi characterizes the current value of x. Fuzzification consists of associating a fuzzy vector with the quantity x by passing x through all the membership functions Ai providing grade of membership functions $\mu_i(x), i = 1, 2, \ldots, n$

$$F : R \rightarrow I^n \tag{4.40}$$

$$x \rightarrow \begin{bmatrix} A_1 \\ \vdots \\ A_n \end{bmatrix} = \begin{bmatrix} \mu_1(x) \\ \vdots \\ \mu_n(x) \end{bmatrix}. \tag{4.41}$$

The fuzzification of three different types of MFs is shown in Fig. 4.16. It is to be noted that due to the shape of the MFs the membership values are different for each MF, that is, $\mu_A(x_1) = 1$, $\mu_B(x_1) = 0.75$ and $\mu_C(x_1) = 0.5$ are different. This is again evident that the choice of MF plays an important role in any fuzzy system.

There are other views on fuzzification procedure reported in the literature Díaz-Hermida et al. (2005); Kalaykov (1998); Wang (1997). A promising idea was proposed by Díaz-Hermida et al. (2005) based on probabilistic quantifier fuzzification mechanism. They showed empirically the circumstances in which the new view on crisp representatives of fuzzy sets makes sense for quantification purposes and verified that the proposed quantification model performs better than other probabilistic fuzzification methods.

4.9 Inference Mechanism

Inference is the process of formulating a nonlinear mapping from a given input space to output space. The mapping then provides a basis from which decisions can be made. The process of fuzzy inference involves all input–output membership functions, fuzzy logic operators and if–then rules (Zadeh 1968, 1994).

There are three basic types of fuzzy inference, which have been widely employed in various control applications. The differences between these three fuzzy inferences, also called fuzzy models, lie in the consequents of their fuzzy rules, aggregations and defuzzification procedures. These fuzzy models are

a. Mamdani fuzzy inference
b. Sugeno fuzzy inference
c. Tsukamoto fuzzy inference

4.9.1 Mamdani Fuzzy Inference

The Mamdani type fuzzy modelling was first proposed as the first attempt to control a steam engine and boiler by a set of linguistic control rules obtained from experienced human operator (Mamdani and Assilian 1974). Figure 4.17 is an illustration of a two input-single output Mamdani type fuzzy model.

The choice of T-conorm (e.g. Max) and T-norm (e.g. min or product) operators can be max-min and max-product. Max-min is the most common rule of composition. In max-min rule of composition the inferred output of each rule is a fuzzy set chosen from the minimum firing strength. In max-product rule of composition the inferred output of each rule is a fuzzy set scaled down by its firing strength via algebraic product. A typical rule in Mamdani-type fuzzy model with two-input single-output has the form

$$\text{If } x \text{ is } A \text{ and } y \text{ is } B \text{ then } z \text{ is } C \tag{4.42}$$

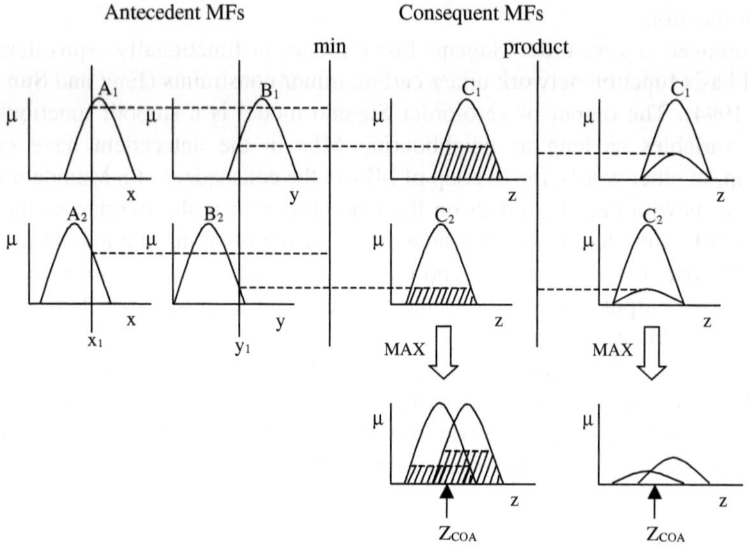

Fig. 4.17 Two-input single-output Mamdani fuzzy model

Where x and y are the input variables and z is the output variable. A, B, and C are the MFs of the corresponding input and output variables. In Mamdani's fuzzy model crisp values are used as inputs and defuzzification (see Sect. 4.10) is used to convert a fuzzy set to a crisp value.

4.9.2 Sugeno Fuzzy Inference

The Sugeno fuzzy model, also known as the TSK fuzzy model, was proposed by Takagi and Sugeno (1985); Sugeno and Kang (1988) in an effort to develop a systematic approach to generate fuzzy rules from a given input–output data set. Figure 4.18 illustrates a two-input single-output Sugeno fuzzy model. A typical fuzzy rule in Sugeno fuzzy model has the form

$$\text{If } x \text{ is } A \text{ and } y \text{ is } B \text{ then } z = f(x, y) \tag{4.43}$$

where A and B are MFs in the antecedent part, while $z = f(x, y)$ is a linear function in the consequent part. Usually $f(x, y)$ is polynomial in the input variables x and y but it can be any function as long as it can appropriately describe the output of the model within the fuzzy region specified by the antecedent of the rule. When $f(x, y)$ is a first-order polynomial, the resulting fuzzy inference system is called a first-order Sugeno fuzzy model which was proposed in Sugeno and Yasukawa (1993); Sugeno and Kang (1988); and Takagi and Sugeno (1985). When $f(x, y)$ is a constant, it is zero-order Sugeno fuzzy model, which can be considered as a special case of Mamdani fuzzy model, in which the consequent of each rule is specified by a fuzzy singleton or by a pre-defuzzified consequent or a special case of Tsukamoto fuzzy model in which the consequent of each rule is specified by an MF of a step function.

Moreover, a zero-order Sugeno fuzzy model is functionally equivalent to a radial basis function network under certain minor constraints (Jang and Sun 1993; Jang 1994). The output of zero-order Sugeno model is a smooth function of its input variables as long as neighbouring MFs in the antecedent have enough overlap. In other words, the overlap of MFs in the consequent of a Mamdani model does not have a decisive effect on the smoothness; it is the overlap of the antecedent MFs that determines the smoothness of the resulting input–output behaviour. The overall output of a Sugeno fuzzy model is obtained via weighted average of the crisp output, thus avoiding the time consuming process of defuzzification required by Mamdani model.

In practice, the weighted average operator is sometimes replaced with the weighted sum operator to reduce computation further, especially in the training of a fuzzy inference system. However, this simplification could lead to the loss of MF linguistic meaning unless the sum of firing strengths is close to unity.

4.9.3 Tsukamoto Fuzzy Inference

In the Tsukamoto fuzzy model, the consequent of each fuzzy If–then rule is represented by a fuzzy set with a monotonic MF (Tsukamoto 1979). As a result, the inferred output of each rule is defined as crisp value included by the rule's firing strength. The overall output is taken as the weighted average of each rule's output. Since each rule infers a crisp output, the Tsukamoto fuzzy model aggregates each rule's output by the method of weighted average and thus avoids the time-consuming process of defuzzification. Figure 4.19 illustrates a two-input single-output Tsukamoto fuzzy model. A typical rule in Tsukamoto-type fuzzy model with two-input single-output has the form same as (4.42).

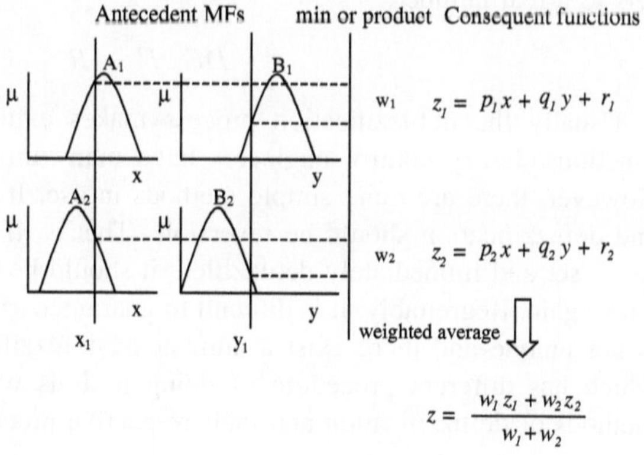

Fig. 4.18 Two-input single-output Sugeno fuzzy model

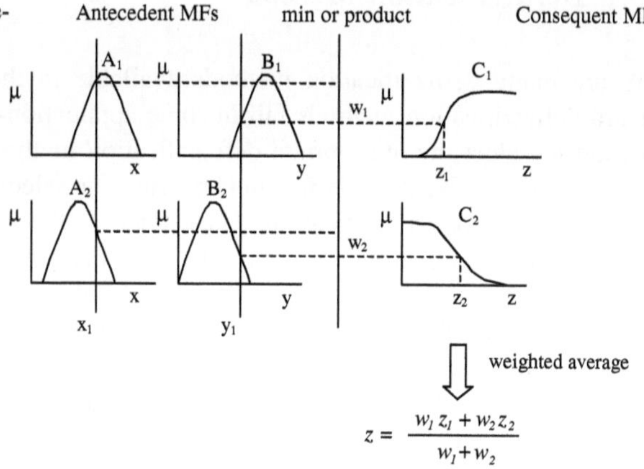

Fig. 4.19 Two-input single-output Tsukamoto fuzzy model

4.10 Defuzzification

Defuzzification is the reverse process of fuzzification. Mathematically, the defuzzification of a fuzzy set is the process of conversion of a fuzzy quantity into a crisp value. This is necessary when a crisp value is to be provided from a fuzzy system to the user. For example, if we develop a fuzzy system for blood pressure control, we will probably want to tell the user of what a blood pressure is expected to be in the next few hours.

Basically, defuzzification is a mapping from a space of fuzzy control actions defined over an output universe of discourse into a space of nonfuzzy (crisp) control actions. In a sense this is the inverse of the fuzzification even though mathematically the maps need not be inverses of one another. In general, defuzzification can be viewed as a map, DF, mapping a fuzzy vector with n fuzzy sets to a real number:

$$DF : I^n \rightarrow R \tag{4.44}$$

Usually the defuzzification process makes explicit use of the membership functions. Fuzzy control engineers have many different ways of defuzzifying. However, there are quite simple methods in use. It is intuitive that fuzzification and defuzzification should be reversible. That is, if a number is fuzzified into a fuzzy set and immediately defuzzified, it should be able to get the same number back again. Regrettably, it is difficult to guarantee as the defuzzification operation is not unique and there exist a number of defuzzification methods and each of which has different procedure of doing it. It is worth discussing the different methods of defuzzification and their respective procedures.

4.10.1 Defuzzification Methods

There are many defuzzification methods available in the literature. Very often standard defuzzification methods fail in some application domain. It is, therefore, important to select the appropriate defuzzification method for a particular application. Unfortunately, there is no standard rule for selecting a particular defuzzification method for an application. The choice of the most appropriate method depends on the application. A good study on the selection of appropriate defuzzification methods has been reported by Runkler (1997). In the following sections, some widely used methods of defuzzification are presented. A classification of defuzzification methods has been proposed by some researchers (Runkler 1997; van Leekwijck and Kerre 1999) as to evaluation of what properties are important for what types of applications and are classified into the following groups:

(a) Area methods: The defuzzification value divides the area under the membership function in two or more or less equal parts.

(b) Distribution methods and derivatives: Conversion of the membership functions into a probability distribution, and computation of the expected value. The main advantage of these approaches is continuity property.

(c) Maxima methods and derivatives: Selection of an element from the core of a fuzzy set as defuzzification value. The main advantage of these approaches is simplicity.

The area methods of defuzzification are mathematically elegant and are widely used in fuzzy control systems. There are a number of methods available under this category, namely:

(1) Bisector of area (BOA)
(2) Centre of gravity (COG) or centre of area (COA)
(3) Centre of sums (COS)

Bisector of Area

Bisector of area Z_{BOA} is a vertical line that divides the area into two equal areas defined as

$$\int_{\alpha}^{z_{BOA}} \mu_A(z)dz = \int_{z_{BOA}}^{\beta} \mu_A(z)dz \tag{4.45}$$

where $\alpha = \min\{z|z \in Z\}$ and $\beta = \max\{z|z \in Z\}$. It sometimes coincident with the centroid line.

Centre of Gravity or Centre of Area

Centre of gravity method of defuzzification is also referred to as centre of area or centroid method in the fuzzy literature. This is the most widely used defuzzification method. The centre of area method finds the centroid of the area under membership function. In the continuous case it is given by the expression as

$$z^* = \frac{\int \mu_c(z) \cdot zdz}{\int \mu_c(z)dz} \tag{4.46}$$

and for a discrete universe with m quantisation levels in the output it is given by

$$z^* = \frac{\sum_{i=1}^{m} \mu_c(z_i) \cdot z_i}{\sum_{i=1}^{m} \mu_c(z_i)} \tag{4.47}$$

Figure 4.20 shows this operation in a graphical way. The value z^* is the centroid of the area, which is the defuzzified value of the combined overlapped consequent fuzzy sets of the rule.

COG method of defuzzification is highly popular and is very often used as a standard defuzzification method in experimental as well as industrial controllers. The calculation needed to carry out the defuzzification using COG is time

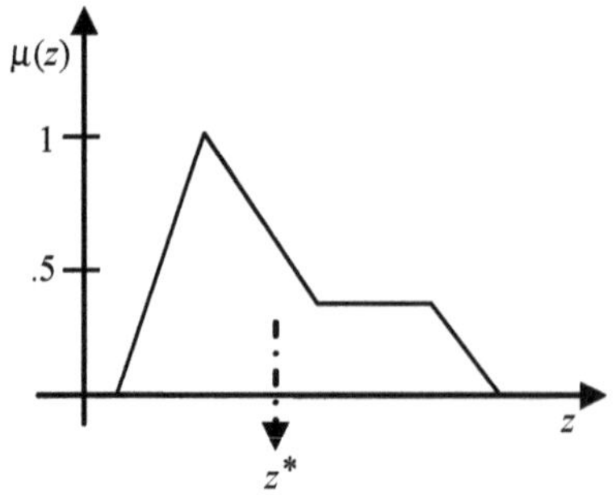

Fig. 4.20 Centre of gravity defuzzification

consuming as it applies numerical integration. It can be seen from Eqs. (4.46) and (4.47) that computation of z^* involves several mathematical operations. This method when used on-line not only requires more memory space to store the results of calculations but also takes more time to produce the output. It becomes more severe when the number of inputs and number of fired rules increase. Therefore it is very important to reduce the computational time and memory space requirements by somehow reducing the mathematical operations involved. Some numerical aspects of the centre of area method of defuzzification are reported by Patel and Mohan (2002) and different methods of performing COG reported by Van Broekhoven and De Baets (2006).

Centre of Sums

This process involves the algebraic sum of individual output fuzzy sets instead of computing the union of the fuzzy sets. Since it calculates the area of individual fuzzy sets, the method is faster than centre of gravity method. The defuzzified value z* is formally given in the discrete case by the expression in (4.48).

$$z^* = \frac{\sum_{i=1}^{m} z_i \cdot \sum_{k=1}^{n} \mu_k(z_i)}{\sum\limits_{i=1}^{m} \cdot \sum\limits_{k=1}^{n} \mu_k(z_i)} \tag{4.48}$$

One drawback of this method is that the overlapping area is added twice. There are two fuzzy sets C_1 and C_2 in Fig. 4.21 and the defuzzified value for the centre of sums method is shown in the Figure. The shaded area is the overlapped area of the fuzzy sets C_1 and C_2 which is calculated twice.

The core of a fuzzy set can be seen as the set of elements that best satisfy the property of the respective fuzzy set. Therefore, it is natural to have the defuzzification procedure select an element of the core of the overall fuzzy output set as defuzzified value (Runkler and Glesner 1994). The class of methods based on this notion is the maxima methods. A number of defuzzification methods are available that can select an element of the core as defuzzified value:

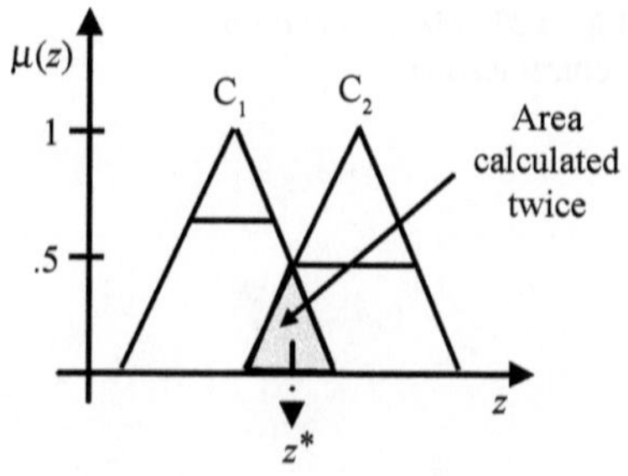

Fig. 4.21 Centre of sums defuzzification

1. Random Choice Of Maxima (RCOM),
2. First Of Maxima (FOM),
3. Last Of Maxima (LOM),
4. Middle Of Maxima (MOM)

However, in general these methods are not well suited for the use in fuzzy controllers because they cannot guarantee the continuity of the controller. Detail description of these methods can be found in van Leekwijck and Kerre (1999).

Van Leekwijck and Kerre (2001) proposed a new defuzzification method called Continuity Focused Choice Of Maxima (CFCOM). The basic idea of the method is to select an element of the core of the overall fuzzy output set in a way that the continuity of the fuzzy controller is assured. The CFCOM method uses a specific formula to select an element from the core, so that the continuity of the controller is guaranteed if the constraints for input terms, output terms, and rules are fulfilled.

There are a number of defuzzification methods reported in the literature that in fact first convert the membership function into a probability distribution and then compute the expected crisp value. These methods are grouped as distribution methods. The widely used methods under this category are:

(1) Centre of gravity (COG)
(2) Mean of maxima (MeOM)
(3) Basic defuzzification distributions (BADD)
(4) Generalized level set defuzzification (GLSD)
(5) Indexed centre of gravity (ICOG)
(6) Semi-linear defuzzification (SLIDE)
(7) Fuzzy mean (FM)
(8) Weighted fuzzy mean (WFM)
(9) Quality method (QM)
(10) Extended quality method (EQM)

Max-Membership Method

Also, known as the height method, the max-membership method is both simple and quick. This method takes the peak value of each fuzzy set and builds the weighted sum of these peak values. This method is given by the algebraic expression as

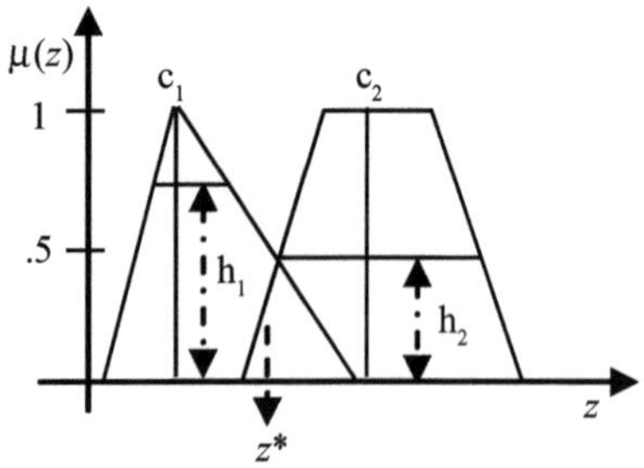

Fig. 4.22 Max-membership defuzzification

$$z^* = \frac{\sum_{k=1}^{m} c_k \cdot h_k}{\sum_{k=1}^{n} h_k} \tag{4.49}$$

Defuzzification using max-membership method is shown in Fig. 4.22. c_k is the peak value of the fuzzy sets and h_k is the height of the clipped fuzzy sets as shown in the Fig. 4.22.

Mean-Max Membership

Also known as the middle of maxima method, a single defuzzified output is generated by the mean or average of all local maxima defined by (4.50).

$$z^* = \frac{\sum_{i=1}^{N} \mu_{\max}(z_i)}{N} \tag{4.50}$$

where $\mu_{max}(z_i)$ is the maximum membership value and N is the number of times the membership function reaches the maximum support value. Figure 4.23 shows the two maxima a and b. The defuzzified output z^* is calculated from the mean of the two values as follows.

$$z^* = \frac{(a+b)}{2} \tag{4.51}$$

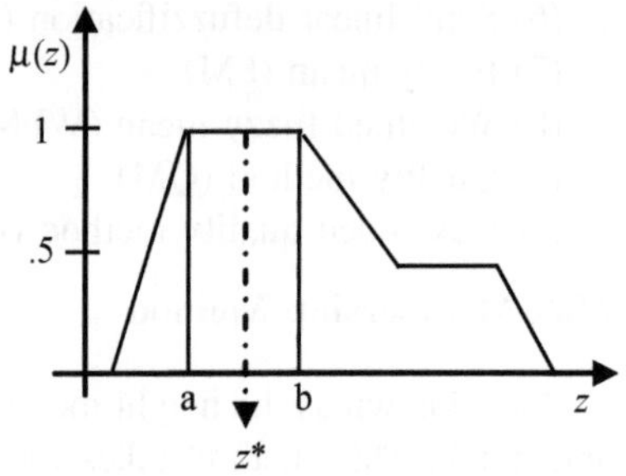

Fig. 4.23 Mean-max defuzzification

Mean of Maxima

Mean of maxima Z_{MeOM} is the average of the maximizing z at which the membership function reaches a maximum μ^*. Mathematically expressed as

$$z_{MeOM} = \frac{\int_{z'} z dz}{\int_{z'} dz} \tag{4.52}$$

where $Z' = \{z | \mu_A(z) = \mu^*\}$. In particular, if $\mu_A(z)$ has a single maximum at $z = z^*$, then $Z_{MeOM} = z^*$. Moreover, if $\mu_A(z)$ reaches its maximum whenever $z \in [z_{left}, z_{right}]$, then $z_{MeOM} = (z_{left} + z_{right})/2$.

It has been reported by Yager and Filev (1994) that the MeOM method of defuzzification produces poor performance at steady-state for fuzzy controllers, and yields a less smooth response curve compared to the COG method.

Weighted Average Method

This method is suitable for symmetrical membership functions. It is given by the algebraic expression as

$$z^* = \frac{\sum \mu_c(z) \cdot z'}{\sum \mu_c(z)} \tag{4.53}$$

where Σ denotes an algebraic sum. This is shown in Fig. 4.24. In the Figure, there are two trapezoidal membership functions A and B. $\mu_A(z) = 1$ is the weight for $z' = a$ and $\mu_B(x) = 0.5$ is the weight for $z' = b$. The defuzzified value of the two clipped trapezoidal MFs can be calculated using the Eq. (4.54):

$$z^* = \frac{\{a(1) + b(0.5)\}}{(1 + 0.5)} \tag{4.54}$$

There are a number of methods reported in the literature, which do not belong to the any of the groups discussed above. These methods are discussed briefly in the following section.

Fig. 4.24 Weighted average defuzzification

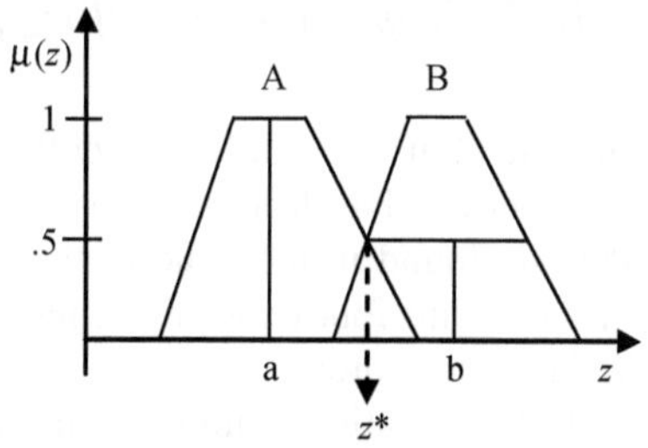

Constraint Decision Defuzzification (CDD)

CDD method of defuzzification proposed by Runkler and Glesner (1994) adds an extra fuzzy component to the fuzzy system that cooperates with a basic defuzzification operator such as MOM discussed earlier. The parameters of this method consist of the fuzzy constraints and the basic defuzzification method. The fuzzy constraints are related only to the defuzzification process itself. For example, the defuzzification value should be big or the defuzzification value should be near the COA.

Fuzzy Clustering Defuzzification (FCD)

Genther et al. (1994) proposed the FCD method of defuzzification based on fuzzy clustering. Fuzzy clustering partitions a set of input data into a number of clusters. Every element gets a membership degree in every cluster. The fuzzy C-Means algorithm computes the centre of each cluster and gives it a membership degree. The FCD method selects the centre with the greatest membership degree as defuzzification value. The parameter in this method is the number of groups that the fuzzy C-Means has to put the input data in. Genther et al. shows that the choice of the number of groups has a decisive influence on the defuzzified value.

Many researchers have been working on the choice of defuzzification methods and their performance in fuzzy systems, especially in fuzzy controllers. A set of criteria is to be found to evaluate the defuzzification methods and identify the properties of the methods that determine the suitability for specific applications. As for example, continuity and computational efficiency are of utmost importance for fuzzy controllers, but they are far less critical for decision support systems.

Some researchers have formulated a set of reasoning and structural basis for choice of defuzzification methods. Among the well-known methods, Yager and Filev (1994) have contributed to the understanding of the process of defuzzification from the perspective of invariant transformations between different uncertainty paradigms. Similarly, Roychowdhury and Wang (1994) have attempted to understand the problem of defuzzification from the scope of optimal selection of an element from a fuzzy set.

4.10.2 Properties of Defuzzification

In general, it is assumed that defuzzification procedure transforms a fuzzy set into a numeric value. Runkler and Glesner (1993) identified a set of thirteen features that are found in most the of the defuzzification methods. These features can be grouped into four core properties of defuzzification (Roychowdhury and Pedrycz 2001). These are:

Defuzzification computes one numeric value: This implies that the defuzzification operation is always injective. Therefore, two fuzzy sets can have the same defuzzified value. It is also assumed that the defuzzified value is always within the support set.

Defuzzified value determined by MF: Very often, MF is critical in determining the defuzzified value. Concentration of a fuzzy set monotonically leads to the normal of a fuzzy set. Similarly, the dilation operator monotonically leads the defuzzified value away from the of the fuzzy set. Neither scaling nor translation of fuzzy sets affects the MF. Therefore, defuzzified values do not get scaled or translated.

Defuzzified value of two triangular MFs contained within the bounds of individual defuzzified values: If fuzzy set $C_f = T(A_f, B_f)$ where A_f and B_f are fuzzy sets and T is the T norm, $Def(A_f) \leq Def(C_f) \leq Def(B_f)$, and so it is true for T-conorm (T^*) $C_f = T^*(A_f, B_f)$.

Defuzzified value falls in the permitted zone in the case of prohibitive information: It is not unusual in any specific applications where a strange fuzzy set may be inferred from the inference mechanism. Standard defuzzification mechanisms do not work on fuzzy inference generated two peaked fuzzy sets.

4.10.3 Analysis of Defuzzification Methods

Very often standard defuzzification methods fail in some application domain. It is, therefore, important to select the appropriate defuzzification method for a particular application. Unfortunately, there is no standard rule for selecting a particular defuzzification method for an application. The choice of the most appropriate method depends on the application. A good study on the selection of appropriate defuzzification methods has been reported by Runkler (1997).

The defuzzification is defined as a mapping from fuzzy set B in $Z \subset R$ (which is the output of the fuzzy inference mechanism) to crisp point $z^* \in Z$. In other words, defuzzification is the process of specifying a point in Z that best represents the fuzzy set B. There are a number of choices in determining the representative point $z^* \in Z$. In a fuzzy controller, the final control output depends not only on the rule-base but also on the chosen inference mechanism, e.g. Mamdani, Takagi–Sugeno or Tsukamoto-type and defuzzification method. Choice of appropriate defuzzification method is an important performance factor of any fuzzy controller. Very often the fuzzy control designer uses different symmetric and/or asymmetric shaped output fuzzy MFs with a view to represent data distribution. Eventually, the existing defuzzification methods cause the fat shape dominance phenomenon, that is, the fat shaped MFs dominate the thin shaped MFs. The phenomenon escalates even further if the fuzzy rules are somehow inconsistent. Therefore, the following criteria should be taken into consideration while choosing a defuzzification method:

- Plausibility: The point $z^* \in Z$ should represent the fuzzy set B from an intuitive point of view, that is, it should lie approximately in the middle of the support set of B or should have a high degree of membership in B.

- Computational complexity: Computational time for any defuzzification procedure should be fairly short so that it can be applied to any real-time application. This is particularly important for fuzzy control application, which requires on-line calculation of the crisp output.
- Continuity: A small change in B should result in a small change in $z^* \in Z$. This is especially important for stability analysis of fuzzy control systems.

There have been various studies reported on defuzzification methods in the literature (Driankov et al. 1993). An empirical study of the performance of defuzzification methods applied to different fuzzy controllers has been reported by Lancaster and Wierman (2003). They investigated standard methods such as true centre of gravity, fast centre of gravity and mean of maxima and found these methods have some advantages over the other methods. They also developed some new methods such as plateau average, weighted plateau average, sparus, capitis, and clivosus. These methods are not discussed here further but interested readers are referred to Lancaster and Wierman (2003); van Broekhoven and De Baets (2004, 2006, 2008). A very comprehensive review of defuzzification methods can also be found in Van Leekwijck and Kerre (1999); Roychowdhury and Pedrycz (2001). A comparative analysis of different defuzzification methods is given in (Driankov et al. 1993). Good theoretical analysis on defuzzification process and problems has been reported in Yager and Filev (1994); Wang (1997).

4.11 Summary

Fuzzy logic is one of the tools for intelligent control. This chapter introduces different aspects of fuzzy logic, membership functions, fuzzy models, rule-base, fuzzification, inferencing, defuzzification and finally approaches to fuzzy modelling and control. For any fuzzy system design, membership function shape plays an important role. Choice of fuzzy models such as Mamdani-, Takagi-Sugeno- or Tsukamoto-type mainly depends on the type of application. Rule-base construction is crucial and mainly need to depend on an expert knowledge. Choice of defuzzification is also important in designing efficient fuzzy system or controller.

The fuzzy mathematics discussed in this chapter will be applied to design and development of fuzzy controller for the flexible arm in the following chapters.

References

Ang KK, Quek C, Pasquier M (2003) POPFNN-CRI(S): Pseudo outer product based fuzzy neural network using the compositional rule of inference and singleton fuzzifier. IEEE Trans Syst Man Cybern Part B-Cybern 33:838–849

Bezdek JC (1981) Pattern recognition with fuzzy objective function algorithm. Plenum Press, Berlin

Bezdek JC, Keller J, Krisnapuram, Pal NR (1999) Fuzzy models and algorithms for pattern recognition and image processing. Kluwer Academic Publisher, Berlin

Boverie S, Demaya B, Titli A (1991) Fuzzy logic control compared with other automatic control approaches. In: Proceedings of the 30th conference on decision and control, vol 2. Briton, UK, pp 1212–1216

Brachman RJ, Levesque HJ (2004) Knowledge representation and reasoning. Morgan Kaufmann, Massachusetts

Castilo O, Melin P (2008) Type-2 fuzzy logic: theory and applications, studies in fuzziness and soft computing, vol 223. Springer, Heidelberg

Denna M, Mauri G, Zanaboni AM (1999) Learning fuzzy rules with Tabu search—an application to control. IEEE Trans Fuzzy Syst 7(2):295–318

Díaz-Hermida F, Losada DE, Bugarín A, Barro S (2005) A probabilistic quantifier fuzzification mechanism: the model and its evaluation for information retrieval. IEEE Trans Fuzzy Syst 13(5):688–700

de Oliveira JV, Pedrycz W (eds) (2007) Advances in fuzzy clustering and its applications. Wiley, Chichester

Drainkov D, Hellendorn H, Reinfrank M (1993) An introduction to fuzzy control, Springer-Verlag, New York

Espinosa J, Vandewalle J, Wertz V (2005) Fuzzy logic identification and predictive control. Springer, London

Genther H, Runkler TA, Glesner M (1994) Defuzzification based on fuzzy clustering. In: Proceedings of the IEEE international conference on fuzzy systems, Orlando, pp 1645–1648

Gilachet S, Foulloy L (1995) Fuzzy controllers: synthesis and equivalences. IEEE Trans Fuzzy Syst 3(2):140–148

Guillaume S (2001) Designing fuzzy inference systems from data: an interpretability-oriented review. IEEE Trans Fuzzy Syst 9(3):426–443

Hu B-G, Mann GKI, Gosine RG (2001) A systematic study of fuzzy PID controllers—function-based evaluation approach. IEEE Trans Fuzzy Syst 9(5):699–712

Huang H, Pasquier M, Quek C (2006) Optimally evolving irregular-shaped membership functions for fuzzy systems 2006 IEEE congress on evolutionary computation. Sheraton Vancouver Wall Centre Hotel, Vancouver, BC, Canada, 16–21 July, pp 3309–3316

Jang JSR (1994) Structure determination in fuzzy modelling: a fuzzy CART approach. In: Proceedings of international conference on fuzzy systems, Orlando, FL, pp 480–485

Jang J-SR, Sun C-T (1993) Functional equivalence between radial basis function networks and fuzzy inference systems. IEEE Trans Neural Netw 4(1):156–159

Jang J-SR, Sun C-T, Mizutani E (1997) Neuro-fuzzy and soft computing. Prentice Hall, Upper Saddle River

Kalaykov I (1998) Fuzzy controllers can be extremely fast. In: Proceedings of the national conference on automation and informatics, symposium on intelligent control, vol 2. Sofia, Bulgaria, pp 31–36

Klir G, Yuan B (1995) Fuzzy sets and fuzzy logic: theory and applications. Prentice Hall, New Jersy

Kohonen TK (1989) Self-organization and associative memory, 3rd edn. Springer, New York

Kosko B (1992) Neural networks and fuzzy systems. Prentice-Hall, Englewood Cliffs

Kovacic Z, Bogdan S (2006) Fuzzy controller design: theory and application. CRC Press, Taylor and Francis Group

Lancaster SS, Wierman MJ (2003) Empirical study of defuzzification. In: Proceedings of the international conference of the North American fuzzy information processing society, pp 121–126

Liska J, Melsheimer SS (1994) Complete design of fuzzy logic systems using genetic algorithms. In: Proceedings of the third IEEE conference on fuzzy systems, IEEE world congress on computational intelligence, vol 2, pp 1377–1382

Mamdani EH, Assilian S (1974) Application of fuzzy algorithms for control of simple dynamic plant. Proc. IEEE 121:1585–1588

Margaliot M, Langholz G (2000) New approaches to fuzzy modelling and control: design and analysis, vol 38. World Scientific Publishing Company Inc, Singapore

Mendel JM (2001) Uncertain rule-based fuzzy logic systems: introduction and new directions. Prentice Hall, NJ

Murata, T., Ishibuchi, H. and Gen, M. (1999) Construction of fuzzy classification systems using multiple fuzzy rule tables. In: Proceedings on SMC '99 Conference, vol 4, pp 875–880

Nefti S (2002) New algorithm for simplification of rule base generated by automated fuzzy modelling. IEEE Int Conf Syst Man Cybern, Vol 2, pp 190–195

Nie JH, Lee TH (1996) Rule-based modelling: fast construction and optimal manipulation. IEEE Trans Syst Man Cybern—Part A: Syst Hum 26(6):728–738

Pal NR, Bezdek JC, Tsao ECK (1995) Generalized clustering networks and Kohonen self-organizing scheme. IEEE Trans Neural Netw 6:521–521

Patel AV, Mohan BM (2002) Some numerical aspects of center of area defuzzication method. Fuzzy Sets Syst 132:401–409

Pedrycz W, Gomide F (1998) An introduction to fuzzy sets: analysis and design. The MIT Press, Cambridge

Pluger NJ, Yen J, Langari R (1992) A defuzzification strategy for fuzzy logic controllers employing prohibitive information in command formulation. In: Proceedings of the first IEEE conference on fuzzy systems, San Diego, CA, pp 717–723

Roychowdhury S, Pedrycz W (2001) A survey of defuzzification strategies. Int J Intell Syst 16:679–695

Roychowdhury S, Wang BH (1994) Cooperative neighbours in defuzzification. Fuzzy Sets Syst 78:37–49

Runker TA (1997) Selection of appropriate defuzzification methods using application specific properties. IEEE Trans Fuzzy Syst 5(1):72–79

Runkler TA, Glesner M (1993) A set of axioms for defuzzification strategies—towards a theory of rational defuzzification operators. In: Proceedings of the 2nd IEEE international conference on fuzzy systems, San Francisco, CA, pp 1161–1166

Runkler TA, Glesner M (1994) Defuzzification and ranking in the context of membership value semantics, rule modality, and measurement theory. In: Proceedings of the first European congress on fuzzy and intelligent technologies, Aachen

Russo M (1998) FuGeNeSys-a fuzzy genetic neural system for fuzzy modelling. IEEE Trans Fuzzy Syst 6:373–388

Setnes M, Babuska R, Verbruggen HB (1998) Rule-based modelling: precision and transparency. IEEE Trans Syst Man and Cybern: Part C Appl Rev 28(1):165–169

Sugeno M, Kang GT (1988) Structure identification of fuzzy model. Fuzzy Sets Syst 28:15–33

Sugeno M, Yasukawa T (1993) A fuzzy-logic-based approach to qualitative modelling. IEEE Trans Fuzzy Syst 1(1):7–31

Takagi H, Hayashi I (1991) NN-driven fuzzy reasoning. Int J Approx Reasoning 5(3):191–212

Takagi T, Sugeno M (1985) Fuzzy identification of systems and its application to modeling and control. IEEE Trans Syst Man Cybern 15:116–132

Tsukamoto Y (1979) An approach to fuzzy reasoning method. In: Gupta MM, Ragade RK, Yager RR (eds) Advances in fuzzy set theory and applications. North-Holland, Amsterdam, pp 137–149

Tung WL, Quek C (2002) DIC: a novel discrete incremental clustering technique for derivation of fuzzy membership functions. PRICAI: Trends Artif Intell. Lect Notes Comput Sci 2417:485–491

Van Broekhoven E. and De Baets B (2004) A comparison of three methods for computing the centre of gravity defuzzification. In: Proceedings of the 2004 IEEE international conference on fuzzy systems, pp 1537–1542

Van Broekhoven E, De Baets B (2006) Fast and accurate center of gravity defuzzification of fuzzy system outputs defined on trapezoidal fuzzy partitions. Fuzzy Sets Syst 157:904–918

Van Broekhoven E, de Baets B (2008) Monotone Mamdani–Assilian models under mean of maxima defuzzification. Fuzzy Sets Syst 159:2819–2844

Van Leekwijck WV, Kerre EE (1999) Defuzzification: criteria and classification. Fuzzy Sets Syst 108(2):159–178
Van Leekwijck W, Kerre EE (2001) Continuity focused choice of maxima: yet another defuzzification method. Fuzzy Sets Syst 122:303–314
Wang L-X (1997) A course in fuzzy systems and control. Prentice-Hall International, Inc, NJ
Wong C-C, Lin N-S (1997) Rule extraction for fuzzy modelling. Fuzzy Sets Syst 88:23–30
Yager RR, Filev D (1994) Essentials of fuzzy modelling and control. Wiley, NY, Chichester
Zadeh LA (1965) Fuzzy sets. Inf Control 8:338–353
Zadeh LA (1968) Fuzzy algorithms. Inf Control 12:94–102
Zadeh LA (1972) A fuzzy-set-theoretic interpretation of linguistic hedges. J Cybern 2:4–34
Zadeh LA (1973) Outline of a new approach to the analysis of complex systems and decision process. IEEE Trans Syst Man Cybern 3:28–44
Zadeh LA (1975a) Fuzzy logic and approximate reasoning. Syntheses 30:407–428
Zadeh LA (1975b) The concept of a linguistic variable and its application to approximate reasoning-I. Inf Sci 8:199–249
Zadeh LA (1976) The linguistic approach and its application to decision analysis. In: Ho YC, Mitter SK (eds) Directions in large scale systems, pp 339–370
Zadeh LA (1979) A theory of approximate reasoning. In: Hayes J, Michie D, Mikulich LI (eds) Machine intelligence 9. Halstead Press, New York, pp 149–194
Zadeh LA (1994) The role of fuzzy logic in modelling, identification and control. Model Ident Control 15(3):191–203
Zadeh LA (1999) From computing with numbers to computing with words—from manipulation of measurements to manipulation of perceptions. IEEE Trans Circuits Syst—I: Fundam Theory Appl 45(1):105–119
Zadeh LA (2000) Outline of a computational theory of perceptions based on computing with words. In: Sinha NK, Gupta MM, Zadeh LA (eds) Soft computing and intelligent systems: theory and applications. Academic Press, London, pp 3–22
Zadeh LA (2001) A new direction in AI—toward a computational theory of perceptions. AI Mag 22(1):73–84
Zadeh LA (2002) Toward a perception-based theory of probabilistic reasoning with imprecise probabilities. J Stat Plann Infer 105:233–264
Zadeh LA (2008) Is there a need for fuzzy logic? Inf Sci 178(13):2751–2779

Chapter 5
Fuzzy Control

5.1 Introduction

Traditional mathematical techniques of system modelling and control have significant limitations. In many cases it is difficult to describe the system behaviour by a set of mathematical equations when the system is nonlinear and partially known or unknown. Moreover, there are a lot of uncertainties and unpredictable dynamics that do not let the system model to be described mathematically. Such uncertainties and unpredictable behaviour in complicated and ill-defined systems can be modelled using the linguistic approach proposed by Zadeh as a model of human thinking, which introduced fuzziness into systems theory (Zadeh 1965, 1973). Recently, an emergence of results using intelligent control technologies incorporating fuzzy logic, which avoids the need for a mathematical model of such a highly non-linear system, is noted. Among the few available control approaches, the fuzzy logic control method is characterised by its ease of incorporation of human expertise, easy to understand and design.

Most of the classical controller design methodologies such as Nyquist, Bode, state-space, optimal control, root locus, H_∞, and μ-analysis are based on assumptions that the process is linear and stationary and hence is represented by a finite dimensional constant coefficient linear model. These methods do not suit complex systems well because few of those represent uncertainty and incompleteness in process knowledge or complexity in design. But the fact is the real world is too complex. In particular, many industrial processes are highly nonlinear and complex. As the complexity of a system increases, quantitative analysis and precision become difficult. However, many processes that are nonlinear, uncertain, incomplete or non-stationary are controlled by skilled human operators successfully. Without requiring any precise mathematical description or knowledge about the process, the expert human operator models the process in a heuristic or experiential manner. It is evident that such human experiential knowledge is becoming more and more important in control systems design. This experiential perspective in controller design requires the acquisition of heuristic and qualitative, rather than quantitative, knowledge or expertise from the human operator. During the past several years, fuzzy control has

N. Siddique, *Intelligent Control*, Studies in Computational Intelligence 517,
DOI: 10.1007/978-3-319-02135-5_5,

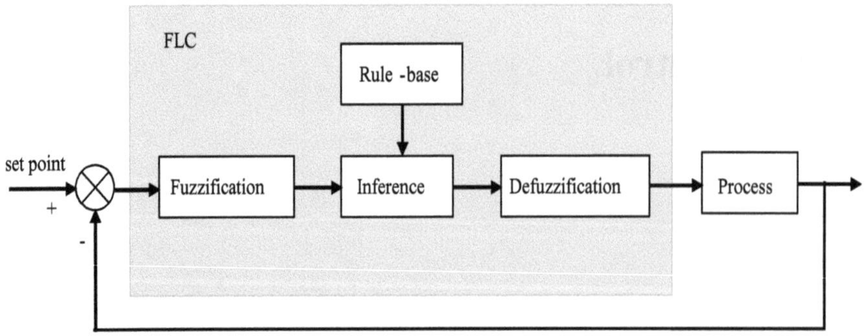

Fig. 5.1 Configuration of a fuzzy logic controller

emerged as one of the most active and powerful areas for research in the application of such complex and real world systems using fuzzy set theory (Zadeh 1965, 1994). The control of complex nonlinear systems has been approached in recent years using fuzzy logic techniques. A generic fuzzy logic controller (FLC) has the basic configuration illustrated in Fig. 5.1.

The seminal ideas of fuzzy logic applied to systems modelling and control can be found in the early papers of Zadeh in the 1960s and 1970s (Zadeh 1968, 1973). There are many interpretations of fuzzy modelling. For instance, a fuzzy set is a fuzzy model of human concept. In other words, fuzzy models consist of linguistic explanations about the system behaviour and deals with fuzzy modelling of a plant for control (Takagi and Sugeno 1985; Zadeh 1973, 1994; Kovacic and Bogdan 2006; Shin and Xu 2009). Just as modern control theory, a fuzzy controller can be designed based on a fuzzy model of a plant if the fuzzy model can be identified (Sugeno and Yasukawa 1993). A detail study on the mathematical foundation for fuzzy logic control has been made in Chap. 4.

5.1.1 *Fuzzification for Control*

In fuzzy control applications, the observed data are usually crisp. Since the data manipulation in an FLC is based on fuzzy sets described by a membership function, fuzzification is necessary. Therefore, fuzzification defines the mapping from observed input data to fuzzy sets in certain input universe of discourse. This process consists of associating to each fuzzy set a membership function (MF). These MFs can be thought of as maps from the real numbers to the interval $I = [0, 1]$. If there are n fuzzy sets associated with a given quantity $x \in R$ such that n maps $F_i : R \to I,\ i = 1, \ldots, n$ are defined. They determine to what extent the label associated with fuzzy set i characterizes the current value of x and produces a membership value $\mu_i(x) \in I$. Practically, the value of $\mu_i(x)$ depends on the shape of MFs. There are different types of MFs used in fuzzy control literature but there is no strict rule for choosing a certain type of MF. The choice of MFs mainly depends on the application domain and ease of computation. The most common choices of MFs are triangular, trapezoidal, Gaussian and bell shaped membership functions. A detail discussion on MFs is presented in Chap. 4.

5.1.2 Inference Mechanism for Control

Inference is the process of formulating a nonlinear mapping from a given input space to output space. The mapping then provides a basis from which decisions can be made. The process of fuzzy inference involves all the input and output membership functions, fuzzy logic operators, and if–then rules.

The different types of fuzzy inference mechanisms have been discussed in Chap. 4. These have been widely employed in various control applications. The differences between these three fuzzy inferences (also called fuzzy models) lie in the consequents of their fuzzy rules, aggregations and defuzzification procedures. The widely known fuzzy models used in control applications are

(a) Mamdani fuzzy models
(b) Takagi–Sugeno fuzzy models
(c) Tsukamoto fuzzy models

The most widely used fuzzy models are Mamdani and Takagi–Sugeno fuzzy models. It is due to the fact that Mamdani-type fuzzy model (Mamdani and Assilian 1974) is easy to apply without much a priori information about the system whereas Takagi–Sugeno-type fuzzy model (Takagi and Sugeno 1985) requires estimation of consequent parameters from available data. Tsukamoto fuzzy model (Tsukamoto 1979) somehow did not attract lots of applications though the defuzzification procedure is very straightforward using the monotone functions.

Another important aspect of fuzzy control is the choice of inference mechanism. There is no strict rule for choice of specific fuzzy inference. It is mainly depends on the preference of the designer, availability of a priori information and the application domain. Applicability of stability analysis should another criterion for the choice of the inference mechanisms. In this respect, Takagi–Sugeno-type fuzzy models have distinct advantages over the other two types of models. A wide variety of stability analysis techniques, presented in detail in Chap. 9, is easily applicable to Takagi-Sunego type fuzzy controllers. Literature is abounding in this area. In this chapter, Mamdani-type fuzzy model will be explored as to demonstrate the simplicity and ease of developing fuzzy controller for highly nonlinear system. As a matter of fact, Mamdani-type fuzzy model has some distinct disadvantages. It cannot distinguish specific information from the input space and requires the antecedent of the rules span over the whole input space (Yager and Filev 1994). This is how the Mamdani-type fuzzy controller needs to undergo an optimisation procedure. A second issue with the Mamdani-type fuzzy controller is the analysis of stability, which needs further investigation.

5.1.3 Rule-Base for Control

The rule-base of a fuzzy system describes in qualitative terms how an output behaves when subjected to various inputs. Fuzzy rule-base and its construction for general purpose fuzzy systems have been discussed in detail in Chap. 4. In case of a fuzzy controller, the inputs may be error, change of error or sum of error and the output may be the control action or change of control action. The construction of rule-base appears to be limited by the elucidation of the heuristic rules for control. The common approach is to partition the input space for defining the rule-base. For example, the rule-base in Table 5.1 shows partitioning of the input and output space comprising error e, change of error Δe and control variable u for a PD-like fuzzy controller or Δu for a PI-like fuzzy controller. Inputs and output are partitioned into five fuzzy sets defined as PB $\hat{=}$ Positive Big, PS $\hat{=}$ Positive Small, ZO $\hat{=}$ Zero, NS $\hat{=}$ Negative Small and NB $\hat{=}$ Negative Big.

The rule-base in Table 5.1 shows some distinct features. The negative linguistic values of the control variable are placed below the diagonal and the positive linguistic values are placed above the diagonal. Zero control values are placed along the diagonal. Most rule-bases have skew-symmetric property, namely $u_{ij} = -u_{ji}$. The linguistic values of the control variable increase with increasing distance from the diagonal line. If the quantization level of the independent variables is halved, then the boundaries of the control regions look like staircase shapes. As they become infinitesimal, the boundaries become straight lines as shown in Fig. 5.2. The common characteristics of the rule-base for fuzzy logic controller reflect the partitioning of the input space defined by the antecedent of the rules, i.e. fuzzy values of error e and change of error Δe. As it can be seen from the rule-base in Table 5.1, it created five distinct bands of regions in the input space.

Table 5.1 Rule-base with error and change of error

	Change of error Δe				
Error e	PB	PS	ZO	NS	NB
NB	ZO	PS	PB	PB	PB
NS	NS	ZO	PS	PB	PB
ZO	NB	NS	ZO	PS	PB
PS	NB	NB	NS	ZO	PS
PB	NB	NB	NB	NS	ZO

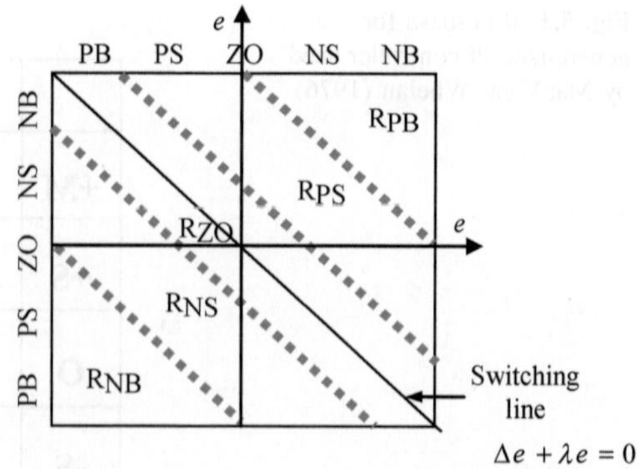

Fig. 5.2 Band-wise partitioning of input space into regions R_{NB}, R_{NS}, R_{ZO}, R_{PS}, and R_{PB}

For increasing fuzziness (i.e. higher levels of completeness) of the input fuzzy sets (i.e. MFs) an idealised multi-band partitioning of the input space with smooth borders is possible as illustrated in Fig. 5.2.

It is to be noted that the magnitude of the control input $|u|$ is approximately proportional to the distance from the main diagonal line. It can also be called switching line as the control inputs above and below it have opposite signs. The switching line is defined by:

$$\Delta e + \lambda e = 0 \tag{5.1}$$

As is most often the case, the construction of fuzzy rules is a heuristic approach as demonstrated in Table 5.1 and Fig. 5.2 and is based on judgments of experienced operators. All these heuristics and operators' experience can steer to formulate some workable meta-rules such as:

1. If both $e(k)$ and $\Delta e(k)$ at time instant k are zero, then maintain present control setting.
2. If conditions are such that $e(k)$ will go to zero at a satisfactory rate, then maintain present control setting.
3. If $e(k)$ is not self-correcting, then control action $\Delta u(k)$ is not zero and depends on the sign and magnitude (small, medium, large, etc.) of $e(k)$ and $\Delta e(k)$.

Using these meta-rules, Mac Vicar-Whelan (1976) devised the production rules for the fuzzy PI controller as shown in Fig. 5.3. The controller variables are quantized into fuzzy sets of several levels such as M = medium, S = small and O = zero. Based on values assumed by $e(k)$ and $\Delta e(k)$ the control action $\Delta u(k)$ is given by the corresponding table entry.

As quantization levels of the control and measurement variables become infinitely fine, the Mac Vicar-Whelan controller approaches a deterministic PI or PD controller (Tang and Mulholland 1987).

For finite universes of the inputs, the rule table can be obtained in the form of a lookup table. For instance, by sampling the universes of the input variables and apply the reasoning, lookup table entries can be calculated off-line for a predefined

Fig. 5.3 Rule-base for generalized PI controller used by MacVicar-Whelan (1976)

e \ Δe	-M	-S	O	+S	+M
+M	O	-S	-M	-M	-L
+S	+S	O	-S	-S	-M
O	+M	+S	O	-S	-M
-S	+M	+S	+S	O	-S
-M	+L	+M	+M	+S	O

Fig. 5.4 Rule-base as lookup table

e(k) \ Δe(k)	-M	-S	O	+S	+M
+M	0	-1	-3	-3	-4.5
+S	+1	0	-1	-1	-3
O	+3	+1	0	-1	-3
-S	+3	+1	+1	0	-1
-M	+4.5	+3	+3	+1	0

rule-base and membership functions. Lookup table is a significant simplification of the internal representation of the fuzzy controller. For example, the rule table in Fig. 5.3 will look something like Fig. 5.4. The computation of the crisp output u inferred by the fuzzy controller for some actual inputs will be very straightforward.

5.1.4 Defuzzification for Control

Basically, defuzzification is a mapping from a space of fuzzy control actions defined over an output universe of discourse into a space of non-fuzzy (crisp) control actions. In a sense this is the inverse of the fuzzification even though mathematically the maps need not be inverses of one another.

Usually the defuzzification process makes explicit use of the MFs of different shapes discussed in Chap. 4. There is a variety of methods for defuzzifying a fuzzy set A (describe by MF). A classification of different defuzzification methods and their computational procedures have been discussed in detail in Chap. 4. Unfortunately, there is no strict rule for the choice of specific defuzzification method in control applications. Different defuzzification method use different computation procedure. Computation needed to carry out the defuzzification operations is, in general, time consuming. Therefore, it is important to consider the computational time for any defuzzification procedure to be applied to control problems, which should be fairly short so that it can be applied to any real-time application. This is particularly important for fuzzy control application, which requires on-line calculation of the crisp output.

Many researchers have been working on the choice of defuzzification methods for fuzzy controllers and their impact on the performance. For example, continuity and computational efficiency are of utmost importance for fuzzy controller design. Some guidelines are discussed in Chap. 4 for evaluation of the defuzzification methods and identify the properties of the methods that determine the suitability for fuzzy controllers.

5.2 Theoretical Analysis of Fuzzy Controllers

The limitations of the conventional mathematical model-based control theory are well known and a brief discussion is presented in the introduction. The basic principles of feedback control using experience, intuition and practical skills have been known for centuries. The seminal works by Zadeh (1973, 1975a, b, 1994) introduced the idea of formulating control algorithms in terms of human perception, experience, linguistic variables and approximate reasoning.

Let $x = (x_1, \ldots, x_n)$ be a vector of process state variables, y be the process output variable and u be the process input variable or control variable. The conventional closed-loop model, when linearised around the set-point, is given by

$$x(k+1) = A \cdot x(k) + b^T \cdot u(k) \tag{5.2}$$

$$y(k) = c^T \cdot x(k) \tag{5.3}$$

$$u(k) = k \cdot y(k) \tag{5.4}$$

where A is the process matrix, b and c are vectors and k is a scalar. The state equations can be written as

$$x(k+1) = A \cdot x(k) + b^T \cdot u(k) \tag{5.5}$$

$$u(k) = k \cdot c^T \cdot x(k) \tag{5.6}$$

The fuzzy counter part of the above model can be described using linguistic variables. Let the linguistic variable x_i (e.g. error, change of error etc.) have the term set X_i (e.g. NB, NS etc.) described by membership functions denoted by $\tilde{X}_i$. Thus, the linguistically defined process state vector is denoted by $\tilde{X} = (\tilde{X}_1, \ldots, \tilde{X}_n)$. Similarly, u takes linguistically defined values U with membership functions $\tilde{U}$. Then the fuzzy model of the closed-loop system can be described as

$$\tilde{X}(k+1) = \tilde{X}(k) \times \tilde{U}(k) \circ \tilde{A} \tag{5.7}$$

$$\tilde{U}(k) = \tilde{X}(k) \circ \tilde{K} \tag{5.8}$$

where $\tilde{A}$ is a fuzzy relation on $X \times U \times X$, $\circ$ is the composition operation and $\tilde{K}$ is the controller which is a fuzzy relation on $X \times U$ representing the meaning of a set of If–Then rules of the form

$$\textit{If } x_1 \textit{ is } X_i \textit{ and} \ldots x_n \textit{ is } X_j \textit{ Then } u \textit{ is } U_k \tag{5.9}$$

Based on the principle described in Eqs. (5.7)–(5.9), many research efforts have been put to design and develop fuzzy controllers. Li and Gatland (1995, 1996) proposed a more systematic design method for PD and PI-type FLC's. For PID (proportional-integral-derivative)-type FLC, they also presented a simplified rule generation method using two two-dimensional (2-D) spaces instead of a three-dimensional space. Palm (1992, 1994) proposed a sliding mode fuzzy controller which generates the absolute value of switching magnitude in the sliding mode control law using the error and the change-of-error. Most of researches use two input variables in the rule antecedent regardless of the complexity of the controlled plants.

5.2.1 Consideration of Process Variables

The general form of control law is defined as:

$$u(k) = f(e(k), e(k-1), \ldots, e(k-n), u(k-1), \ldots, u(k-n)) \tag{5.10}$$

where $u(k)$ is the control input, $e(k)$ is error defined as $e(k) = y_d - y(k)$, y_d is the desired output, $y(k)$ is the actual output and $f(.)$ is a nonlinear function that describes the control law. The parameter n defines the order of the controller. Different variants of control algorithms can be derived from the control law (5.10) such as proportional (P), differential (D), integral (I), proportional plus integral (PI), proportional plus differential (PD) and proportional, integral plus differential (PID) controllers.

A fuzzy system is characterized by a set of linguistic statements based on expert knowledge. The expert knowledge is usually in the form of if–then rules, which are implemented by fuzzy conditional statements. The collection of fuzzy rules that are expressed as fuzzy conditional statements forms the rule base or the rule set of an FLC. The design parameters of the rule-base include

- Choice of process state and control output variables
- Choice of the content of the rule antecedent and the rule-consequent
- Choice of term-sets for the process state and control output variables
- Derivation of the set of rules

If one has made the choice of designing a P-, PD-, PI-, or PID-like fuzzy logic controller, this already implies the choice of process state and control output variables as well as the content of the rule antecedent and rule consequent for each of the rules. The process state variables representing the contents of the rule antecedent (if part of the rule) are selected as follows

- Error denoted by $e(k)$
- Change of error denoted by $\Delta e(k)$
- Sum of error denoted by Σe

The control output (process input) variables representing the content of the rule-consequent (then part of the rule) are selected as follows

- Control output denoted by $u(k)$
- Change of control output $\Delta u(k)$

By analogy with the conventional controller they are defined as

- $$e(k) = y_d - y(k)$$
- $$\Delta e(k) = e(k) - e(k-1)$$
- $$\sum_{k=1}^{n} e(k) = \sum_{k=1}^{n-1} e(k) + e(k)$$
- $$\Delta u(k) = u(k) - u(k-1)$$
- $$u(k) = u(k-1) + \Delta u(k)$$

where y_d is the desired output or set-point, $y(k)$ is the process output, k is the sampling time and n is the maximum sample number.

Most works in fuzzy control field use the error $e(k)$ and the change-of-error $\Delta e(k)$ as input variables regardless of the complexity of controlled plants. Also, either control input $u(k)$ (in PD-type) or incremental control input $\Delta u(k)$ (in PI-type) is typically used as output variables representing the rule consequent

(Choi et al. 1999, 2000). These conventional FLCs are naturally based on the concept of linear proportional-derivative (PD) or proportional-integral (PI) control scheme. Such FLCs with two-inputs and single output are sufficient for simple second order plants. However, in the cases of complex and higher order plants, all possible process states are required as input variables for implementing state feedback FLCs which demands a huge number of control rules and much effort to construct the rule-base. Therefore, only the error and the change-of-error or error and the sum of error are used as input variables for most of the FLC implementation.

5.2.2 Types of Fuzzy Controllers

Depending on the combination of input variables such the error, change-of-error and sum of error, there are a variety of FLC implementations possible.

P-like FLC
The conventional proportional (P)-like controller is described as

$$u = k_p \cdot e(k) \tag{5.11}$$

where k_p is the proportional gain coefficient. The rule for P-like controller is given in the symbolic form as

$$\text{If } e \text{ is } A_i \text{ then } u \text{ is } B_j \tag{5.12}$$

where A_i and B_j, i, j = 1, 2, ... n, are the linguistic variables. Function of the control output for such single-input and single output (SISO) system is then a curve as shown in Fig. 5.5. Figure 5.5 demonstrates the control output for n = 4, i.e. 4 MFs for each input and output.

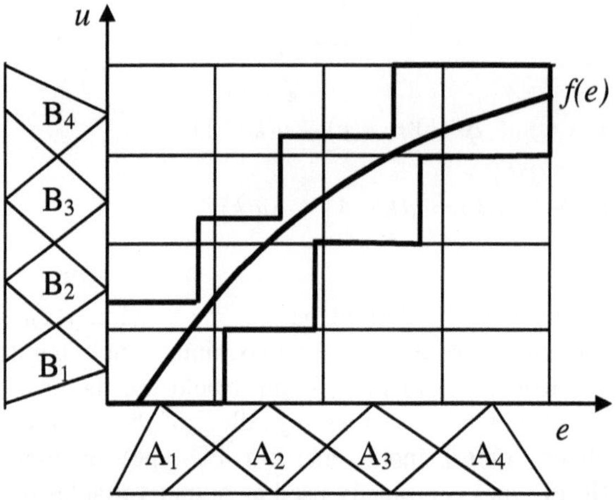

Fig. 5.5 Function of control output for SISO systems

PD-like FLC

The general equation of proportional-derivative (PD)-like controller is given as

$$u(k) = k_p \cdot e(k) + k_d \cdot \Delta e(k) \tag{5.13}$$

where k_p and k_d are the proportional and differential gain coefficients. The PD-like FLC consists of rules of the form:

$$\text{If } e \text{ is } A_i \text{ and } \Delta e \text{ is } B_j \text{Then } u \text{ is } C_k \tag{5.14}$$

where A_i, B_j and C_k are the linguistic values for the inputs and output and $i = 1, \ldots, n_1, j = 1, \ldots, n_2$, and $k = 1, \ldots, m$. The number of rules will be $n_1 \times n_2$. PD-like FLC ensure simplicity, fast response and has good dynamic properties.

PI-like FLC

A conventional proportional-integral (PI)-controller is described as

$$u = k_p e + k_I \int edt \tag{5.15}$$

where k_p and k_I are the proportional and the integral gain coefficients. Taking the derivative with respect to time of Eq. (4.15) yields

$$\dot{u} = k_p \cdot \dot{e} + k_I \cdot e \tag{5.16}$$

which can be written as

$$\Delta u(k) = k_p \cdot \Delta e(k) + k_I \cdot e(k) \tag{5.17}$$

This yields incremental PI-like controller equation. The PI-like FLC rule base accordingly consists of rules of the form:

$$\text{If } e \text{ is } A_i \text{ and } \Delta e \text{ is } B_j \text{ Then } \Delta u \text{ is } C_k \tag{5.18}$$

In this case, to obtain the value of the control output $u(k)$, the change of control output $\Delta u(k)$ is added to $u(k-1)$ such that

$$u(k) = \Delta u(k) + u(k-1) \tag{5.19}$$

Another way to express the PI-controller is the absolute integral PI-like controller as

$$u(k) = k_p \cdot e(k) + k_I \cdot \sum e(k) \tag{5.20}$$

where $\sum e(k)$ is the sum of error, k_p and k_i are the proportional and integral gain coefficients.

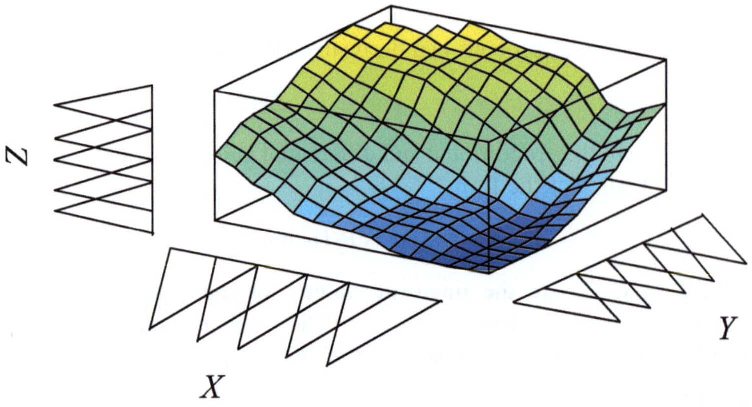

Fig. 5.6 Control surface of a two-input single-output FLC

The absolute PI-like FLC consists of rules of the form:

$$\text{If } e \text{ is } A_i \text{ and } \sum e \text{ is } B_j \text{ Then } u \text{ is } C_k \tag{5.21}$$

where A_i, B_j and C_k are the linguistic variables. Due to the integral term in the controller, PI-like FLC is slow and takes long time to settle though resolves the static error.

Controller output forms a surface for PD- and PI-like controllers. The control surface of a two-input and single-output (MISO) system is shown in Fig. 5.6, where X and Y represent inputs and Z represents the controller output.

PID-like FLC

Conventional PID controllers are still the most widely adopted method in industry for various control applications, due to their performance, simple structure, ease of design, and low cost in implementation. However, PID controllers might not perform satisfactorily if the system to be controlled is of highly nonlinear or if it involves uncertainties. Fuzzy control has been successful in handling nonlinearities and uncertainties through use of fuzzy set theory. It has been shown that many fuzzy PID controllers are nonlinear PID controllers and perform better than conventional PID controllers in most cases (Chen 1996; Farinwata et al. 2000). It is thus believed that by combining these two techniques together a better control system can be achieved. Hu et al. (1999, 2001) suggest that if a fuzzy controller is designed to generate control actions within PID concepts like a conventional PID controller, then it is called the fuzzy PID controller.

Therefore, a further option to obtain a better performance in respect of rise time, settling time, overshoot and steady-state error is to develop a proportional-integral-derivative (PID)-like FLC. The basic idea of a PID controller is to choose the control law by considering the error e, change of error Δe and integral of error $\sum e = \int_0^t e dt$, and thus giving the controller as

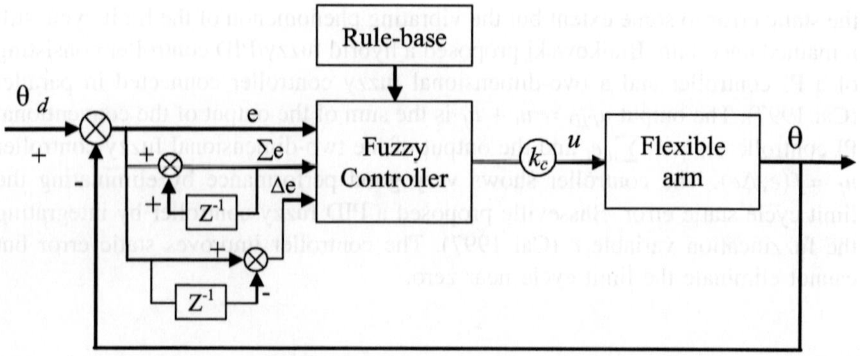

Fig. 5.7 Fuzzy PID-like controller

$$u_{PID} = k_P \cdot e + k_d \cdot \Delta e + k_I \cdot \int_0^t e \cdot dt \tag{5.22}$$

The controller in discrete time is written as

$$u_{PID} = k_P \cdot e + k_d \cdot \Delta e + k_I \cdot \sum e \tag{5.23}$$

The fuzzy control rule corresponding to the PID-controller has the form

$$\text{If } e \text{ is } A_i \text{ and } \Delta e \text{ is } B_j \text{ and } \sum e \text{ is } C_k \text{ Then } u \text{ is } D_l \tag{5.24}$$

where $i = 1, \ldots, n_1, j = 1, \ldots, n_2, k = 1, \ldots, n_3$ and $l = 1, \ldots, m$. A_i B_j C_k and D_l are the linguistic labels of the input and output variables. The fuzzy controller can be constructed using e, Δe and $\sum e$ as inputs and torque u as output. Figure 5.7 shows the block diagram of a typical 3-input single-output fuzzy PID-like controller.

Theoretically, the number of rules to cover all possible input variations for a three-term fuzzy controller is $n_1 \times n_2 \times n_3$, where n_1, n_2 and n_3 are the number of linguistic labels of the three input variables. For $n_1 = n_2 = n_3 = 5$, there will be 125 rules, which will be a huge time consuming computation.

An alternative form of the PID-like FLC is defined by

$$\Delta u(k) = K_1 e(k) + K_2 e(k-1) + K_3 e(k-2) \tag{5.25}$$

where K_1, K_2 and K_3 are the gain coefficients. Further extension of the input variables of the FLC is theoretically possible but it leads to an increasing complexity of the structure of the controller as it becomes reasonably difficult for an expert to determine the control rules considering the second and higher order differences.

There are several methods proposed for the implementation of PID-like FLCs by many researchers in pursuit of performance improvement. Braae and Rutherford (1978, 1979) developed a fuzzy controller where an integration unit is placed before the fuzzification unit and after the defuzzification unit. The FLC improved the static error to some extent but the vibrating phenomenon of the limit cycle still remained uncertain. Bialkowski proposed a hybrid fuzzy/PID controller consisting of a PI controller and a two-dimensional fuzzy controller connected in parallel (Cai 1997). The output $u_{PID} = u_i + u_f$ is the sum of the output of the conventional PI controller $u_i = k_I \sum_i e_i$ and the output of the two-dimensional fuzzy controller $u_f = f(e, \Delta e)$. The controller shows very good performance by eliminating the limit cycle static error. Basseville proposed a PID fuzzy controller by integrating the fuzzification variable e (Cai 1997). The controller improves static error but cannot eliminate the limit cycle near zero.

5.3 Fuzzy Controller for Flexible Arm

There has been little work reported on the application of fuzzy logic controllers to flexible arms. The first of this kind was reported by Kubica and Wang using fuzzy logic control for fast moving single-link apparatus (Kubica and Wang 1993). A rule-base fuzzy control in combination with a model based scheme is used to control the tip position by Rattan et al. (1994) considering the presence of joint friction and changes in the payload. It is a computer based simulated system controlled by a PD controller. Similar sort of controller was developed by Liu and Lewis (1994). The control scheme is composed of a feedback linearisation inner-loop and fuzzy logic outer-loop. A reduced-order computed torque control is used to linearise the whole system in the inner-loop and then a 33-rule-based fuzzy controller is used to command the rigid modes to track the desired trajectories. A methodology for fuzzy logic controller for very flexible-link arm is proposed and the design parameters are analysed in Lee et al. (1994). It is observed in this investigation that error and rate of change of error as inputs gives the best results. Meressi (1995) proposed a hybrid controller where the overall controller has a linear and a nonlinear part. The linear part is designed using optimal technique to achieve a reasonable transient response while the nonlinear part is controlled by a fuzzy controller to enhance the performance of the linear part by decreasing both the rise time and the settling time (Meressi 1995). Vukovich and Lee (1999) showed in their experiments with fuzzy logic controller for a single-link arm that the result of the fuzzy controller is slightly better than that of the PD controller. A rule-based supervisory control for a two-link flexible arm has been reported in (Garcia-Benetiz et al. 1993). A fuzzy learning control for a two-link flexible arm is found in (Moudgal et al. 1994a, b, 1995). Even though the controller was able to achieve adequate performance for varying configurations, its performance degrades when there is a payload at the endpoint.

Fuzzy controllers are more robust than PID controllers, since they can cover wider ranges of operating conditions than PID controllers. Significant research efforts have been made on controllers design of the fuzzy-PID type (Carvajal et al. 2000; Hu et al. 2001; Li et al. 2005; Sooraksa et al. 2002; Tang et al. 2001a, b) including its simplified versions of fuzzy-PD (Mohan and Patel 2002) and fuzzy-PI (Tang et al. 2001a, b) control for different applications. The conventional approach to FLC design is to generate a fuzzy rule set based upon the system states such as error, change of error or sum of error, thus producing a two-input single-output PD-, or PI-type or three-input single-output PID-type control rule base. PI-type FLCs are most common and practically followed by the PD-type FLCs. The performance of PI-type FLCs is known to be quite satisfactory for linear first-order systems (Lee 1993; Mudi and Pal 1999). But, as with conventional PI-controllers, the performance of PI-type FLCs for higher order systems, and for systems with integrating elements or large dead time, and also for non-linear systems may be very poor due to large overshoot and excessive oscillation. Such systems may be ultimately uncontrollable (Lee 1993). On the other hand, good performance is achieved with PD-type fuzzy controller during the transient state. Generally speaking, the PD-type fuzzy controller will show a rapid response at the transient state. However, at the steady state, elements of error and change of error are possibly too small and the control signal, through fuzzy inference, becomes zero. The zero control signals will cause steady-state error or oscillations at the steady state (Chao and Teng 1997; Chung et al. 1998). Much of the research on fuzzy control of flexible arm has considered only simulation examples with two inputs and one output. Most of them have used error and rate of change of error as inputs. Moudgal et al. used the error and acceleration as inputs to the FLC (Moudgal et al. 1994a, b) and implemented the scheme with a two-link arm.

Current literature includes various efforts on modelling and control of flexible arms, both from theoretical and experimental points of view. Numerous researchers have investigated a variety of techniques for representing flexible and rigid dynamic models of such mechanisms as discussed in Chap. 3. These models are derived on the basis of different assumptions (small deflection for example) and mode shape functions. There are many uncertainties or unmodelled dynamics (payload changes for examples) in the system as well. Hence, even if a relatively accurate model of the flexible arm can be developed (as described in Chap. 3), it is very often too complex to use this model in a controller development. Thus, two choices present themselves for development of controllers.

1. Make further effort to deal with the nonlinear mathematical models
2. Find ways of non-conventional techniques that do not require mathematical model

Different model-based conventional control techniques for flexible arm, their difficulties and limitations were discussed in Chap. 3. Among the few non-conventional control approaches that do not require mathematical models, fuzzy logic, neural networks and hierarchical schemes are characterised by their ease of incorporation of experiential knowledge. Nonlinearity can be constructed as a fuzzy limit of analytic control theory (Lee et al. 1994) so that it is natural to follow the second choice mentioned above. The prospects and advantages of fuzzy control were described in Sect. 5.1. Fuzzy logic controllers have been applied to various complex industrial possesses, which are characterised as being highly nonlinear from analytical point of view (Sugeno 1985). But there have been little reported applications of fuzzy logic controllers to flexible arm, which are being addressed in this chapter. The objective of this chapter is to develop a fuzzy logic controller for the flexible arm described in Chap.2.

The PD-, PI-, and PID-type fuzzy controllers are investigated in terms of performance, input–output selection, rule-base construction and minimization. In fact a PID-type fuzzy controller would be a better choice, but unfortunately it requires a time consuming huge rule-base to process. A trade off should be found between PD-, PI-, and PID-type fuzzy controllers. In the following sections, a comparative assessment of the performance of each type of controller with different parameters and rule-bases is carried out.

5.3.1 Input–Output Selection

The fuzzy controller to be designed will be used to control the flexible-arm described earlier in Chap.2. Rather than going for development of a mathematical model of the system with available states such as hub angle denoted as θ, hub velocity denoted as $\dot{\theta}$ and end-point acceleration denoted as α, a fuzzy model using the derived states, namely, the hub angle error e, change of hub angle error Δe, sum of hub angle error Σe and input torque u at each discrete-time step during the control process, is developed.

The states hub angle θ, hub velocity $\dot{\theta}$ and end point acceleration α can be measured directly from the system whereas error e, change of error Δe, sum of Σe can be derived from the hub angle (θ) as

$$e = \theta_d - \theta \tag{5.26}$$

$$\Delta e = e(k) - e(k-1) \tag{5.27}$$

$$\sum e(k) = \sum e(k-1) + e(k) \tag{5.28}$$

$$u(k) = u(k-1) + \Delta u(k) \tag{5.29}$$

where θ is the measured hub angle and θ_d is the desired hub angle.

It is very common to normalise the universe of discourse. The normalised universes of discourse for error e, change of error Δe, sum of error Σe and input torque u are $E = \left[-a_e^*, +a_e^*\right]$, $\Delta E = \left[-a_{\Delta e}^*, +a_{\Delta e}^*\right]$, $\Sigma E = \left[-a_{\Sigma e}^*, +a_{\Sigma e}^*\right]$ and $U = \left[-a_u^*, +a_u^*\right]$ respectively. Due to some physical constraints of the experimental flexible arm the normalised universes of discourse for the error, change of error, sum of error and torque input were set as $E = [-36, +36]$, $\Delta E = [-25, +25]$, $\Sigma E = [-150, +150]$ and $U = [-3, +3]$ respectively.

5.4 PD-Like Fuzzy Logic Controller

Though the FLC exhibits superior applicability to the traditional PID controllers and is highly robust (Tang and Mulholland 1987; Chou and Lu 1993), PI-like and PD-like FLCs possess mainly the same characteristics as traditional PI and PD controllers, respectively. That is, the PD-like FLC adds damping and reliably predicts large overshoots, but does not improve the steady-state response (Chao and Teng 1997; Chung et al. 1998). PD control can reliably predict and correct large overshoots, but the derivative control will affect the steady-state error of a system.

A PD-like FLC can be developed by using either an error and change of error or an error and velocity model as

$$u = k_p.e + k_d.\Delta e \tag{5.30}$$

$$u = k_P\,.e + k_d\,.v \tag{5.31}$$

where k_p and k_d are the proportional and the differential gain coefficients and e is the error, Δe is the change of error and v is the hub velocity. In the next section two different types of PD-like FLC will be developed and analysed. These are:

- PD-like FLC with error and change of error as inputs
- PD-like FLC with error and velocity as inputs

Hub velocity v is measured using a tachometer from the arm system rather than derived it from hub angle as rate of change of error.

5.4.1 PD-Like Fuzzy Controller with Error and Change of Error

In this type of FLC, it is assumed that no mathematical model for the flexible-arm is available except two states, namely, the hub angle error and change of error. Only hub angle θ is measured from the system and the error and change of error are derived from θ. The hub angle error and change of error are defined as

$$e(k) = \theta_d - \theta(k) \tag{5.32}$$

$$\Delta e(k) = e(k) - e(k-1) \tag{5.33}$$

where θ_d is the desired hub angle, e is the error and Δe is the change in angle error. Figure 5.8 shows the block diagram of the PD-like FLC with error and change of error as inputs.

Triangular membership functions are chosen for inputs and output. The membership functions for hub angle error, change of hub angle error, and torque input are shown in Fig. 5.9. The universe of discourse for the hub angle error,

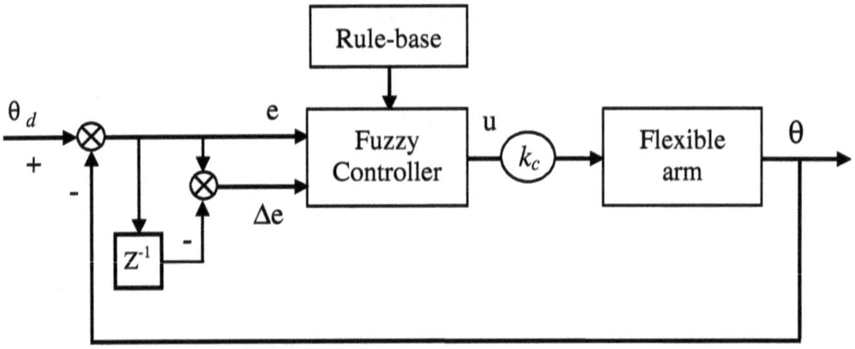

Fig. 5.8 PD-like FLC with hub angle error and change of hub angle error

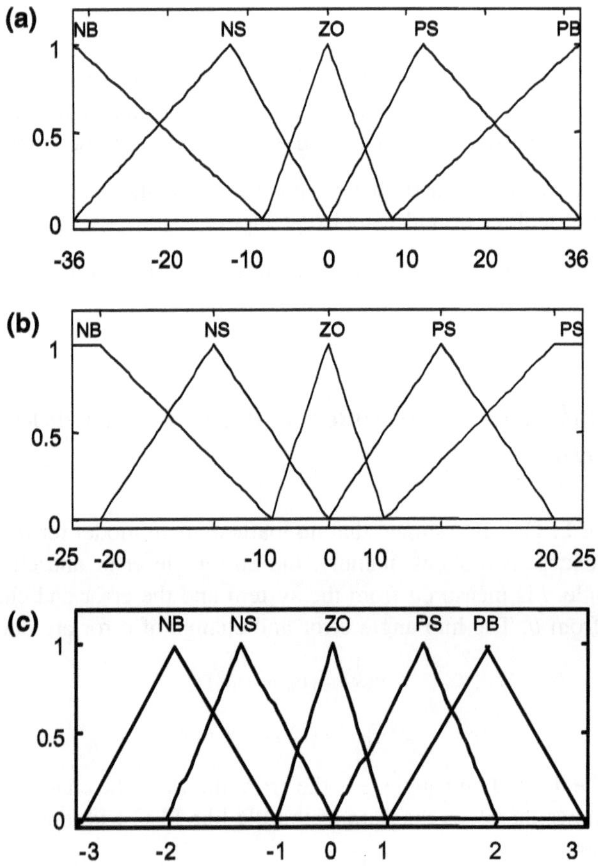

Fig. 5.9 Membership functions for inputs and output. Hub angle error (**a**); Change of hub-angle error (**b**); Torque input (**c**)

Table 5.2 Rule-base for PD-like FLC with angle error and change of angle error

Hub angle error	Change of error				
	PB	PS	ZO	NS	NB
NB	ZO	PS	PB	PB	PB
NS	NS	ZO	PS	PS	PB
ZO	NS	ZO	ZO	ZO	PS
PS	NB	NS	NS	ZO	PS
PB	NB	NB	NB	NS	ZO

change in hub angle error are chosen as [−36, +36] degree, and [−25, +25]. The universe of discourse of the output, i.e., input torque is chosen as [−3, +3] volts. To construct a rule base, the hub angle error, change of angle error and torque input are partitioned into five primary fuzzy sets as

$$\begin{aligned} \text{Hub angle error E} &= \{\text{NB, NS, ZO, PS, PB}\} \\ \text{Change of angle error C} &= \{\text{NB, NS, ZO, PS, PB}\} \\ \text{Torque U} &= \{\text{NB, NS, ZO, PS, PB}\} \end{aligned}$$

where E, C and U are the universes of discourse for hub angle error, change of hub angle error and torque input respectively. The *n*th rule of the rule base for the FLC, with error and change of error as inputs, is as

$$\mathrm{R}_n \ : \ \mathrm{IF}(e \text{ is } E_i) \text{ and } (\Delta e \text{ is } C_j) \text{ THEN } (u \text{ is } U_k) \tag{5.34}$$

where R_n, $n = 1, 2, \ldots, N_{\max}$ is the *n*th fuzzy rule, E_i, C_j, and U_k, for $i, j, k = 1, 2, \ldots, 5$ are the primary fuzzy sets. The five primary fuzzy sets, i.e. $\mathrm{MF_S}$ are NB, NS, ZO, PS, PB defined earlier. The rule base is shown in Table 5.2.

The membership functions defined in Fig. 5.9 and the rule-base defined in Table 5.2 form the control surface of the controller, which is shown in Fig. 5.10. The controller is applied to the single-link arm presented in Chap. 2. The performance of the system is shown in Fig. 5.11. For a demanded hub angle of 36°, it reached a maximum overshoot of 50°. The PD-like FLC shows rapid response at transient state, i.e. a rise time of 17 time units and a settling time of 44 time units. The performance of the PD-like FLC is very promising in respect of rise time, maximum overshoot and settling time but it shows a significant amount of steady state error of 2.56°.

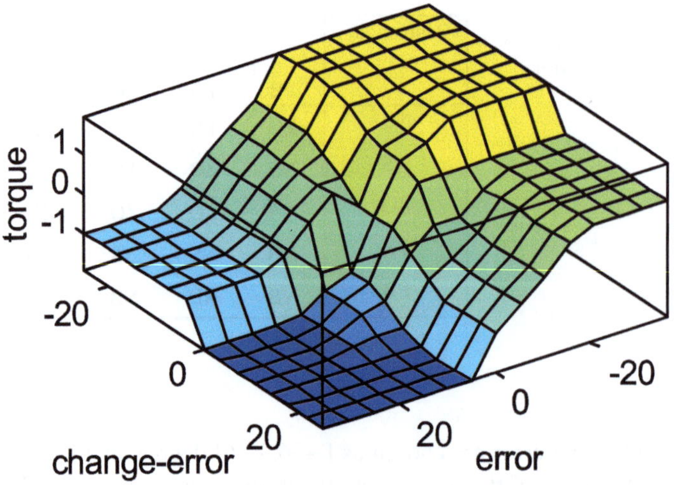

Fig. 5.10 Control surface of the controller with hub-angle error and change of hub-angle error

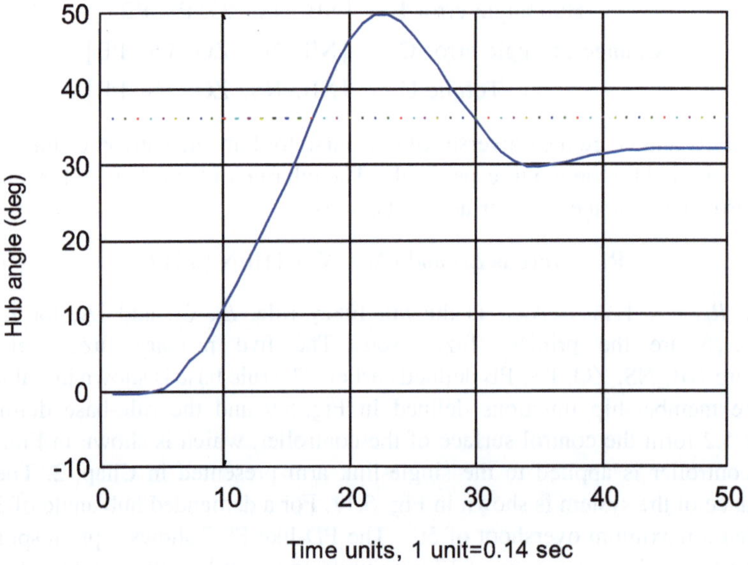

Fig. 5.11 Hub angle with FLC with hub-angle error and change of hub-angle error

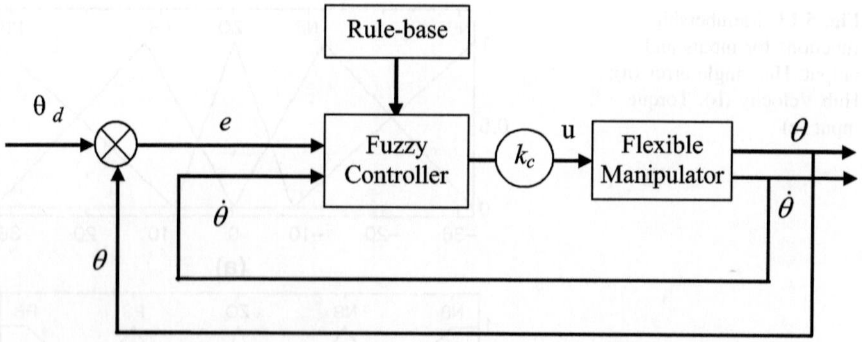

Fig. 5.12 Block diagram of a PD-like FLC with hub angle error and hub velocity

5.4.2 PD-Like Fuzzy Controller with Error and Velocity

In this type of FLC, two states are available, namely, the hub angle error and hub velocity, defined as

$$e = \theta_d - \theta \tag{5.35}$$

$$v = \dot{\theta} = \frac{\theta(k) - \theta(k-1)}{\Delta t} \tag{5.36}$$

where θ_d is the desired hub angle, θ is the measured hub angle, e is the angle error and v is the velocity. This type of implementation is also known as collocated PD-type controller mentioned in Chap. 3. The block diagram of the PD-like FLC with error and velocity is shown in Fig. 5.12. In this type of FLC, the hub velocity is measured from the system instead of deriving it from Eq. (5.36). Triangular membership functions are chosen for inputs and output. The membership functions for hub angle error, hub velocity and torque input are shown in Fig. 5.13. The universes of discourse for the hub angle error and velocity are chosen as [−36, +36] degree and [−33, 153] degree/s. The universe of discourse of the output is chosen as [−3, +3] volts. To construct a rule base, the hub angle error, hub velocity and torque input are partitioned into five primary fuzzy sets as

$$\begin{aligned} \text{Hub angle error E} &= \{\text{NB, NS, ZO, PS, PB}\} \\ \text{Hub Velocity V} &= \{\text{NB, NS, ZO, PS, PB}\} \\ \text{Torque U} &= \{\text{NB, NS, ZO, PS, PB}\} \end{aligned}$$

where E, V and U are the universes of discourse for hub angle error, hub velocity and torque input respectively. The nth rule of the rule base for the FLC, with error and velocity as inputs, is as

$$\text{R}_n : \text{IF}(e \text{ is } E_i) \text{ and } (v \text{ is } V_j) \text{ THEN } (u \text{ is } U_k) \tag{5.37}$$

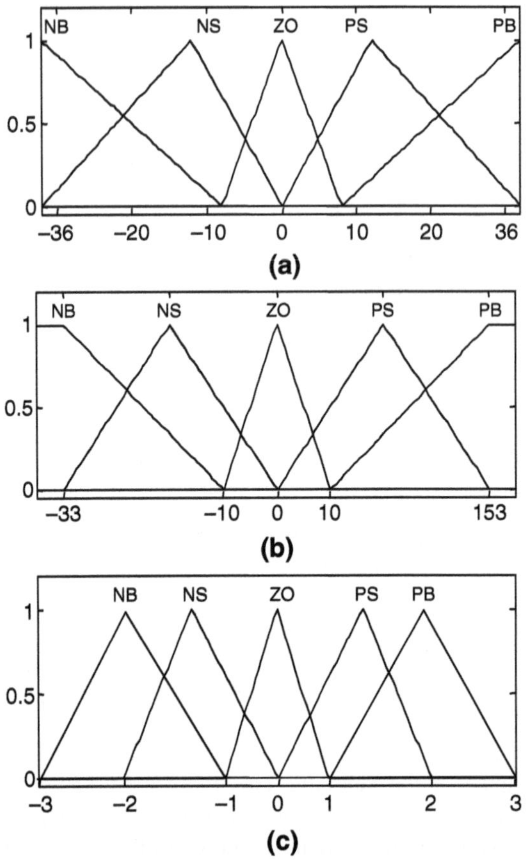

Fig. 5.13 Membership functions for inputs and output. Hub angle error (**a**); Hub Velocity (**b**); Torque input (**c**)

Table 5.3 FLC rule-base with hub-angle error and hub velocity

Angle error	Velocity				
	PB	PS	ZO	NS	NB
NB	ZO	PS	PB	PB	PB
NS	NS	ZO	ZO	PS	PB
ZO	NS	ZO	ZO	ZO	PS
PS	NB	NS	ZO	ZO	PS
PB	NB	NB	NB	NS	ZO

where R_n, $n = 1, 2, \ldots, N_{max}$, is the nth fuzzy rule, E_i, V_j, and U_k, for $i, j, k = 1, 2, \ldots, 5$ are the primary fuzzy sets. The rule base is shown in Table 5.3.

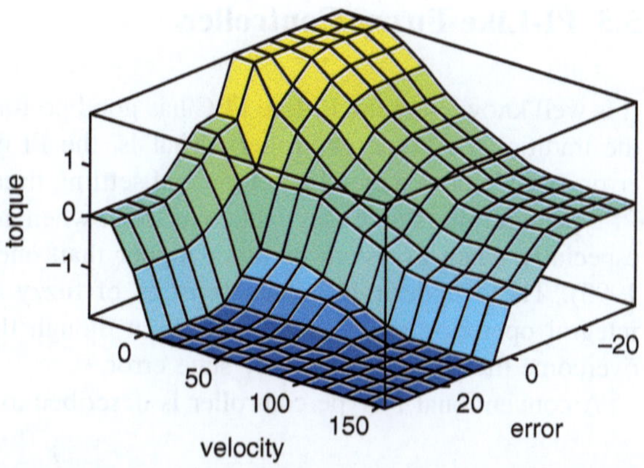

Fig. 5.14 Control surface of the controller with hub-angle error and hub velocity

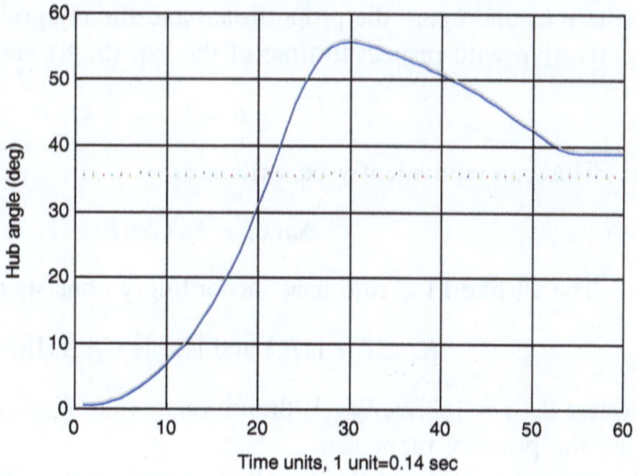

Fig. 5.15 Hub angle with FLC with hub-angle error and hub velocity

The control surface of the controller with error and velocity is shown in Fig. 5.14. The controller was implemented on the single-link arm. The performance of the system thus achieved is shown in Fig. 5.15. It is easily noticeable that rise time and overshoot are larger than that with PD-like FLC with hub angle error and change of hub angle error discussed in Sect. 5.4.1. For a demanded hub angle of 36°, it reached a maximum overshoot of 56.06°. It has slower rise time of 21 time units and slower settling time of 55 time units as well as a steady state error of −3.07°. PD-like FLC with hub angle error and change of error shows better performance in respect of rise time, maximum overshoot, settling time and steady state error than PD-like FLC with hub angle error and velocity.

5.5 PI-Like Fuzzy Controller

It is well known that the PI-like FLC has good performance at the steady state like the traditional PI-type controllers. That is, the PI-like FLC reduces steady-state error, but yields penalized rise time and settling time (Chao and Teng 1997). The PI-type controllers give inevitable overshoot when attempt to reduce the rise time, especially when a system of order higher than one is under consideration (Lee 1993). These undesirable characteristics of fuzzy PI controllers are caused by integral operation of the controller, even though the integrator is introduced to overcome the problem of steady state error.

A conventional PI-type controller is described as

$$u = k_P \cdot e + k_I \cdot \int edt \tag{5.38}$$

where k_P and k_I are the proportional and the integral gain coefficients. Taking the derivative with respect to time of the Eq. (5.38) yields

$$\dot{u} = k_P \cdot \dot{e} + k_I \cdot e \tag{5.39}$$

This can equivalently be written as

$$\Delta u(k) = k_P \cdot \delta e(k) + k_I \cdot e(k) \tag{5.40}$$

The PI-like FLC rule base, accordingly consists of rules of the form:

$$R_n : \text{If}(\, e \text{ is } E_i) \text{ and } (\Delta e \text{ is } C_j) \text{ THEN } (\Delta u \text{ is } CU_k) \tag{5.41}$$

where $R_n, n = 1, 2, ..., N_{max}$ is the nth fuzzy rule, E_i, C_j and CU_k, for $i, j, k = 1, 2, ..., 5$ are the primary fuzzy sets.

This type of controller is called an incremental PI-like FLC. The inputs are the same as a PD-like FLC with error and change of error except the control input is incremented at each time. Actually, the rules of fuzzy controller are designed with phase plane in mind, in which the fuzzy controllers drive a system into the so-called sliding mode. The tracking boundaries in the phase plane, however, are related not with incremental control input but with control input itself, which is calculated as

$$u(k) = \Delta u(k) + u(k-1) \tag{5.42}$$

To select the maximum variation of the incremental control input Δu giving satisfactory rise time and maximum overshoot is not so easy as in the case where the control input itself is to be determined (Lee 1993). One natural approach to overcome such difficult situation is to adopt the rate of change of error. Such controller may be called as PID fuzzy controller, which will be addressed later. Furthermore, a primary objective of this chapter is to investigate the performance of the PD- and PI-like FLCs with different inputs such as error (e), change of error (Δe) and sum of error (Σe) and hence this type of controller is not investigated and analysed further. Rather an

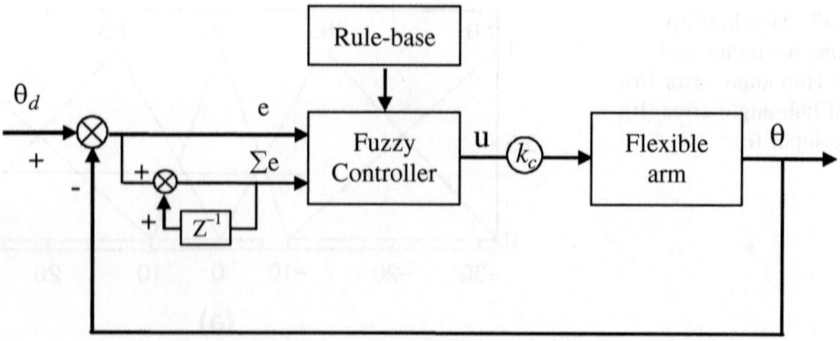

Fig. 5.16 Block diagram of a PI-type FLC

absolute PI-type controller is investigated. In an absolute PI-like FLC, error and sum of error are used as inputs and it is expressed as

$$u(k) = k_P \cdot e(k) + k_I \cdot \sum e(k) \quad (5.43)$$

where $\sum e(k)$ is the sum of error.

The block diagram of the absolute PI-like FLC is shown in Fig. 5.16. In this type, the hub angle is measured from the system and the sum of hub angle error is derived from the hub angle error. Triangular membership functions are chosen for inputs and output. The membership functions for hub angle error, sum of hub angle error and torque input are shown in Fig. 5.17. The universes of discourse for the hub angle error and sum of hub-angle error are chosen as [−36, +36] degree and [−150, +150] degree/s. The universe of discourse of the output is chosen as [−3, +3] volts.

To construct a rule base, the hub angle error, sum of hub angle error and torque input are partitioned into five primary fuzzy sets as

$$\begin{aligned}\text{Hub angle error E} &= \{\text{NB, NS, ZO, PS, PB}\}\\ \text{Sum of hubangle S} &= \{\text{NB, NS, ZO, PS, PB}\}\\ \text{Torque U} &= \{\text{NB, NS, ZO, PS, PB}\}\end{aligned}$$

where *E*, *S* and *U* are the universes of discourse for hub-angle error, sum of hub-angle error and torque input respectively. The *n*th rule of the rule base for this PI-type FLC is as

$$\mathrm{R}_n : \text{IF}(e\,\text{is}\,E_i)\text{ and }(s\,\text{is}\,S_j)\text{ THEN }(u\,\text{is}\,U_k) \quad (5.44)$$

where R_n, $n = 1, 2, \ldots, N_{\max}$ is the *n*th fuzzy rule, E_i, S_j, and U_k, for $i, j, k = 1, 2, \ldots, 5$ are the primary fuzzy sets. The rule base for the PI-like FLC is shown in Table 5.4.

A difficulty arises from deciding on the number of time units to go back in calculating the sum in Eq. (5.43). Even the literature on conventional control theory tends to be somewhat vague on this point, and many texts use an indefinite integral type of notation when representing the integral term, though obviously it is

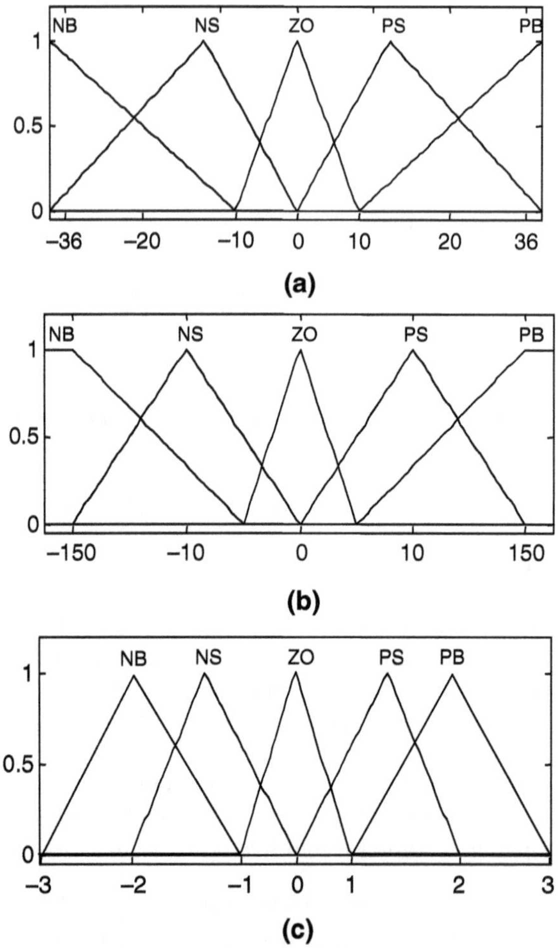

Fig. 5.17 Membership functions for inputs and output. Hub angle error (**a**); Sum of hub-angle error (**b**); Torque input (**c**)

Table 5.4 Rule-base for PI-like of FLC with angle error and sum of angle error

Hub angle error	Sum of error				
	PB	PS	ZO	NS	NB
NB	ZO	PS	PB	PB	PB
NS	NS	ZO	ZO	PS	PB
ZO	NS	ZO	ZO	ZO	PS
PS	NB	NS	ZO	ZO	PS
PB	NB	NB	NB	NS	ZO

Fig. 5.18 Control surface of the controller with error and sum of error

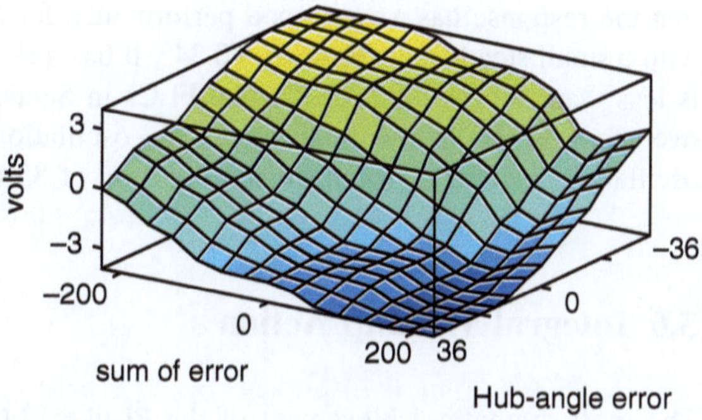

Fig. 5.19 Hub angle with PI-like FLC

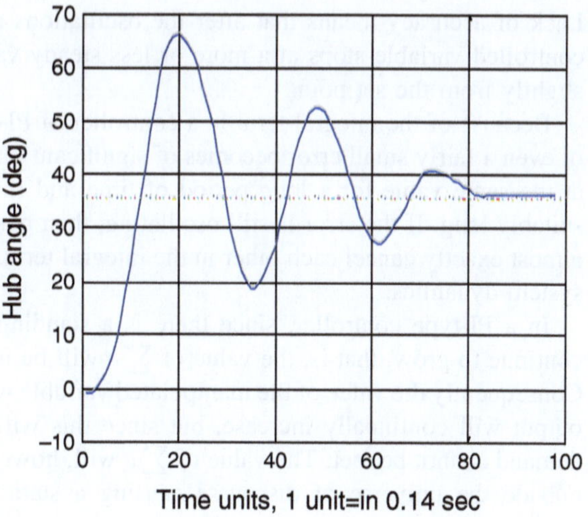

not to be taken literally. The reason for this vagueness may be that traditionally in conventional control, the integral term is approximated by analogue circuitry, and the integral limits cannot easily be stated precisely anyway (Lewis 1997).

Experience with the system suggested using 10 time units to indicate recent tendencies in the error, and experimentation demonstrated that this works very well. It was also convenient to work with an average rather than a sum so that the base value can be easily compared with the current error. Thus, the $\sum e$ base value is calculated as

$$\sum e(0) = \sum_{k=-9}^{0} e(k) \tag{5.45}$$

The control surface of the controller with hub angle error and sum of hub angle error is shown in Fig. 5.18.

The controller was implemented on the single-link arm. The response of the absolute PI-like FLC for the flexible-link arm is shown in Fig. 5.19. It can be seen that the response has a very good performance for a demanded hub angle of 36° with a small steady state error of −0.34°. It has a rise time of 12 time units, which is less than the rise time of PD-like FLCs in Sects. 5.4.1 and 5.4.2 and a larger overshoot of 66.45° with an excessive oscillation around the set point. The oscillations caused a prolonged settling-time of 85 time units.

5.6 Integral Windup Action

The most dramatic achievement of the PI-like FLC is to allow major improvements in steady-state accuracy with almost no negative impact on relative stability. Lack of accuracy means that after the oscillations die out in the system and the controlled variable stops at a more or less steady value, this value still may vary slightly from the set point.

Because of the integral term in a controller of PI-type or PID-type, the integral of even a fairly small error becomes of significant magnitude if the error continues at the same value for a long period of time and if the interval of the integral is suitably long. If the error is still oscillating, then positive and negative overshoots almost exactly cancel each other in the integral term, so that it has no effect on the system dynamics.

In a PI-type controller, since there is a standing error, the integral term will continue to grow, that is, the value of $\sum e$ will be increased at each sample time. Consequently the value of the manipulated variable will increase and the demanded output will continually increase, but since this will already be a maximum, the demand cannot be met. The value of $\sum e$ will, however, still be large and will fall outside the universe of discourse causing a static output. The effect is called integral wind-up action and results in the controller having a very poor response.

Many techniques have been developed for dealing with the problems of integral wind-up and the main ones are:

- Fixed limits on the integral term;
- Stop summation on saturation;
- Integral subtraction;
- Use of velocity algorithm;
- Analytical methods.

Fixed Limits

A maximum and minimum value for the integral summation is fixed and if the term exceeds this value it is reset to the maximum or minimum as appropriate, that is, if $s(n)_{max}$ then $u(n)_{max}$ and if $s(n)_{min}$ then $u(n)_{min}$.

Stop Summation

The value of the integral sum is frozen when the control actuator saturates and the integrator value remains constant while the actuator is in saturation. The scheme can be implemented either by freezing the summation term when the manipulated variable falls outside the range $u(n)_{min}$ and $u(n)_{max}$ or by the use of a digital input signal from the actuator which indicates that it is at a limit.

The stop summation gives a better response if the integral term is unfrozen once the sign of the error changes. The sign of the error will change before the actuator comes out of saturation.

Integral Subtraction

The idea behind this method is that the integral value is decreased by an amount proportional to the difference between the calculated value of the manipulated variable and the allowable maximum value. The integral summation $s(n) = s(n-1) + e(n)$ is replaced by $s(n) = s(n-1) + e(n) - K[u(n) - u_{\max}]$. The integral sum is thus decreased by the excess actuation and increased by the error. The rate of decrease is dependent on the choice of the parameter K. If K is not properly chosen then a continual saturation/desaturation oscillation can occur.

Velocity algorithm and analytical approach are not suitable to prevent integral wind-up for the fuzzy PI or PID controller and are not investigated in this study. Implementation of these algorithms can be found in (Bennet 1994).

In this study, the integral summation is calculated as follows

1. Calculate $\sum_{k=1}^{N} e(k) = \sum e(k-1) + e(k)$
2. If $k = N$, stop summation
3. Set $\sum e = 0$

A difficulty arises from deciding on the value of N. Even the literature on conventional control theory tends to be somewhat vague on this point. Experiment with the system shows that for $N = 10$ works very well. The integral windup action for different values of N is investigated for absolute PI-like FLC in this section. Figure 5.20a shows the integral wind up action for integral sum of 12 and 16. Neither of them settled within this time interval. The overshoot with integral sum 16 is higher (72.12°) than the overshoot with integral sum 12 (61.73°). Figure 5.20b shows the integral wind up action for integral sum 10 and 14. As can be seen from the figure that the controller with integral sum 10 achieved the set point at 82 time units with an overshoot of 66.34° where as the controller with integral sum 14 did not settle within the time interval and it has larger overshoot of 71.23°.

5.7 PID-Like Fuzzy Controller

Generally, PD-like two-term fuzzy controllers usually cannot eliminate steady state error and PI-like two-term fuzzy controller can eliminate steady state error but it has slower response due to the integral control variable. These characteristics have been studied and verified in Sects. 5.4 and 5.5 for flexible arm. In order to meet the design criteria of fast rise time, minimum overshoot, shorter settling time and zero steady state error, a further option is to develop a PID-like FLC which enables fast rise time, smaller overshoot and settling time from PD part and

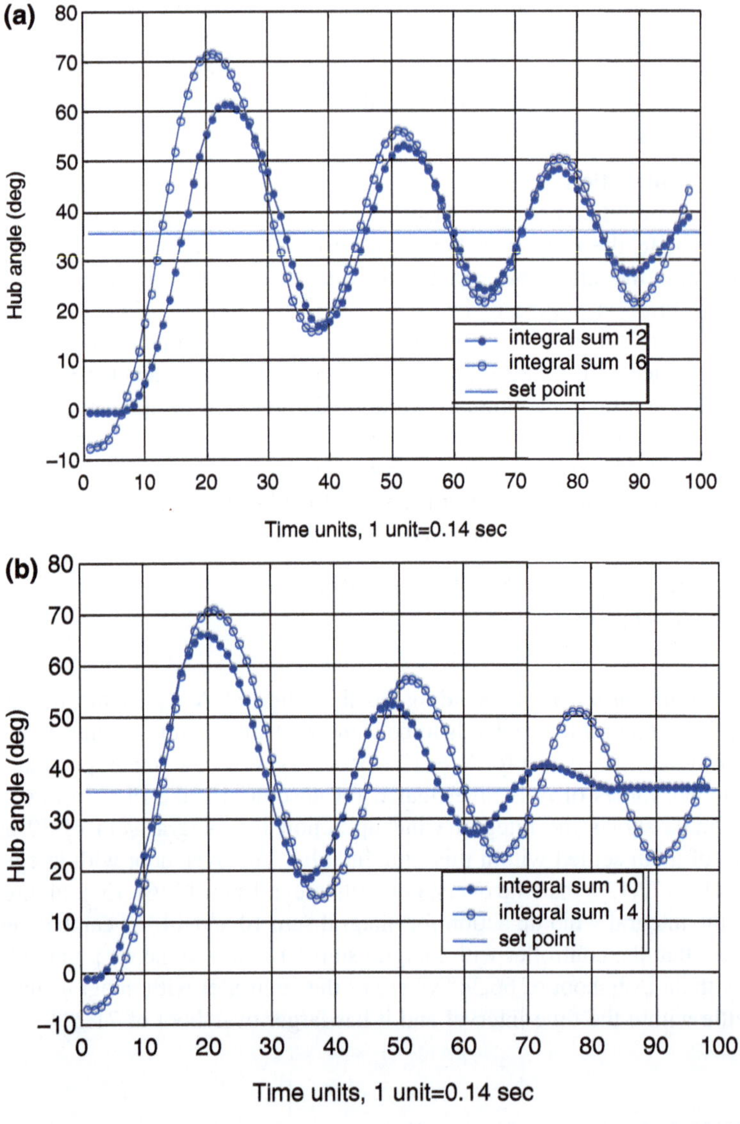

Fig. 5.20 Integral wind up action in PI-type FLC. Integral wind up action with $N > 10$(**a**); Integral wind up action with N around 10 (**b**)

minimum steady state error from PI part of the PID controller. The generic fuzzy PID controller is a four-dimensional (three input-one output) fuzzy system. The basic idea of a PID controller is to choose the control law by considering the error e, change of error Δe and integral of error or sum of error $\sum e$, and thus giving the controller as

$$u_{PID} = k_P \cdot e + k_D \cdot \Delta e + k_I \cdot \sum e \tag{5.46}$$

The fuzzy control rule corresponding to the PID-controller has the form

$$R_{PID}^{(n)} : \ if\ e\ is\ E_i\ and\ \Delta e\ is\ CE_j\ and \sum e\ is\ SE_k\ then\ u\ is\ U_l \tag{5.47}$$

The 3-input single-output PID-like FLC for flexible arm is shown in Fig. 5.7. Theoretically, the number of rules to cover all possible input variations for a three-term fuzzy controller is $n_1 \times n_2 \times n_3$, where n_1, n_2, and n_3 are the number of linguistic labels of the three input variables. In particular, if $n_1 = n_2 = n_3 = 7$, then the number of rules $R = 7 \times 7 \times 7 = 343$. In practical applications the design and implementation of such a large rule base is a tedious task, and it will take a substantial amount of memory space and reasoning time. Because of a long reasoning time the response of such a generic PID-like FLC will be too slow and hence not suitable for flexible arm where a fast response is desired.

A variety of approaches have been made to overcome the problems of PID-like fuzzy controllers in (Tzafestas and Papanikolopoulos 1990; Brehm 1994). Kwok et al. have considered a novel means of decomposing a PID controller into a fuzzy PD controller in parallel with various types of fuzzy gains, fuzzy integrators, fuzzy PI controller and deterministic integral control (Kwok et al. 1990, 1991). The various PID configurations are shown in Fig. 5.21a–f. For a process whose steady state gain is known or can be measured easily as k_p, then integral action is not necessary. If k_p is not known, integral action is necessary. This can be achieved by placing a conventional integral controller in parallel with the fuzzy PD controller. These hybrid types of PID controllers are not true fuzzy PID controllers as they include deterministic controls as well. A detailed description of these kinds of decompositions can be found in (Harris et al. 1993).

5.8 PD-PI-Type Fuzzy Controller

A variety of approaches have been made to overcome the problems of fuzzy PID controllers in (Tzafestas and Papanikolopoulos 1990; Brehm 1994). Kwok et al. have considered a novel means of decomposing a fuzzy PID controller into a fuzzy PD controller in parallel with various types of fuzzy gains, fuzzy integrators, fuzzy PI controller and deterministic integral control (Kwok et al. 1990, 1991). For a process whose steady state gain is known or can be measured easily as k_p, then integral action is not necessary. If k_p is not known, the integral action is necessary. This can be achieved by placing a conventional integral controller in parallel with the fuzzy PD controller. These hybrid types of PID controllers are not true fuzzy PID controllers as they include deterministic controls as well. A detailed description of these kinds of decompositions can be found in (Harris et al. 1993).

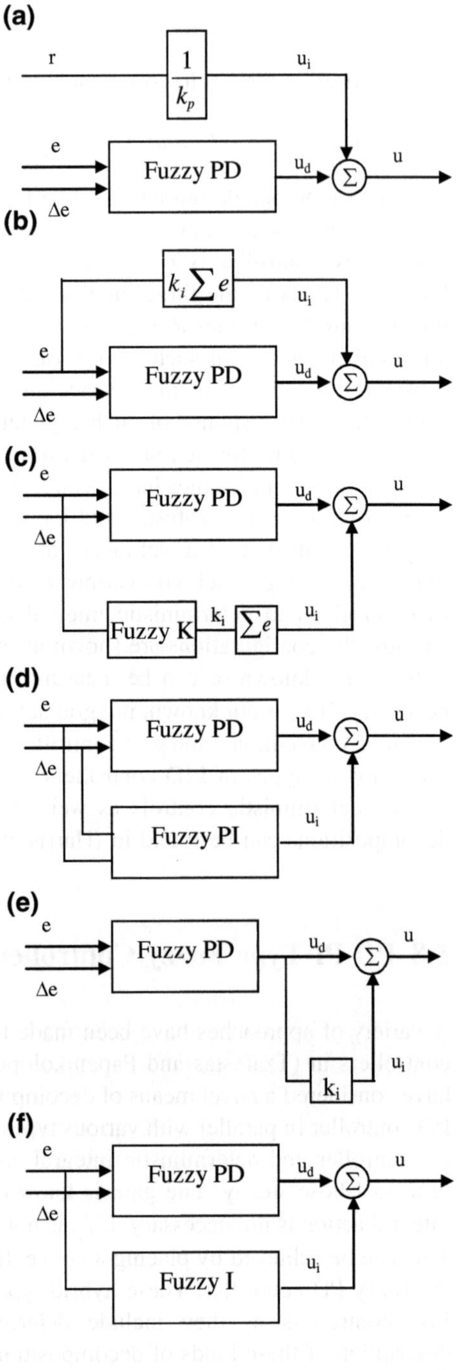

Fig. 5.21 Different implementations of fuzzy PID controller

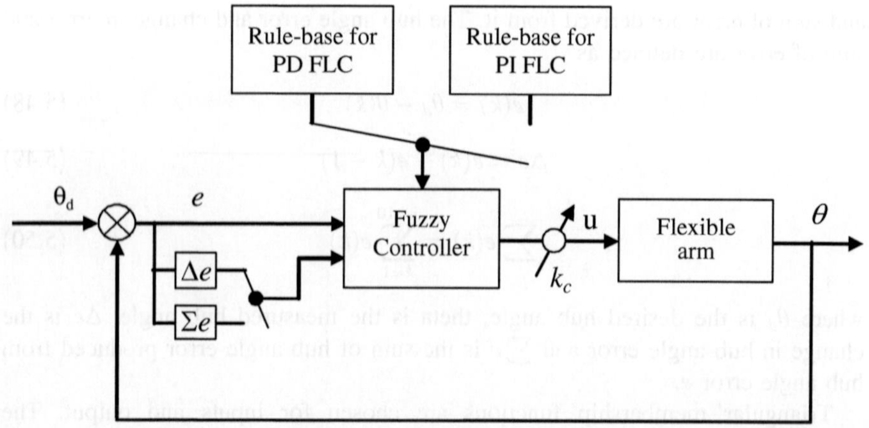

Fig. 5.22 Block diagram of a PD-PI-like FLC system

To overcome the problems of PD-, PI-type controllers described in Sects. 5.4 and 5.5, a number of approaches have been proposed. Steady state error in PD-like fuzzy controller is reduced by fine tuning the rule bases, performing parameter optimisation and increasing the number of rules (Guerocak and de Lazaro 1994). On the other hand, though the PI-like FLC can solve the steady state error problem, techniques such as scaling factor adjustment, rule modification and membership function shifting are required in order to reduce the rise time and improve the oscillatory behaviour (Maeda and Murakami 1992; Zheng 1992).

A typical method for rule reduction in a fuzzy PID-type controller is to divide the three-term PID controller into two separate fuzzy PD and fuzzy PI parts (Kwok et al. 1990; Zhang and Mizumoto 1994; Chen and Linkens 1998). This hybrid PD-like and PI-like fuzzy controller with n linguistic labels in each input variable require only $n \times n + n \times n = 2n^2$ rules, e.g. for $n = 5$ there will be $5 \times 5 + 5 \times 5 = 50$ rules, which is significantly smaller than n^3 rules (e.g. $5 \times 5 \times 5 = 125$) required by a generic PID-like fuzzy controller. This is the number of rules processed during execution of the controller consuming a significant amount of processing time and memory space. A further reduction is possible if the controller is switched from PD- to absolute PI-like after a certain period of time. In that case only one set of rules, $n \times n$ rules for each type of controller, will be executed at a time and thus the executed rules in a controller rule base will be reduced to only 25 rules for 5 linguistic labels in each input variable. Having been impressed with this idea, a switching type FLC is developed for the flexible arm where a PD-like FLC is executed first and then switched to a PI-like FLC. The block diagram of this switching PD-PI-like fuzzy controller is shown in Fig. 5.22.

The state variables used in PD-PI-like FLC are the same as in PD-like with hub-angle error and change of hub-angle error and absolute PI-like FLC namely, the hub-angle error, sum of hub-angle error and torque input. Actually the hub angle is measured from the system and the other states namely the error, change in error and sum of error are derived from it. The hub angle error and change in error and sum of error are defined as

$$e(k) = \theta_d - \theta(k) \tag{5.48}$$

$$\Delta e = e(k) - e(k-1) \tag{5.49}$$

$$\sum e(k) = \sum_{k=1}^{n=10} e(k) \tag{5.50}$$

where θ_d is the desired hub angle, theta is the measured hub angle, Δe is the change in hub-angle error and $\sum e$ is the sum of hub-angle error produced from hub angle error e.

Triangular membership functions are chosen for inputs and output. The membership functions for hub-angle error, change of angle error, sum of error and torque input are the same as it is in PD- and PI-like controller shown in Figs. 5.2 and 5.12 respectively. The universes of discourse for the hub-angle error, change in hub-angle error, sum of hub-angle error and torque input are chosen the same as in PD- and PI-like FLCs. To construct a rule base, the hub-angle error, change of hub-angle error, sum of hub-angle error and torque input are partitioned into 5 primary fuzzy sets as

$$\begin{aligned} \text{Hub-angle error E} &= \{\text{NB, NS, ZO, PS, PB}\} \\ \text{Change of hub-angle error C} &= \{\text{NB, NS, ZO, PS, PB}\} \\ \text{Sum of hub-angle error S} &= \{\text{NB, NS, ZO, PS, PB}\} \\ \text{Torque U} &= \{\text{NB, NS, ZO, PS, PB}\} \end{aligned}$$

where E, C, S and U are the universes of discourse for hub-angle error, change of hub-angle error, sum of hub-angle error and torque input respectively. The nth rule of the rule base for the PD-like FLC is as

$$\mathrm{R}_n: \text{ IF}(e \text{ is } E_i) \text{ and } (\Delta e \text{ is } C_j) \text{ THEN } (u \text{ is } U_k) \tag{5.51}$$

where R_n, $n = 1, 2, \ldots, 25$, is the nth fuzzy rule, E_i, C_j, and U_k, for $i, j, k = 1, 2, \ldots, 5$ are the primary fuzzy sets. The nth rule of the rule base for the PI-like FLC is as

$$\mathrm{R}_n: \text{ IF}(e \text{ is } E_i) \text{ and } \left(\sum e \text{ is } S_j\right) \text{ THEN } (u \text{ is } U_k) \tag{5.52}$$

where R_n, $n = 1, 2, \ldots, 25$, is the nth fuzzy rule, E_i, S_j, and U_k, for $i, j, k = 1, 2, \ldots, 5$ are the primary fuzzy sets. The same rule bases are used for both PD-like and PI-like controllers shown in Tables 5.2 and 5.3. The number of rules processed during execution of the individual controller is 25.

Determination of switching point is important and can result in a frustrating poor performance if chosen at the wrong point. It is obvious that if the controller switched at the point of maximum overshoot of the PD-like FLC, it can yield the best performance. But surprisingly, it does not show a good performance. Experimental investigations showed that a switching point just before or after the point of maximum overshoot gives better result than at the point of maximum overshoot and suggest a trial and error approach to find the switching point around the point of maximum overshoot.

5.9 Some Experimental Results on PD-PI-like FLC

The switching PD-PI-like FLC was applied to the flexible arm in real-time to verify the performance of the controller. The performance of the PD-PI-like FLC was verified for different switching points, control output scaling factors (k_c) and different integral summations Σe. Figure 5.23 shows the performance of the controller for a demanded hub angle of 36° with control output scaling factor $k_c = 86$ and switching point at 20 time units. A higher value of k_c produced a faster rise time of 14 time units causing larger overshoot of 53.63°. An early switching point, say 20 time units, causes a negative overshoot as well, which shows the dominance of the PI-type controller over PD-type. A conclusion can be drawn from this experiment (Fig. 5.23) that an earlier switching time than 20 time units will also cause the maximum overshoot to increase and more oscillations around the set point.

Figure 5.24 shows the hub angle with different switching points at 23 time units, 24 time units and 25 time units with scaling factor $k_c = 76$ for a set point of 36°. A switching point at 23 time units, indicated by solid line with dots, shows slightly larger overshoot and settling time and causes steady state error to increase. A switching point at 24 time units and 25 time units, indicated by solid line with circles and solid line with plus (+) respectively, yield better performance of the controller. These two curves have the same rise time of 15 time units, the same

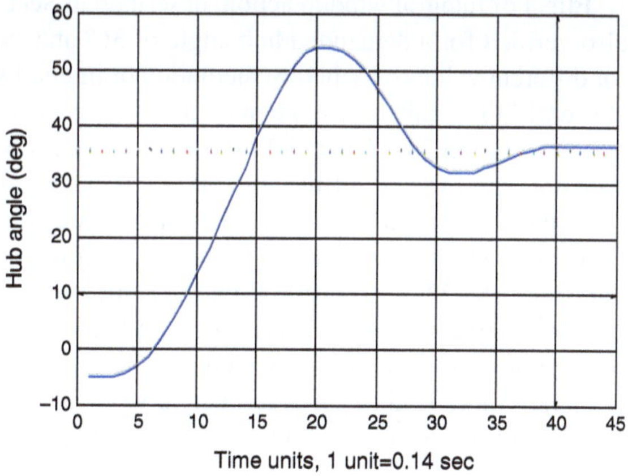

Fig. 5.23 Hub angle at switching point 20 time units and scaling factor $k_c = 86$

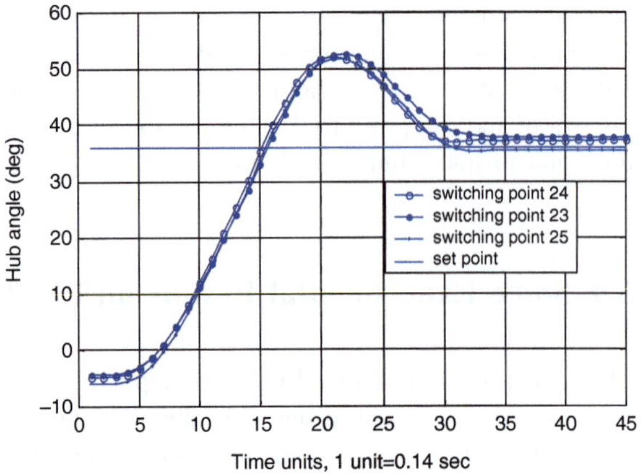

Fig. 5.24 Hub angle at different switching time with scaling factor $k_c = 76$

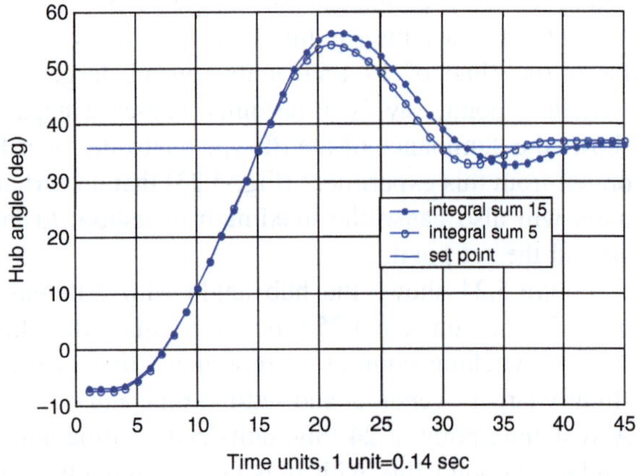

Fig. 5.25 Integral wind up action with $N < 10$ and $N > 10$

overshoot of 52.16° and negligible steady state error. In these experiments scaling factors k_p, k_d and k_i are assumed unity.

Effect of integral windup action, described in Sect. 5.6, on the PD-PI-like FLC is also verified for a demanded hub angle of 36° and switching point at 17 time units for different values of N in the calculation of integral summation. Switching point is chosen 17 time units to see the integral effect on the overshoot, which was observed earlier in Fig. 5.24. Figure 5.25 shows the integral windup action for $N = 15$ (line with dot) and for $N = 5$ (line with circle). It can be seen from the figure that in both cases there are larger overshoot and a smaller negative overshoot. The overshoot is little larger (56.18°) and settling time is longer (41 time units) for $N = 15$ than the overshoot (54.23°) and settling time (37 time units) for $N = 5$.

The integral windup actions shown in Fig. 5.26 are for values of $N = 10$ (line with circle) and $N = 11$ (line with dot). The overshoot and settling time for $N = 10$ are 55.23° and 31 time units respectively whereas the overshoot and settling time for $N = 11$ are 55.17° and 37 time units respectively. The experiment shows that

Fig. 5.26 Integral wind-up action with $N = 10$, $N = 11$

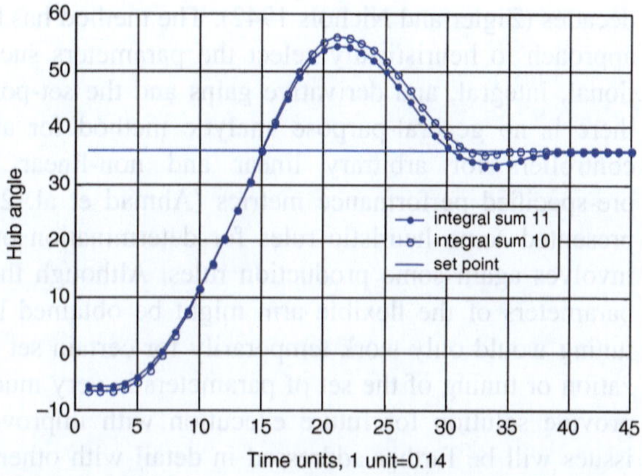

for $N = 10$ gives the best result in respect of overshoot and settling time. The control output-scaling factor k_c was set 76 in the experiments shown in Figs. 5.25 and 5.26.

5.10 Choice of Scaling Factors

The use of normalised domains (i.e. universes of discourse) requires a scale transformation, which maps the physical values of the input, output or any other process state variables into normalised domains. A denormalisation is also required, which maps the normalised values of the control output variables into their physical domains. Therefore, choice of scaling factors is important that performs the scale transformation. These scaling factors are similar to gain coefficients in conventional controller and are very critical to the controller performance and stability. Incorrect choice of these scaling factors can cause instability, excessive oscillation and deteriorated damping effects (Driankov et al. 1993; Li 1997; Palm 1995). The switching PD-PI-like fuzzy controller can be represented as a function of the scaling factors as:

$$k_c u(k) = f\left(k_p, e(k), k_d, \Delta e(k), k_I, \Sigma e(k)\right) \tag{5.53}$$

where k_p, k_d, k_I and k_c are the scaling factors for e, Δe, Σe and u respectively.

There are basically two widely practiced approaches to the determination of the scaling factors:

(i) Analytical
(ii) Heuristic

The most of the industrial controllers are of PI- and PID-type and these controllers mainly relied on the Ziegler-Nichols tuning rules for the past seven decades (Zigler and Nichols 1942). The method has been known as fairly accurate approach to heuristically select the parameters such as the controller's proportional, integral, and derivative gains and the set-point weighting. Unfortunately, there is no general-purpose analytic method for automatically tuning of fuzzy controllers for arbitrary linear and non-linear systems that can optimize pre-specified performance metrics (Ahmad et al. 2012). Driankov et al. (1993) presented some heuristic rules for determination of the scaling factors but this involves again some production rules. Although the initial value of the control parameters of the flexible arm might be obtained heuristically but the heuristic tuning would only work temporarily for certain set of available data. An optimization or tuning of the set of parameters is very much demanded such that it will provide solution for future execution with improved system performance. The issues will be further addressed in detail with other different technologies in the next few chapters.

5.11 Summary

An investigation into the development of PD-, PI-, and PID-like FLC has been carried out. It has been demonstrated that the system response with PD-like FLC with error and change of error as inputs exhibit positive and negative overshoot, fast rise time and small settling time. With error and velocity as inputs, on the other hand, a response with a relatively larger positive overshoot, slower rise time and settling time and noticeable steady state error is achieved. Whereas, with PI-like FLC with error and sum of error as inputs, the steady state error in the system is significantly improved at the expense of relatively larger overshoot and settling time.

A switching PD-PI-like FLC has been proposed on the basis of exploiting advantages of each of the PD- and PI-like FLCs. It has been shown that switching PD-PI-like FLC achieves improved performance and has some advantages over the PD-, PI- and PID-like FLCs. Firstly, it improves the steady state error caused by PD-like FLC and reduces the rise time and settling time caused by PI-like FLC. Secondly, it achieves a performance similar to a PID-like controller without using a huge rule-base. With this strategy the number of rules during execution of the FLC is reduced from n^3 to only n^2. In the case of a generic PID-like FLC, this has amounted to a reduction from 125 rules to 25, i.e. about 80 % reduction of the rule-base. In the case of a hybrid PD- and PI-like FLC, this has amounted to a reduction from 50 rules to 25, i.e. about 50 % reduction of the rule-base.

Integral windup action is an important issue, which causes a static output degrading the system performance. This was investigated with the proposed PD-PI-like FLC. It has been demonstrated that the adopted approach in calculating the sum of error in PI-like FLC resulted improvement in the system performance.

Performance of the PD-PI-like FLC crucially depends on the value of switching point. Since a PD-like FLC ensures a minimised overshoot, the switching point should be chosen at a point after achieving the maximum overshoot so that the PI-like FLC is dominated by the PD-like FLC. The situation is reverse in an earlier switching, which causes the overshoot and oscillation around set point to increase. The control output-scaling factor (k_c) has a significant effect on the performance of the FLC like any other controllers. A heuristic selection is adopted in choosing the value of the scaling factor.

References

Ahmad S, Siddique NH, Tokhi O (2012) Evolutionary tuning of modular fuzzy controller for two-wheeled wheelchair. Int J Comput Intell Appl 11(2):1–23

Bennet S (1994) Real time computer control: an introduction, 2nd edn. Prentice Hall, New York, London

Braae M, Rutherford DA (1978) Fuzzy relations in a control setting. Kybernetes 7:185–188

Braae M, Rutherford DA (1979) Theoretical and linguistic aspects of the fuzzy controller. Automatica 15:553–577

Brehm T (1994) Hybrid fuzzy logic PID controller. In: Proceedings of 3rd IEEE conference on fuzzy systems. 3:1682–1687

Cai Z-X (1997) Intelligent control: principles, techniques and applications. World Scientific, Singapore, NJ, London, Hong Kong

Carvajal J, Chen G, Ogmen H (2000) Fuzzy PID controller: design, performance evaluation, and stability analysis. Inf Sci 123:249–270

Chao C-T, Teng C-C (1997) A PD-like self-tuning fuzzy controller without steady-state error. Fuzzy Sets Syst 87:141–154

Chen G (1996) Conventional and fuzzy PID controllers: an overview. Int J Intell Control Syst 1:235–246

Chen M, Linkens DA (1998) A hybrid neuro-fuzzy PID controller. Fuzzy Sets Syst 99:27–36

Choi B-J, Kwak S-W, Kim BK (1999) Design of a single-input fuzzy logic controller and its properties. Fuzzy Sets Syst 106(8):299–308

Choi B-J, Kwak S-W, Kim BK (2000) Design and stability analysis of single-input fuzzy logic controller. IEEE Trans Syst Man Cybern B Cybern 30(2):303–309

Chou CH, Lu HC (1993) Real-time fuzzy controller design for hydraulic servo system. Int J Comput Ind 22:129–142

Chung H-Y, Chen B-C, Lin J-J (1998) A PI-type fuzzy controller with self-tuning scaling factors. Fuzzy Sets Syst 93:23–28

Driankov D, Hellendorn H, Reinfrank M (1993) An Introduction to Fuzzy Control. Springer-Verlag, New York

Farinwata SS, Filev D, Langari R (2000) Fuzzy control synthesis and analysis. Wiley, Chichester

Garcia-Benetiz E, Yurkovich S, Passino KM (1993) Rule-based supervisory control of a two-link flexible manipulator. J Intell Rob Syst 7:195–213

Guerocak HB, de Lazaro AS (1994) A fine tuning method for fuzzy logic rule bases. Fuzzy Sets Syst 67:147–161

Harris CJ, Moore CG, Brown M (1993) Intelligent control: aspects of fuzzy logic and neural nets, (World scientific series vol. 6). World Scientific, Singapore, NJ, London, Hong Kong

Hu BG, Mann GKI, Gosine RG (1999) New methodology for analytical and optimal design of fuzzy PID controllers. IEEE Trans Fuzzy Syst 7(5):521–539

Hu B-G, Mann GKI, Gosine RG (2001) A systematic study of fuzzy PID controllers—function-based evaluation approach. IEEE Trans Fuzzy Syst 9(5):699–712

Kovacic Z, Bogdan S (2006) Fuzzy controller design: theory and application. CRC Press, Taylor and Francis Group

Kubica K, Wang D (1993) A fuzzy control strategy for a flexible-link robot, IEEE International Conference on Robotics and Automation, vol 2. Atlanta, pp 236–241, 2–6 May

Kwok DP, Tam D, Li CK, Wang P (1990) Linguistic PID controllers. In: Proceedings of 11th IFAC world congress, Tallinn, USSR, pp 192–197

Kwok DP, Tam D, Li CK, Wang P (1991) Analysis and design of fuzzy PID control systems. In: Proceedings of IEE control'91 conference, vol 2. Heriot Watt University, Edinburg, pp 955–960

Lee J (1993) On methods for improving performance of PI-type fuzzy logic controllers. IEEE Trans Fuzzy Syst 1(1):298–301

Lee JX, Vukovich G, Sasaidek JZ (1994) Fuzzy control of a flexible link manipulator. In: Proceeding of American control conference; Baltimore, Maryland, USA, pp 568–574

Lewis HW (1997) The foundation of fuzzy control. Plenum Press, New York, London

Li H-X (1997) A comparative design and tuning for conventional fuzzy control. IEEE Trans Syst Man Cybern B Cybern 27(5):884–889

Li H-X, Gatland HB (1995) A new methodology for designing a fuzzy logic controller. IEEE Trans Syst Man Cybern 25:505–512

Li H-X, Gatland HB (1996) Conventional fuzzy control and its enhancement. IEEE Trans Syst Man Cybern B Cybern 26:791–797

Li H-X, Zhang L, Cai K-Y, Chen G (2005) An improved robust fuzzy-PID controller with optimal fuzzy reasoning. IEEE Trans Syst Man Cybern B Cybern 35(6):1283–1294

Liu K, Lewis FL (1994) Hybrid feedback linearization/fuzzy logic control of a flexible-link manipulator. J Intell Fuzzy Syst 2:325–336

MacVicar-Whelan PJ (1976) Fuzzy sets for man-machine interaction. Int J Man Mach Stud 8:687–697

Mamdani EH, Assilian S (1974) Application of fuzzy algorithms for control of simple dynamic plant. Proc IEEE 121:1585–1588

Maeda M, Murakami S (1992) A self-tuning fuzzy controller. Fuzzy Sets Syst 51:29–40

Meressi T (1995) Non-linear and fuzzy logic vibration control of flexible manipulators. In: Proceedings of the American control conference, Seattle, Washington, USA, pp 4238–4242

Mohan BM, Patel AV (2002) Analytical structures and analysis of the simplest fuzzy PD controllers. IEEE Trans Syst Man Cybern B 32(2):239–248

Moudgal VG, Kwong WA, Passino KM, Yurkovich S (1994a) Fuzzy learning control for a flexible-link robot. In: Proceedings of American control conference, Maryland, pp 563–567

Moudgal VG, Yurkovich S, Passino KM (1994b) Expert supervisory control for a two-link flexible robot. In: Proceedings of IEEE international conference on robotics and automation, San Diego, 8–13 May, pp 3296–3301

Moudgal VG, Kwong WA, Passino KM, Yurkovich S (1995) Fuzzy learning control of a flexible-link robot. IEEE Trans Fuzzy Syst 3(2):199–210

Mudi RK, Pal NR (1999) A robust self-tuning scheme for PI- and PD-type fuzzy controllers. IEEE Trans Fuzzy Syst 7(1):2–16

Palm R (1992) Sliding mode fuzzy control. Int Control Fuzzy Syst 1992:519–526

Palm R (1994) Robust control by fuzzy sliding mode. Automatica 30(9):1429–1437

Palm R (1995) Scaling of fuzzy controllers using cross-correlation. IEEE Trans Fuzzy Syst 3(1):116–123

Pedrycz W, Gudwin R, Gomide F (1997) Nonlinear context adaptation in the calibration of fuzzy sets. Fuzzy Sets Syst 88(1):91–97

Rattan KS, Chiu B, Feliu V, Brown HB Jr (1994) Rule-based fuzzy control of a single-link flexible manipulator in the presence of joint friction and load changes. In: Proceedings of the American control conference, Seattle, Washington, pp 2749–2750

Shin YC, Xu C (2009) Intelligent systems: modelling, optimistion and control. CRC Press, Taylor and Francis Group

Sooraksa P, Pattaradej T, Chen G (2002) Design and implementation of fuzzy PID controller for handlebar of a bicycle robot. Integr Comput Aided Eng 9(4):319–331

Sugeno M (1985) An Introductory Survey of Fuzzy Control. Inf Sci 36:59–83

Sugeno M, Yasukawa T (1993) A fuzzy-logic-based approach to qualitative modelling. IEEE Trans Fuzzy Syst 1(1):7–31

Takagi T, Sugeno M (1985) Fuzzy identification of systems and its application to modeling and control. IEEE Trans Syst Man Cybern 15:116–132

Tang KL, Mulholland RJ (1987) Comparing fuzzy logic with classical controller designs. IEEE Trans Syst Man Cybern 17:1085–1087

Tang KS, Man KF, Chen G, Kwong S (2001a) An optimal fuzzy PID controller. IEEE Trans Ind Electron 48(4):757–765

Tang WM, Chen G, Lu RD (2001b) A modified fuzzy PI controller for a flexible-joint robot arm with uncertainties. Int J Fuzzy Sets Syst 118:109–119

Tsukamoto Y (1979) An approach to fuzzy reasoning model. Adv Fuzzy Set Theory Appl 137:149 (Amsterdam, North-Holland)

Tzafestas S, Papanikolopoulos NP (1990) Incremental fuzzy expert PID control. IEEE Trans Ind Electron 37:365–371

Vukovich G, Lee JX (1999) Experiments of fuzzy versus PD controls of flexible link manipulator. In: Proceeding of 14th world congress of IFAC, Beijing, China, pp 397–402

Yager RR, Filev D (1994) Essentials of fuzzy modelling and control. Wiley, NY, Chichester

Zadeh LA (1965) Fuzzy Sets. Inf Control 8:338–353

Zadeh LA (1968) Fuzzy algorithms. Inf Control 12:94–102

Zadeh LA (1973) Outline of a new approach to the analysis of complex systems and decision process. IEEE Trans Syst Man Cybern 3:28–44

Zadeh LA (1975a) Fuzzy logic and approximate reasoning. Syntheses 30:407–428

Zadeh LA (1975b) The concept of a linguistic variable and its application to approximate reasoning-I. Inf Sci 8:199–249

Zadeh LA (1994) The role of fuzzy logic in modelling. Ident Control Model Ident Control 15(3):191–203

Zadeh LA (2008) Is there a need for fuzzy logic? Inf Sci 178(13):2751–2779

Zhang Z, Mizumoto M (1994) On rule self-generating for fuzzy control, International Journal of Intelligent Systems. 9(12):1047–1057

Zheng L (1992) A practical guide to tune of proportional and integral (PI) like fuzzy controller. In: Proceedings of 1st IEEE international conference on fuzzy systems, San Diego, CA, pp 633–641

Ziegler JG, Nichols NB (1942) Optimum settings for automatic controllers. Trans ASME 64(1942):759–768

Chapter 6
Evolutionary-Fuzzy Control

6.1 Introduction

Fuzzy logic controllers (FLC) have found many successful applications in engineering and industrial process control (Kovacic and Bogdan 2006; Chan et al. 2008; Siddique and Adeli 2013). Despite all these successes, FLCs experience a deficiency in knowledge acquisition and rely to a great extent on empirical and heuristic knowledge. One of the most important considerations in designing FLCs is construction of the membership functions (MF) for each linguistic term as well as the rule-base. In most existing applications, the fuzzy rules are generated by an expert in the area, especially for the control problems with only a few inputs. The correct choice of MFs is by no means trivial but plays a crucial role in the success of an application. Previously, generation of MFs had been a task mainly done either iteratively by trial and error or by human experts. With an increasing number of inputs and linguistic variables, the possible number of rules for the system increases exponentially, which makes it difficult for experts to define a complete set of rules and associated MFs for a good system performance. The pitfall of fuzzy logic system is that it cannot learn from experience. Even if an initial rule-base and MFs are known a priori, tuning of scaling parameters is of importance in improving the performance of the FLCs. When acquired knowledge is not enough for the systems to be controlled, some kind of learning or adaptation is essential. Researchers have been trying to employ learning and adaptation mechanisms that start from an empty or a randomly generated knowledge and learn towards an optimal knowledge. Incorporation of any learning or knowledge acquisition mechanism to fuzzy controllers would be preferable.

There are many methods reported for tuning and learning of fuzzy controllers in the literature (Adeli and Hung 1995; Cordon et al. 2001; Jang et al. 1997; Siddique and Adeli 2013; Vidyasagar 2002). However, methods such as neural-fuzzy and evolutionary-fuzzy approach are the most widely used by researchers. Neural network based tuning and learning of fuzzy controllers has found huge applications in intelligent control (Chan and Kazabov 2004; Chen and Linkens 1998; Siddique and Tokhi 2006), especially for two reasons: learning capability of neural networks can easily be incorporated as a supportive mechanism for fuzzy systems

N. Siddique, *Intelligent Control*, Studies in Computational Intelligence 517,
DOI: 10.1007/978-3-319-02135-5_6,

(or controllers) and neural networks can be hybridised with fuzzy controllers in a straightforward way. The detailed treatment of the two topics can be found in (Jang et al. 1997; Siddique and Adeli 2013). Efforts have been made to automate the construction of rule-bases and define the MFs in various ways using neural networks (Lin and Lee 1991, 1992, 1993, 1995; Nauck and Kruse 1992, 1993). Neural network-based tuning and learning of fuzzy controllers will be addressed in the next chapter. Though popular, neural-fuzzy approach poses a potential problem of applying to designing controllers, i.e. the neural-fuzzy approach essentially demands some kind of experiential data, which may not be available always.

An alternative to learning mechanism in the absence or unavailability of experiential data would be adaptation, whereby the fuzzy controller starts from a randomly generated knowledge from a human expert. The general advantage of fuzzy system is that experts can design fuzzy systems or FLC based on heuristic or no a priori information about the system to be controlled. Very often the expert knowledge represents partial, incomplete or incorrect description of the system as well. The construction of MFs and rule-base using expert knowledge sometimes does not reflect the actual data distribution of the system, which results in poor performance. It is also often the case that the extracted rules may be independent of the MFs, there may be redundant or useless rules that will never fire, there may be inappropriate number of MFs, insufficient overlap between MFs, or inappropriate choice of MF-type. It is especially true for a complex nonlinear system, e.g., flexible robotic arm system, and system with large number of input variables. In these cases, an optimal performance for the FLC cannot be guaranteed. Adaptation, readjustment or optimisation of the MFs and the rule-base are essential for the improvement of performance and robustness of operation of the fuzzy system over the entire data range in changing operating conditions.

Therefore, the design of a fuzzy controller or system can be formulated as a search problem in a high-dimensional space where each point in the space represents a rule set, membership function and the corresponding controller's performance, that is, the performance of the controller forms a hyper-surface in the space according to some given performance criteria. Thus, finding the optimal location of this hyper-surface is a search problem, which is equivalent to developing the optimal fuzzy system design (Shi et al. 1999). The following features characterize the hyper-surface:

- The number of possible fuzzy sets for each variable is unbounded, and it makes the hyper-surface infinitely large.
- The hyper-surface is non-differentiable since changes in the number of fuzzy sets are discrete and can have discontinuous effect on the fuzzy controller's performance, which restricts application of any derivative based optimisation algorithms.
- The hyper-surface is complex and noisy since the mapping from a fuzzy rule set to its performance is indirect and dependent on the evaluation method used.
- Different fuzzy rules and MFs may have similar performance making the hyper-surface multi-modal.

- The hyper-surface is deceptive since similar fuzzy rule sets and MFs may have quite different performance.

These characteristics of a fuzzy controller make evolutionary algorithms a more suitable method for searching the hyper-surface over many other search or optimisation methods such as Tabu-search, simulated annealing, and hill climbing search. A generic block diagram of the evolutionary algorithm (EA) based fuzzy controller design is shown in Fig. 6.1.

It is important in designing of the EA based fuzzy controller to decide which parts of the fuzzy controller or system are subject to optimisation, i.e. different blocks in Fig. 6.1. It is also important to distinguish between tuning and learning of fuzzy control components. Tuning deals with optimisation of an existing fuzzy controller or system with initially defined MFs, rule-base and scaling parameters, whereas learning deals with automated design of the fuzzy system carrying out an elaborate search for a set of MFs, rule-base and scaling functions which ensures an optimal performance of a fuzzy system. The tuning or learning can take place in two different forms:

- structure tuning or learning, i.e. rule base tuning or learning and
- parameter tuning or learning, i.e. MFs and scaling factors tuning or learning.

The structure of the fuzzy controllers usually varies for the different type of fuzzy models chosen for the controllers. Different types of fuzzy models are available such as Mamdani-, Sugeno- and Tsukamoto-type, which have been discussed in Chap. 4. Two different forms of tuning or learning are shown using block diagrams in Figs. 6.2 and 6.3.

Evolutionary learning, in this case, is seen as a search or optimisation problem for a randomly generated system. The only requirement is the definition of suitable scalar index that can measure the optimal performance of the fuzzy controller. The

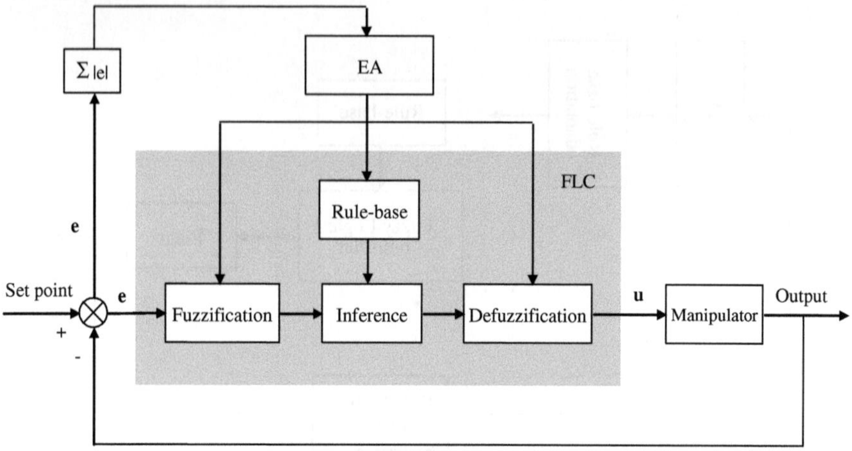

Fig. 6.1 EA-based optimisation of MF parameters and rules

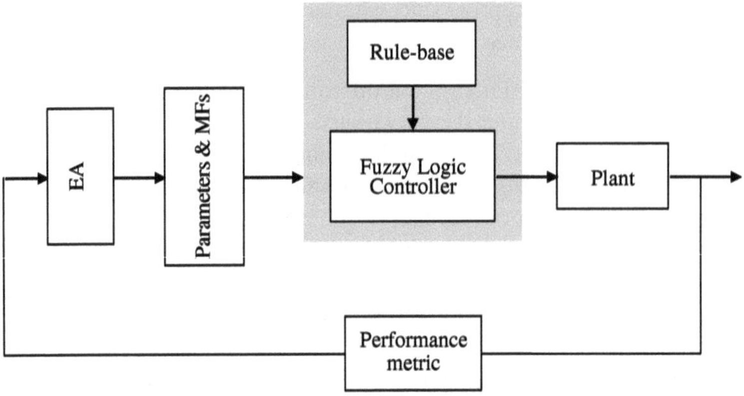

Fig. 6.2 EA based tuning/learning for fuzzy controller parameters and MFs

advantage of the evolutionary learning is the incorporation of a priori knowledge. The a priori knowledge can be in different forms such as linguistic variables, MF parameters, fuzzy rules, and number of rules. There are many approaches reported in the literature for evolutionary learning of fuzzy controllers. These approaches are mainly used for three levels of learning fuzzy rules:

- Michigan approach—where each chromosome in the population represents a single rule and a rule set is represented by the entire population.
- Pittsburgh approach—where each chromosome in the population represents the entire rule set, maintains a population of candidate rule sets and applies genetic operators (selection, crossover and mutation) to produce new generations of rule sets.
- Iterative rule learning approach—known as the third approach to reduce the dimension of the search space by encoding individual rules like the chromosome

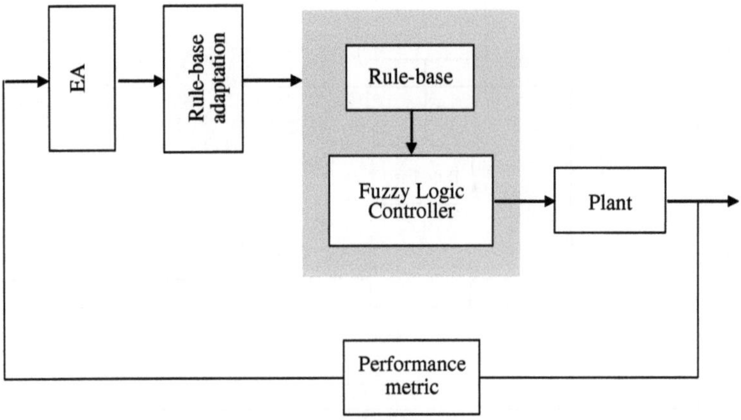

Fig. 6.3 EA based tuning/learning for fuzzy controller rule-base

in Michigan approach and a new rule is adopted and added to the rule-base in an iterative way during execution of the EA.

Three main strategies, including Pittsburgh-type, Michigan-type, and the iterative rule learning genetic fuzzy systems, focus on generating and learning fuzzy rules in genetic fuzzy systems. First, the Pittsburgh-type genetic fuzzy system (Stewart et al. 1997) was characterized by using a fuzzy system as an individual in genetic operators. Second, the Michigan-type genetic fuzzy system was used for generating fuzzy rules in (Ishibuchi et al. 1999), where each fuzzy rule was treated as an individual. Thus, the rule generation methods in (Ishibuchi et al. 1999) were referred to as fuzzy classifier systems. Third, the iterative rule learning genetic fuzzy system (Castillo et al. 2001; Cordon et al. 1999; Gonzalez and Perez 1999) was adopted to search one adequate rule set for each iteration of the learning process. Detail discussion of the three approaches is beyond scope of this book. Interested readers are referred to (Siddique and Adeli 2013; Cordon et al. 2001). Moreover, an initial design approach for fuzzy controllers, especially for Mamdani-type fuzzy controller for a non-linear system such as flexible arm, has been discussed in Chap. 5. Therefore, issues of evolutionary tuning of fuzzy controllers for both structure and parameters will be addressed in this chapter.

In order to avoid processing of a huge rule-base required by a generic PID-like fuzzy controller, a switching PD-PI-like fuzzy controller has been developed in the previous chapter. Secondly, the design objectives for the PD-PI fuzzy controller were the fast rise time, smaller overshoot and minimized steady state error as if it were a generic PID-type controller. This encounters a number of problems such as designing the MFs for error and change of error and rule-base for a PD part of the FLC and designing MFs for error and sum of error and rule-base for PI part of the FLC. Switching point of PD-PI fuzzy controller is determined empirically by trial and error, which requires few trials and has been investigated in Chap. 5. Therefore, determination of switching point is not included in the EA-based optimisation procedure carried out in this chapter. Among the remaining problems to be resolved in the fuzzy controller design are optimisation of MFs of each linguistic terms and optimisation of control rules for both PD and PI parts of the controller while keeping the scaling factors as it is.

Evolutionary algorithms are general purpose search and optimisation algorithms that can be deployed for tuning fuzzy controllers, i.e. rule-base and parameters of MFs (Hoffmann 2001; Siarry and Guely 1998; Zajaczkowski and Verma 2009). In most of the cases, either the rule-base is fixed and the parameters of the MFs are adjusted or MFs are fixed and the rule-base is optimised by evolutionary algorithms. Some researchers have optimised the rule-base, MFs, scaling factors and controller parameters, which seem somewhat redundant.

A systematic and automated approach of designing fuzzy controller (especially the PD-PI like fuzzy controller developed in Chap. 5) using EA with suitable parameter settings will be investigated further. Therefore, a brief overview of EAs and different algorithms are presented first followed by the application of genetic algorithm (GA), a variant of EA discussed later, in the design of fuzzy controllers.

6.2 Overview of Evolutionary Algorithms

As mentioned earlier that evolutionary algorithms are general purpose search and optimisation algorithms based on Darwin's (1859) theory of evolution. Evolutionary systems have first been viewed as optimization processes in the 1930s. The fundamental idea of evolution as a computational process gained momentum in the late 1950s nearly a century after Darwin's theory of evolution. Fraser (1957) was the first to conduct a simulation of genetic systems representing organisms by binary strings. Box (1957) proposed an evolutionary operation to optimizing industrial production. Friedberg (1958) proposed an approach to evolve computer programs. The fundamental works of (Fogel 1962) in evolutionary programming, (Holland 1962) in genetic algorithms, (Rechenberg 1965) and (Schwefel 1968) in evolution strategies had great influences on the development of evolutionary algorithms and computation as a general concept for problem solving and as a powerful tool for optimization. Since the development years of 1960s, the field evolved into three main branches (De Jong 2006): evolution strategies, evolutionary programming, and genetic algorithms. There were significant contributions to the field by many people. Among them are De Jong (1975), Goldberg (1989) and Fogel (1995) are few to name. The 1990s have seen another set of development in the evolutionary algorithms such as Koza (1992) developed genetic programming, Reynolds (1994, 1999) developed cultural algorithms and Storn and Price (1997) developed differential evolution. Evolutionary algorithms have now found a wide spread of applications in almost all branches of science and engineering.

The process of evolution can be modelled algorithmically and simulated on a computer (Fogel 1998). In the simplest form the model can be expressed as

$$\Pi(g+1) = \Psi[\Phi[\Pi(g)]] \tag{6.1}$$

Fig. 6.4 Basic principle of evolutionary algorithm

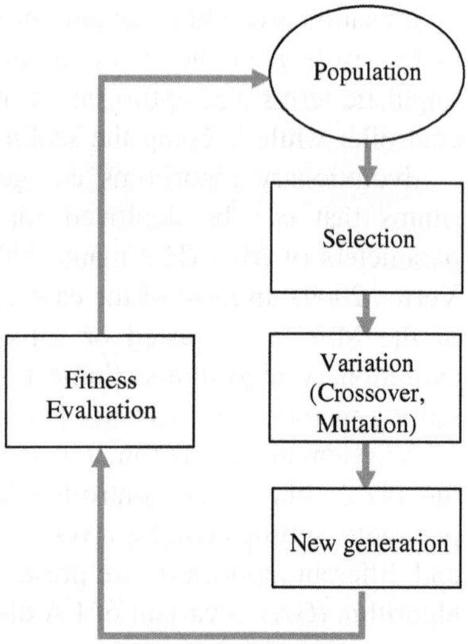

The operations of random variation Φ and selection Ψ are applied on a population of $\Pi(g)$ at generation g to evolve to a new population of $\Pi(g+1)$ in the next generation. The variations with higher fitness have better chance of being survived in the struggle for existence, leading to a process of continual adaptation (Darwin 1859). The basic principle of evolutionary algorithm is shown in Fig. 6.4. Application of the variations and selection operation over generations drives the population towards a particular optimal point in a search space.

Based on the representation of the population, variation operation, selection mechanism and the way of fitness evaluation under the same basic principle of evolutionary algorithm, there are a variety of evolutionary algorithms widely in use:

- Evolutionary Programming,
- Evolution Strategies,
- Genetic Algorithms,
- Genetic Programming,
- Differential Evolution,
- Cultural Algorithm.

6.2.1 Evolutionary Programming

Evolutionary Programming (EP), originally developed by Lawrence Fogel in the 1960s, is a stochastic optimisation strategy (Fogel 1962). EP was then further developed by David Fogel in the 1990s (Fogel 1991, 1992). The basic difference of EP from other EAs in that it emphasises the development of behavioural models rather than the genetic models. EP is derived from simulation of adaptive behaviour in evolution. That is, EP considers phenotypic[1] evolution. The evolutionary process consists of finding a set of optimal behaviours from a space of observable behaviours. For this purpose, the fitness function measures the behaviour error of an individual with respect to the environment of that individual.

Hwang (1999) and Kang et al. (2000) proposed EP based design of an optimal fuzzy rule base for modelling and control that simultaneously evolve the structure and the parameter of fuzzy rule base for a given task.

6.2.2 Evolution Strategies

Evolution strategies (ES) were developed as a method to solve parameter optimisation problems by Rechenberg in the 1960s (Rechenberg 1965) and further developed by (Schwefel 1968). Evolution-strategic optimisation is based on the hypothesis that during the biological evolution the laws of heredity have been developed for fastest phylogenetic adaptation. ES imitate, in contrast to the GAs, the

[1] The phenotype describes the outward appearance of an individual.

effects of genetic procedures on the phenotype. The presumption for coding the variables in the ES is the realization of a sufficient strong causality, i.e. small changes of the cause must create small changes of the effect. The theory states that evolutionary progress takes place only within a very narrow band of the mutation step size.

Bäck and Kursawe (1994) gave a brief overview on the application of ES in the design of fuzzy controllers and showed the advantages of ES over genetic algorithms. Flexibility, completeness and consistency are essential for fuzzy systems to exhibit an excellent performance and compactness is crucial when the number of input variables increases. Jin et al. (1999) proposed a systematic design paradigm for flexible, complete, consistent and compact fuzzy systems using evolution strategies. The structure of the fuzzy rules, which determines the compactness of the fuzzy systems, is evolved along with the parameters of the fuzzy systems.

6.2.3 Genetic Programming

Friedberg (1958) and Friedberg et al. (1959) were among the first to evolve computer programs. The word 'evolution' was not used at that time though author intent to simulate evolution. Dunham and North pursued this line of research within IBM through the 1970s and 1980s up until early 1990s (Dunham et al. 1974). It is John R. Koza who applied genetic algorithm (GA) approach to perform an automatic derivation of equations, logical rules or program functions (Koza 1992). He first used the term genetic programming (GP) where rather than representing the solution to the problem as a string of parameters as in a conventional GA, he used a tree encoding scheme or structure. The leaves of the tree, called *terminals*, represent input variables or numerical constants. Their values are passed to *nodes*, at the junctions of branches in the tree, which perform some arithmetical or program function before passing on the result further towards the root of the tree.

Alba et al. (1999) and Tunsted and Jamshidi (1996) proposed the use of GP paradigm to evolve fuzzy rule-bases represented as type-constrained syntactic trees. The models have been verified on fuzzy controllers, which showed good parameterization of the algorithm leading to near-optimal solutions. Bastian (2000) showed how GP can be utilised in identification of fuzzy models. The approach showed the use of genetic programming to identify the input variables, the rule base and the involved membership functions of a fuzzy model. Hoffman and Nelles (2001) applied GP for structure identification of a neuro-fuzzy model where a Takagi-Sugeno-Kang fuzzy model describes the underlying rules and GP then identifies the optimal partitioning of the input space.

6.2.4 Differential Evolution

Differential Evolution (DE) is a population-based direct search algorithm which has been mainly used to solve continuous optimization problems (Storn 1995,

1999; Storn and Price 1997). DE was developed by Kenneth Price in an attempt to solve the Chebyshev polynomial fitting problem that had been posed to him by Rainer Storn. This was done by modifying genetic annealing originally developed by (Price 1994) to use floating-point encoding scheme. The main difference between DE and other EAs is that DE uses differences of two randomly selected individuals (parameter vectors) as the source of perturbing the vector population rather than probability function as an evolution strategy. DE performs mutation based on the distribution of the solutions in the current population first and then applies crossover operator to generate offspring. In this way, search directions and possible step sizes depend on the location of the individuals selected to calculate the mutation values.

Vakula and Sudha (2012) employed DE algorithm in a study to systematically tune the optimal parameters of a fuzzy logic controller.

6.2.5 Cultural Algorithm

Culture is the sum total of the learned behaviour of a population that is generally considered to be the tradition of that population and transmitted from generation to generation. Some social researchers suggested that culture might be symbolically encoded and transmitted within and between populations as an inheritance mechanism. Using this idea, Robert Reynolds developed a computational model (Reynolds 1994, 1999) called cultural algorithm (CA). A cultural algorithm is dual-inheritance mechanism where the population space represents the genetic traits and belief space represents the cultural traits. These behavioural traits are passed from generation to generation using several socially motivated operators.

Lin et al. (2009) proposed a CA-PSO (Particle Swarm Optimisation) based algorithm for the optimisation of the structural parameters of functional link fuzzy network and applied for prediction applications.

6.2.6 Genetic Algorithm

What is known as Genetic Algorithm (GA) today is the most widely applied and well-known evolutionary algorithm. The whole tribute goes to John Holland whose extensive work in the field during 1960s and 1970s made GA a widely popular optimisation methodology. In GA, the individuals are represented by means of string similar to the way genetic information is coded in organisms as chromosomes (Holland 1975).

Genetic algorithms are exploratory search and optimisation methods that were devised on the principles of Darwinian evolution and population genetics, which was first introduced by (Holland 1975). Unlike other optimisation techniques, GA does not require mathematical descriptions of the optimisation problem, but instead relies on a cost function in order to assess the fitness of a particular solution to the problem in question (Goldberg et al. 1989; Michalewicz 1994). A genetic adaptive plan can be defined as a quadruple as

$$\Lambda = \{\Sigma, \Pi^N, \Phi, \Omega\} \tag{6.2}$$

where Σ is the coding format, Π^N is population of size N, Φ is fitness re-scaling algorithm and $\Omega = [\omega_1, \omega_2, \ldots, \omega_m]$ is the set of genetic operators.

The most common genetic operators are reproduction ω_r, crossover ω_c, and mutation ω_m. Genetic plan refers to the process through which successive populations are generated using evaluation, selection, mating and deletion. Let Ψ be a probability distribution over Π which is derived from the fitness of each trial. A genetic plan can then be formally expressed as the mapping $\Lambda : (\Psi \times \Pi \times \Omega) \rightarrow \Pi'$. An investigation has been done on efficacy of some of the more common GA techniques with a view to derive a genetic plan suited for fuzzy learning.

The general structure of the genetic algorithm is as follows

Algorithm: $t = 0$

```
initialise [P(t)]
evaluate [P(t)]
do while (not termination-condition)
  {
    P'(t) ← reproduce [P(t)]
    evaluate[P'(t)]
    P(t) ← select[P'(t)]
    P(t + 1) ← P(t)
    t = t + 1
  }
enddo
```

In this algorithm, $P(t)$ denotes population of n individuals at generation t. An offspring population $P'(t)$ of size m is generated by means of reproduction operator such as crossover and mutation from a mating pool. The genetic operators such as crossover, mutation and selection will be discussed later in the relevant section.

The different evolutionary algorithms discussed above have their distinct advantages of application over the others in respective domains. For example, EP has specific advantages when applied to hardware systems, GP is specifically suitable for automatic programming due to its tree-structured chromosome representation and ES is found advantageous in many applications due to its self-adaptation of the strategy parameters. DE is specifically suitable for solving continuous optimization problems. CA seems suitable for exploiting the information of specific belief space. GA, in general, has the most versatile representation mechanism and set of genetic operators that suit for a wide variety of optimisation applications.

6.3 Evolutionary Fuzzy Control

In evolutionary fuzzy control, EA collaborate a fuzzy controller to tune, optimize or learning the parameters, MFs and rule-base of the fuzzy controller. To assess the performance of the fuzzy controller, usually a plant is embedded within the loop. Due to computational effort and time required for EA, it is obviously an off-line approach.

Evolutionary approach is mainly a search or optimization procedure for finding optimality either in the set of all design parameters or only a subset of parameters. Therefore, the design of a fuzzy system can be formulated as a search problem in a high-dimensional space where each point in the space represents a rule set, MFs, scaling parameters and the corresponding system performance, that is, the performance of the system forms a hyper-surface in the space according to given performance criteria. The hyper-surface is infinitely large, non-differentiable, complex and noisy, multi-modal and deceptive. Thus, large dimensionality, strong non-linearity, non-differentiability, and noisy and time-varying objective functions are the closely associated factors involved in optimization problems. While applying GA-based optimisation to fuzzy controllers, it is to be distinguished between Takag-Sugeno-type and Mamdani-type fuzzy controllers. The switching type PD-PI FLC for the flexible arm developed in Chap. 5 is a Mamdani-type fuzzy controller, which is to be optimised. This leads to difficult tasks of optimising the MFs, rule-base and the control parameters. Obviously it suggests applying a stochastic method that is capable of searching a high-dimensional search space. There are a variety of stochastic or derivative-free optimization methods that are applicable to the optimization of problem for the FLC such as simulated annealing (Kirkpatrick et al. 1983), tabu-search (Glover 1989), random search method (Matyas 1965), downhill simplex method (Nelder and Mead 1965) and genetic algorithms (GAs) (Bäck 1996; Deb 2008; Eiben and Smith 2007; Goldberg et al. 1989; Michalewicz 1994). In contrast to smart heuristic methods such as simulated annealing, tabu-search, random search and downhill simplex search are local search techniques and use a generate-and-test search manipulating one feasible solution, the GAs work in parallel on a population of potential solutions. GAs are particularly attractive due to its ability to explore an initially unknown search space and to exploit this information to guide subsequent search over generations and identifies useful sub-spaces in which the global minimum is located.

Investigations involving several example applications demonstrated that GAs are capable of optimising major components of a fuzzy controller such as the membership functions as well as rule-bases of fuzzy logic controllers (Kuo and Li 1999). In general, the number of fuzzy rules increases exponentially with increasing number of input variables or linguistic labels. Hence it is very difficult to determine and select which rules in such a large rule space are the most suitable for controlling the process. Secondly, the membership function plays an important role in determining the prescribed control action and the performance of the system. In multi-variable complex processes, the optimisation and selection of membership functions will also be very difficult. There are different arguments on whether the membership functions or rule-bases should be optimised. Based on the research carried out in this area, these can be divided into the following categories: membership function optimisation, rule-base optimisation and other parameters optimisation.

The basic idea is to represent the complete set of membership functions by an individual and to evolve the shape and location of the triangles (or the Gaussian curves). A triangular membership function can be described by its anchor points on the abscissa axis and Gaussian membership functions are characterized by centre m and the width σ. Karr (1991) describes an application to cart-pole balancing system and uses a genetic algorithm to evolve the membership functions of a fuzzy controller. In order to evaluate the fitness of a controller, the system is simulated for a fixed simulation time, repeating the simulation four times for different initial conditions. The resulting optimised fuzzy logic controller turns out to perform by far better than the controller based on membership functions designed by a human expert. These promising results have also been confirmed by an application of the method for the online control of a laboratory pH system with drastically changing system characteristics (Karr and Gentry 1993). In Karr's work, a user needs to declare an exhaustive rule set and then use a GA to design only the membership functions.

Due to the highly complex and nonlinear characteristics of the problem space, uniform distribution of the fuzzy sets is not optimal. Other ways to tackle the nonlinear distribution should be sought. A natural and better way is to employ nonlinear functions in addition to linear function as membership functions. Natural choices are Gaussian functions, sigmoid functions etc. Through inclusion of linear and nonlinear functions, the type of membership function for each fuzzy set will not be predetermined, but instead be evolved during the design process (Shi et al. 1999).

The performance of a fuzzy classification system based on fuzzy if–then rules depends on the choice of fuzzy partition. If a fuzzy partition is too coarse, the performance may be low. If fuzzy partition is too fine, many fuzzy if–then rules cannot be generated because of the lack of training patterns in the corresponding fuzzy subspace. Therefore, the choice of an appropriate fuzzy partition is important and difficult as the same time. To cope with this difficulty, Ishibuchi et al. (1995) introduced the concept of distributed fuzzy if–then rules. They encode all fuzzy if–then rules corresponding to several different fuzzy partitions into tri-value string $\{-1,0,1\}$ and apply GAs to remove the unnecessary rules from fuzzy if–then rules corresponding to the different fuzzy partitions. Since each possible rule for each subspace is coded into the chromosome, the length of the chromosome is very large when the number of input dimensions and/or of different partitions is large.

Akita et al. (2000) proposed an intelligent fire judgment system from the time series of smoke density using genetic algorithms. The system can extract the features of each category automatically. The features make the rules written in if–then form to judge the category of an unknown input data. They also propose a selective elements method for rules generation. In this method, each locus has a real number from 0.0 to 1.0 and the chromosome is independent from the conventional coding method (Akita et al. 2000). Previous works focused on optimising the FLC parameters and reducing the number of rules, while Chin and Qi (1998) used GAs to search for an optimised subset of rules maintaining the controller's performance. GA will eliminate rules that have no significant contribution to improve the system's performance and badly defined and conflicting rules are also eliminated because their existence degrades the performance of the controller. Huang and Huang (1997) used a real valued GA for a grey prediction system. They segmented the search space by introducing two populations at the earlier stage of evolution. Each population is allowed to self evolve for a few generations and merged into a final population by selecting half of the best solutions from each population. Very often there are dummy rules that have no effect on fuzzy inference. They become the potential redundant rules in fuzzy system. Lekova et al. (1998) applied GAs to exclude those redundant rules from a fuzzy system.

Essentially, the problem of dynamically controlling a complex system using fuzzy controllers can be considered as a multi-parameters optimisation problem. In general, the main task for controlling a complex process with FLC is to define a performance response surface, which must be explored by direct search techniques to locate high performance control outputs. Fuzzy control systems are non-linear systems, which have high dimensional, multi-model, discontinuous response surface. The choice of optimisation technique may not be obvious and easy. Even when an appropriate classical optimisation algorithm is available, there are usually various parameters that must be tuned, e.g., the step size in variable metric technique. In much the same manner, a GA can be used to generate the rules, which use membership functions (Chin and Qi 1998). Juang (2005) proposed a GA-based fuzzy system design under reinforcement learning. The precondition part of the fuzzy system is constructed by a clustering-based partitioning. Each individual in the GA population encodes the consequent part parameters of the fuzzy system and is associated with a Q-value. The consequent part is designed by a Q-value based genetic reinforcement learning. The system starts with no fuzzy rules initially and rules are created automatically as it evolves. Chou (2006) proposed a GA based optimal fuzzy controller design where the consequent part of the fuzzy control rule is an index function. The inputs of the controller, after scaling, are utilized by the index function for computing the output linguistic value. The linguistic value is then used to map the suitable fuzzy control actions. This proposed novel fuzzy control rule has crisp input and fuzzified output characteristics. The index function plays a role in mapping the desired fuzzy sets for defuzzification resulting in a controlled hypersurface in the linguistic space formed by the input fuzzy variables. Both linear and nonlinear index functions are verified for controlling systems with different degrees of nonlinearity.

Lee and Takagi (1993) encoded membership functions and all rules into chromosomes, but used a different way to encode the triangular membership functions. They restrict adjacent membership functions to fully overlap and also constrain one membership function to have its centre resting at the lower boundaries of the input range. The drawback of this approach is that the computational efficiency of the fuzzy system is lost using a large number of rules and the robustness decreases with increasing number of rules. Homaifar and McCormick

(1995) proposed a simultaneous design of membership functions and rule sets for fuzzy controller as the membership functions and rule sets are interdependent. The developed methodology was then applied to a cart-centring problem and truck backing system. There are many applications reported in the literature where the chromosomes of the membership functions and the rules are combined together as design objective (Cheong and Lai 2000; Kang et al. 2000; Zhou and Lai 2000). Park et al. (1993) and Khemliche et al. (2002) used combinations of the scaling factor chromosome with the rule chromosome.

Qi and Chin (1997) and Tarng et al. (1996) applied GAs to optimise performance of a fuzzy controller where controller parameters, scaling factors, membership functions and rule structure were encoded into chromosomes. Thus, an integrated optimised fuzzy controller has been obtained for a higher order system.

The use of membership functions, rule structures, scaling parameters and other control parameters into a single chromosome representation for an optimal fuzzy controller design leads to consider multi-objective criteria, which can optimise several performance objectives. In early studies, several objectives were combined into a single objective with weighting. Later days, proper multi-objective approach was applied to parameter identification and structure learning of fuzzy controllers. There are many applications reported in the literature recently where a multi-objective optimisation to fuzzy systems have been applied (Alacala et al. 2009; Antonelli et al. 2012; Celikyilmaz and Turksen 2008; Chaiyaratana and Zalzala 2002; Fazzolari et al. 2013; Gacto et al. 2010; Zhou and Joo 2008). Fazzolari et al. provided an extensive review of the application of multi-objective evolutionary algorithms to design, learning and tuning of fuzzy controllers.

6.4 Merging MFs and Rule-Bases of PD-PI FLC

The advantage of the switching PD-PI fuzzy controller developed in the previous Chapter is that it has a separate PD and PI part and their combined effect is like a PID controller. The PD controller provides a response with fast rise time and minimal peak overshoot and the PI controller has good performance at the steady state that minimises steady state error. A second advantage of the PD-PI fuzzy controller is that with 5 linguistic labels (i.e. MFs) in each input variable it requires two rule-bases each having $5 \times 5 = 25$ rules and thus giving a total of $(5 \times 5) + (5 \times 5) = 50$ rules, which is significantly smaller than that required for a PID controller ($5 \times 5 \times 5 = 125$ rules). A further advantage is that only one set of rules (25 rules) is executed at a time and thus the processing time is less than that of other PID controllers (discussed in Sect. 5.8 of Chap. 5) where a total of 30–50 rules are executed to produce the control input. The block diagram of this switching PD-PI-like fuzzy controller is shown again here in Fig. 6.5.

Data extracted from the experimentation carried out in Chap. 5 for the switching PD-PI fuzzy controller can be split into two separate data sets representing change of error during PD control (i.e. before switching point) and sum of

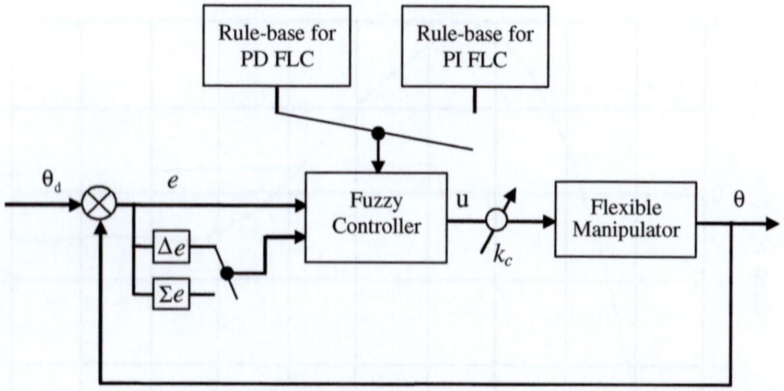

Fig. 6.5 Block diagram of the switching PD-PI-type FLC

error during PI control (i.e. after switching point for the rest of the time). The data set $\Delta e(k)$, $k = 0, 1, \ldots, 25$ is plotted over time instant k in Fig. 6.6. The data set $\Sigma e(k)$, $k > 25, \ldots, T$ is plotted over time instant k in Fig. 6.7. As can be seen from Figs. 6.6 and 6.7 the range of change of error and sum of error are within such a suitable interval so that they can be brought within a common universe of discourse. The two figures are superimposed on one another in Fig. 6.8 to decide on a common universe of discourse. In FLC design, the actual value of the inputs does not matter, rather the MFs for each linguistic variable are important. Therefore, the aim of this Chapter is to unify the MFs for change of error and sum of error so that a further simplification can be achieved in designing an FLC. Now the modified

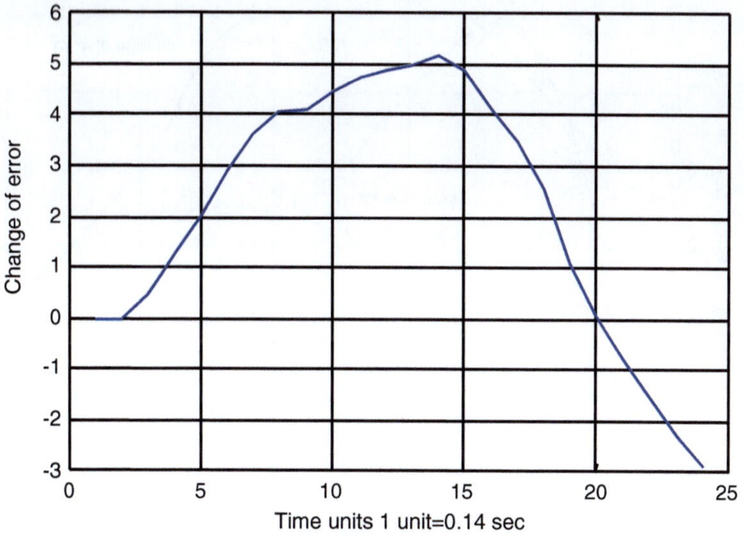

Fig. 6.6 Change of error during PD FLC before switching

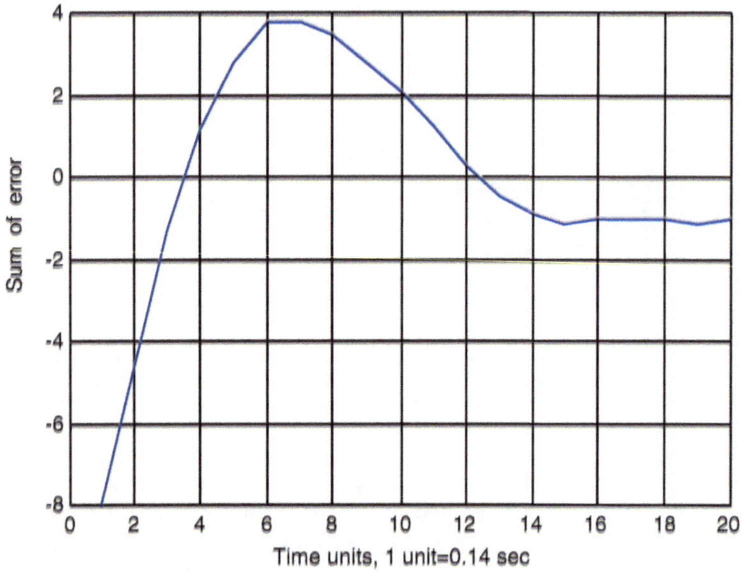

Fig. 6.7 Sum of error after switching to PI FLC

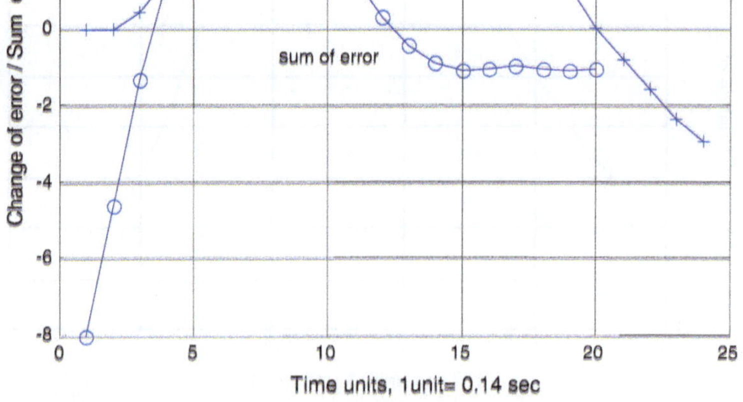

Fig. 6.8 Change and sum of error within a common universe of discourse

common universes of discourse for change error and sum of error are chosen within the interval of [−25, +25].

To construct a rule base, the angle error, change of angle error, sum of error and torque input are partitioned into 5 primary fuzzy sets as

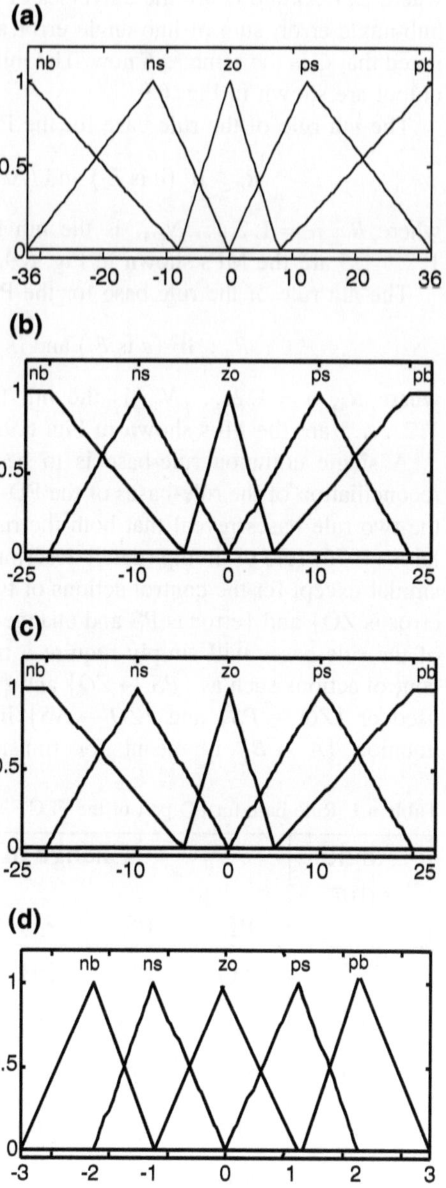

Fig. 6.9 Initial membership functions of inputs and output. Hub-angle error (**a**); Change of hub-angle error (**b**); Sum of hub-angle error (**c**); Torque input (**d**)

Hub angle error $E = \{\text{NB, NS, ZO, PS, PB}\}$
Change of hub angle error $C = \{\text{NB, NS, ZO, PS, PB}\}$
Sum of hub angle error $S = \{\text{NB, NS,ZO, PS, PB}\}$
Torque $U = \{\text{NB, NS,ZO, PS, PB}\}$

where E, C, S and U are the universes of discourse for hub-angle error, change of hub-angle error, sum of hub-angle error and torque input respectively. It is to be noted that C is the same as S now. The initial membership functions for inputs and output are shown in Fig. 6.9.

The nth rule of the rule base for the PD-like FLC is given as

$$R_n : \text{ IF } (e \text{ is } E_i) \text{ and } (\Delta e \text{ is } C_j) \text{ THEN } (u \text{ is } U_k) \tag{6.3}$$

where R_n, $n = 1, 2, \ldots, N_{\max}$ is the nth fuzzy rule, E_i, C_j, and U_k, for $i, j, k = 1, 2, \ldots, 5$ are the MFs shown in Fig. 6.9. The rule base is shown in Table 6.1.

The nth rule of the rule base for the PI-like FLC is given as

$$R_n : \text{ IF } (e \text{ is } E_i) \text{ and } (\text{s is } S_j) \text{ THEN } (u \text{ is } U_k) \tag{6.4}$$

where R_n, $n = 1, 2, \ldots, N_{\max}$is the nth fuzzy rule, E_i, S_j, and U_k, for $i, j, k = 1, 2, \ldots, 5$ are the MFs shown in Fig. 6.9. The rule base is shown in Table 6.2.

A single common rule-base is to be developed for the PD-PI-like FLC by reconciliation of the rule-bases of the PD-part and PI-part. Careful observations of the two rule-bases reveal that both the rule-bases strictly follow MacVicar-Whelan's (1976) rule of thump (detail description is provided in Chap. 5) and they are similar except for the control actions of two rules {error is NS and change/sum of error is ZO} and {error is PS and change/sum of error is ZO}. The reconciliation of the rule-bases will simply require a transformation or adjustment of the two control actions such as $\{PS \rightarrow ZO\}$ and $\{NS \rightarrow ZO\}$ if the rule-base of PD part is used or $\{ZO \rightarrow PS\}$ and $\{ZO \rightarrow NS\}$ if the rule-base of PI part is used. The notation $\{A \rightarrow B\}$ is meant for transformation or adjustment between two

Table 6.1 Rule-base for PD-part of the FLC

Angle error	Change of error				
	PB	PS	ZO	NS	NB
NB	ZO	PS	PB	PB	PB
NS	NS	ZO	PS	PS	PB
ZO	NS	ZO	ZO	ZO	PS
PS	NB	NS	NS	ZO	PS
PB	NB	NB	NB	NS	ZO

Table 6.2 Rule-base for PI-part of the FLC

Angle error	Sum of error				
	PB	PS	ZO	NS	NB
NB	ZO	PS	PB	PB	PB
NS	NS	ZO	ZO	PS	PB
ZO	NS	ZO	ZO	ZO	PS
PS	NB	NS	ZO	ZO	PS
PB	NB	NB	NB	NS	ZO

linguistic terms here. Eventually, the process requires adjustment of three MFs, namely, *NS*, *ZO* and *PS*. The two rules of the PD-part with the required adjustments of the three MFs {*NS*, *ZO*, *PS*} are shown in Eqs. (6.5)–(6.6).

$$\text{IF } (e \text{ is } NS) \text{ and } (\{\Delta e \text{ } or \text{ } \Sigma e\} \text{ is } ZO) \text{ THEN } (u \text{ is } \{PS \rightarrow ZO\}) \quad (6.5)$$

$$\text{IF } (e \text{ is } PS) \text{ and } (\{\Delta e \text{ } or \text{ } \Sigma e\} \text{is } ZO) \text{ THEN } (u\text{is } \{NS \rightarrow ZO\}) \quad (6.6)$$

It will be simple to choose the rule-base of PD-part and see how the adjustments are achieved applying the optimisation procedure. The aim at this stage of development is to optimise the membership functions using the rule-base of the PD-part. Membership functions are generally accommodated within a rule-base during the optimisation process. Therefore, optimising both the membership functions and rule-base is somehow redundant. In this case a drastic improvement of the performance is not the desired anticipation rather a systematic design simplification of the PD-PI-like fuzzy controller is of primary interest.

6.5 Optimising FLC Parameters Using GA

There are a number of implementations of GA found in the literature mainly based on the encoding schemes, chromosome representations and genetic operators. Based on the types of these operators GA has many variants like binary coded GA (Deb 2008), real-coded GA (Goldberg 1991; Eshelman and Schaffer 1993), micro-GA (Krishnakumar 1989), and niching GA (Goldberg and Richardson 1987). In order to ensure convergence properties, elitists GAs are commonly used such as messy-GA (Goldberg et al. 1989) and Eshelman's CHC (Cross-generational elitist selection, Heterogeneous recombination, and Cataclysmic mutation) GA (Eshelman 1991).

Binary-coded GA is the standard genetic algorithm represented by binary strings. The string length should be chosen a priori and is fixed during the run. The difficulty with binary-coded GA is the Hamming cliffs and it can not achieve arbitrary precision of the solution. This suggests a real-coded (or valued) GA, where real parameters are used directly, crossover and mutation operators are applied directly to real parameters and decision variables can be directly used to compute the fitness value. Therefore, a real-valued GA will be used in the optimisation process for the parameters of the triangular membership functions of the inputs and output and the rule-base.

In a typical GA the entire set of parents is replaced by their children during reproduction. This generational replacement technique, however, has some drawbacks. On the one hand, it may be relatively slow in terms of evolution rate, and on the other hand, some of the best individuals may not reproduce and their genes may be lost. One possible solution is to modify the reproduction technique so that it replaces few individuals at a time, i.e. steady-state reproduction. Briefly, this has the following characteristics: i) it creates few children throughout the production, ii) it deletes individuals having poor performance to make room for children, and iii) it inserts randomly selected children into the population.

The input variables error (e) and change/sum of error ($\Delta e/\Sigma e$) and the output variable control input (u) of the fuzzy controller can be partitioned into overlapping sets, which have a linguistic correlation to form membership functions. These MFs are most often triangular in shape but trapezoidal and Gaussian functions can also be used. The membership values control the degree to which the rules are fired, illustrating the interdependent relation between the rule set and the MFs. The GA-base PD-PI-like fuzzy controller with a single rule-base is shown in Fig. 6.10, where the MFs of the error (e), change/sum of error ($\Delta e/\Sigma e$) and the control input (u) will be coded into a suitable representation so that GA can be applied.

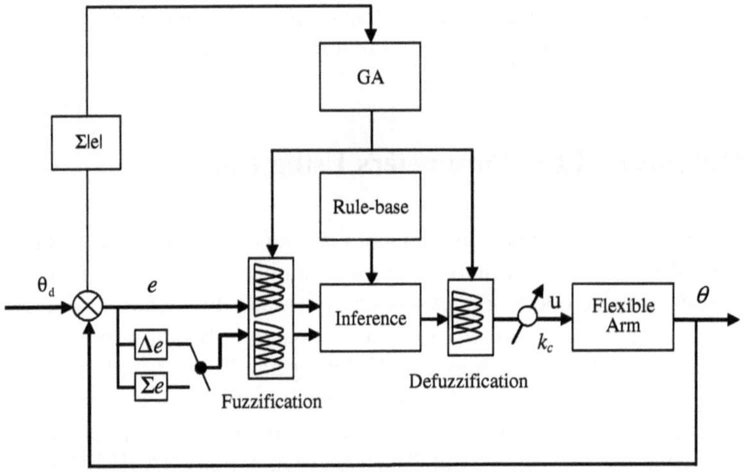

Fig. 6.10 Block diagram of GA-based PD-PI-like FLC

6.5.1 Encoding Scheme

The first thing for the GA is to find a mechanism of encoding genetic information of a population representing an entire search space of a problem domain into chromosomes. Encoding schemes in GA should be such that the representation and the problem space are close together and a set of genetic operations can be performed on them so that genetic information is propagated from generation to generation. A tractable mapping to phenotype should also be ensured to allow fitness values to be calculated at a minimum computation cost. There are many encoding schemes used in GA such as binary, Gray, real valued, hybrid, permutation, value, tree, and grammar coding. A detailed description of the coding schemes can be found in (Siddique and Adeli 2013). Real-valued coding has a number of advantages over the other schemes, which were mentioned earlier. The problem associated with binary coding is that a long string always occupies the computer memory even though only a few bits are actually involved in the crossover and mutation operations. This is particularly the case when a large number of parameters are needed to adjust in the problem and higher precision is required for the final result. To overcome the problem of inefficient use of the computer memory, real-valued chromosome representation for the membership functions is adopted.

6.5.2 Chromosome Representation for MFs

One of the key issues in evolutionary design of fuzzy systems using GAs is the genotype representation, i.e. information encoded into chromosomes. A fuzzy system is specified only when the membership functions associated with each fuzzy set and the rules are determined. To translate membership functions to a representation useful as genetic material, the functions are parameterised with one to four coefficients and each of these coefficients constitutes a gene of the chromosome for genetic algorithms. The most broadly used parameterised membership functions are triangular, trapezoidal, Gaussian, and bell shaped. A detailed description of the different MFs is presented in Chap. 4. These parameterised functions may be classified into two main groups: piecewise linear functions such as triangular and trapezoidal and differentiable functions such as Gaussian and bell shaped. Each coefficient constitutes a gene of the chromosome represented by a real number.

In fuzzy control design, one can frequently assume triangular membership functions for which each membership function can be specified by just a few parameters. In the case of a triangular membership function, it is determined by three parameters: left position, peak and the right position. An overlapping (not more than 50 %) of the MFs is desired to ensure a good performance of the FLC. Therefore, the left and peak position of the next MF is the same as the peak and right position of the previous MF shown in Fig. 6.11a.

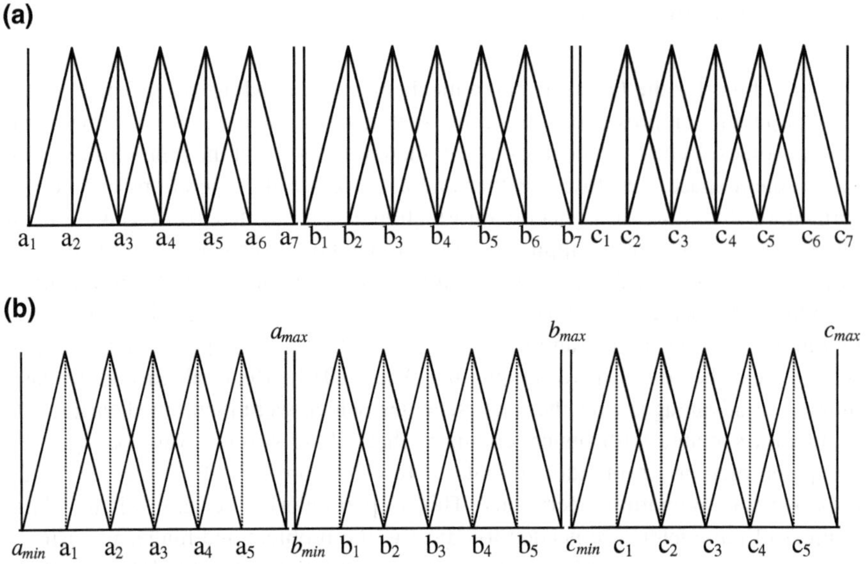

Fig. 6.11 Parameterised membership functions. Parameterised membership functions (**a**); Fixed upper and lower limit of the membership functions (**b**)

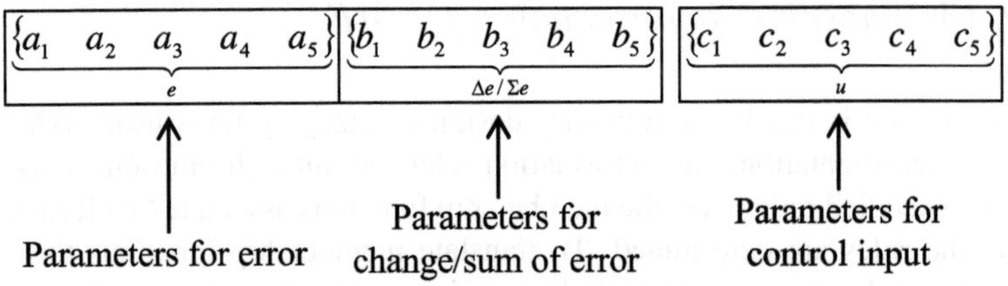

Fig. 6.12 Chromosome representation for membership functions

Seven parameters are needed to define five MFs for each input or output, that is, the five membership functions with each having 3 parameters are *(a1, a2, a3)*, *(a2, a3, a4)*, *(a3, a4, a5)*, *(a4, a5, a6)* and *(a5, a6, a7)* and there are *21* parameters in total for all inputs and output. A reduction of the number of parameters can be achieved by fixing the upper and lower limits of the universe of discourse for each input and output as shown in Fig. 6.11b. Thus, two parameters are reduced per each input and output and the number of parameters is thus reduced to only 15.

Hence, the chromosome for membership functions looks like in Fig. 6.12, where a_is are the parameters for hub angle error, b_is are the parameters for change of error or sum of error and c_is are the parameters for control input.

Fig. 6.13 Encoding of the rule-base

Angle error	Change/sum of error				
	PB ↓ +2	PS ↓ +1	ZO ↓ 0	NS ↓ -1	NB ↓ -2
NB→-2	0	+1	+2	+2	+2
NS→-1	-1	0	+1	+1	+2
ZO→0	-1	0	0	0	+1
PS→+1	-2	-1	-1	0	+1
PB→+2	-2	-2	-2	-1	0

6.5.3 Chromosome Representation for Rule-Base

GAs can be used to optimise the rule-base of an FLC, which is applied to control a system, in this case with two inputs and one output. The linguistic variables can be represented by integer values, for example -2 for NB, -1 for NS, 0 for ZO, +1 for PS and +2 for PB. Applying this coding scheme to the fuzzy rule-base (decision table) represented in Table 6.1, the encoded rule-base shown in Fig. 6.13 is obtained. A chromosome is thus obtained from the decision table by going row-wise and coding each output MF as an integer $\pm n$ (i.e. $-n, \ldots -1, 0, +1, \ldots, +n$), where n is the maximum number required to label the membership functions defined for the output variable of the FLC. In this case, $|n| = 2$. The chromosome of the rule-base is shown in Fig. 6.14.

There will be two different mutation operators for both parts of the chromosome string. The genes in the membership function part of the chromosome will be replaced by a real value whereas genes of the rule-base part of the chromosome will be changed to either up a level or down a level of the integer value to avoid possible large deterioration in performance.

The rule-base consists of 25 parameters and will take long time for an optimisation algorithm to converge to a satisfactory level of performance. As mentioned earlier a design simplification is desired rather than a drastic improvement of the performance. Therefore, the rule-base is not optimised using genetic algorithm in this study rather it is chosen arbitrarily from the PD-part of the FLC discussed in earlier section.

6.5.4 Objective Function

The next important consideration following the chromosome representation is the choice of fitness function. The genotype representation encodes the problem into a

$$\{\underbrace{0 +1 +2 +2 +2}_{1st\ row} | \underbrace{-1 0 +1 +1 +2}_{2nd\ row} | \underbrace{-1 0 0 0 +1}_{3rd\ row} | \underbrace{-2 -1 -1 0 +1}_{4th\ row} | \underbrace{-2 -2 -2 -1 0}_{5th\ row}\}$$

Fig. 6.14 Chromosome representation of the rule-base

string while the fitness function measures the performance of the system. Finding a good fitness measurement is quite important for evolving practical systems using GAs. Unlike traditional gradient-based methods, GAs can be used to evolve systems with any kind of fitness measurement function including those that are non-differentiable and discontinuous. Finding a good fitness measurement can make it easier for the GA to evolve to a useful system. How to define the fitness measurement function for a system to be evolved is problem dependent.

The procedure of evaluating the knowledge base, i.e. membership functions and rule-base, consists of submitting to a simulation model or real system, and returning an assessment value according to a given cost function J subject to minimization. In many cases J is determined as a summation over time of some instantaneous cost rate. As an example, a trial knowledge base can be made to control the model of a process and then sum the errors over the response trajectory. The sum of errors is then directly related to the objective fitness of the trial. The fitness of trial is a measure of the overall worth of a solution, which takes into account the factors of an objective criterion, in this case, the performance of a fuzzy controller implementable with the trial knowledge base. The basic control objective is simply stated as the ability to follow a set point with minimal error. This objective can thus be expressed in terms of minimization of the controller performance indices, which are in common use. These include integral of absolute error (IAE), integral of square error (ISE) and integral of time weighted absolute error (ITAE). Each of these indices has its own merits and demerits. For example, ITAE penalizes errors at large values of time and leads to reduction in steady state errors at the expense of transient errors, while ISE is more suitable for a mathematical analysis criterion (Linkens and Nyongesa 1995a, b). Furthermore, GA is only able to optimise the characteristics explicit in the cost function. In this study, sum of absolute error is used as a measure of performance of the controller.

Assume a controller with multiple inputs and outputs whose overall design effectiveness can be measured by just one output of the overall system such as hub angle error in the case of a flexible arm system. Finally, all membership functions (the rule-base is not used further) can be expressed by some list of m (in this case $m = 15$) parameters, $(p_1, p_2, \ldots, p_m) = p$, where each parameter takes only a finite set of values and can then be specified by the function:

$$J(p) = \int_0^T |e(t)|dt \tag{6.7}$$

Obviously the objective is to minimize $J(p)$ subject to p. In discrete time, it can be written as

$$J(p) = \sum_{k=1}^{n} |e(k)| \tag{6.8}$$

where $e(k)$ is the hub-angle error in the flexible arm system. n is some reasonable number of time units by which the system can be assumed to have settled quite close to a steady state.

6.5.5 Dynamic Crossover

Crossover refers to mating of two parents to produce new chromosomes by blending of genetic information from the parent chromosomes. The analogy carries over to crossover in GAs whereby new solutions are created from the information contained within two (or more) parent solutions. This is the primary mechanism of creating new solutions (i.e. chromosomes) with higher fitness values that survive to next generation. Whether a chromosome will undergo a crossover operation or not is determined by a crossover probability $p_c \in [0, 1]$. The crossover of two individuals parent1 = $\{x_1, \ldots, x_{15}\}$ and parent2 = $\{y_1, \ldots, y_{15}\}$ produces two new chromosomes called offspring's off_1 and off_2 as illustrated in Fig. 6.15.

The above example is a single-point crossover. There are a number of crossover operators used in GA. A detail description of the different crossover operators can be found in (Siddique and Adeli 2013).

Crossover facilitates exploration, while mutation facilitates exploitation of the search space. In general, the probabilities of crossover (p_c) and mutation (p_m) are held constant for the entire run of a GA. An optimal result is not always guaranteed by this approach in many cases. These two parameters can be varied during the run, often starting out by running the GA with a relatively higher value for crossover probability (usually $p_c = 0.86$) and lower value for mutation probability (usually $p_m = 0.01$) and then tapering off the crossover value and increasing the mutation rate toward the end of the run, ending with values of, say, one half and twice the initial values, respectively (Shi et al. 1999; Spears 2000; Yun and Gen 2003).

Since the process to vary these two parameters for obtaining good performance is unknown, it is unclear how to vary the parameters during the run. Normally, they are changed linearly. It is well known to researchers from experience that when the fitness is high, e.g., at the end of the run, low crossover rate and high mutation rate are often preferred (Shi et al. 1999). Also, when the best fitness is stuck at one value for a long time, the system is often stuck at a local minimum in a local neighbourhood, so the system should probably concentrate on exploiting rather than exploring; that is, the crossover rate should be decreased and mutation rate should be increased.

Considering the above analysis into account, Shi et al. (1999) and Yun and Gen (2003) proposed a fuzzy system to adjust the crossover and mutation rates

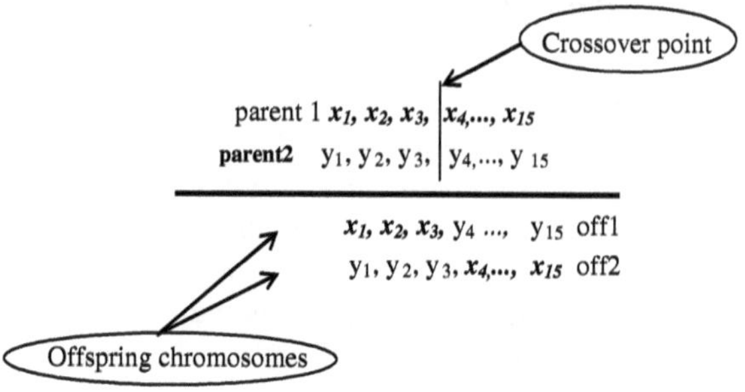

Fig. 6.15 Crossover operation

dynamically using fuzzy rules based on the variance of the fitness of the population. When variance is low, mutation should be emphasized, while when variance is high, crossover should be stressed. McGinley et al. (2008) proposed an adaptive GA (AGA), where individuals that do not undergo crossover are instead subjected to an adaptive rate of mutation. This technique essentially corresponds to splitting the population into two sub-sections: an exploitation (crossover) division and an exploration (adaptive mutation) division. The sizes of these divisions are determined by the population diversity (PD) measure. The adaptive crossover probability (p_c) is defined by the Eq. (6.9):

$$p_c = \left[\left(\frac{PD}{100} * (k_2 - k_1) + k_1\right)\right] \tag{6.9}$$

where $k_1 = 0.4$, $k_2 = 0.8$ and $(0 < PD \leq 100)$.

The p_c varies from the minimum crossover rate $k_1 = 0.4$ to maximum crossover rate $k_2 = 0.8$ based on population diversity PD. PD is calculated by finding the position of the average individual within the problem's search space and summing the Euclidean distances from this average point to the location of every other individual. This measure provides the standard deviation of the population's individuals. The standard deviation is expressed relative to the mean as a coefficient of variation. McGinley et al. (2008) used a population size of 100 in their experiments. But for a small population size of 10 and a maximum generation of 20, this adaptive crossover rate is not promising.

Chromosomes that are created by crossover operation with a fixed crossover rate $p_c$will not improve the population diversity after certain generations for a small size of population. Therefore, a new dynamic crossover probability p_{dc} has been proposed in this study, which will trim down the redundancy of individuals in the population. The dynamic crossover probability is defined as

$$p_{dc} = p_c * \frac{1}{g} \tag{6.10}$$

where $p_c = 0.8$ is the initial crossover probability and g is the current generation number. This means a decreasing crossover probability with increasing generation.

6.5.6 Dynamic Mutation

Mutation refers to randomly changing a gene to produce new genetic material from the parent chromosomes. Due to poor initialisation of population, there may not be enough variety of chromosomes in the population to ensure a good solution or an optimum solution. This is especially true for a small population. A large population ensures diversity. Mutation operation can help increasing the genetic diversity. Mutation operation on an individual chromosome is shown in Fig. 6.16. A gene is randomly selected from the chromosome and changed to a new random

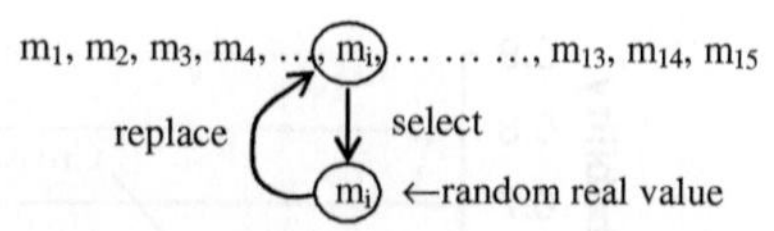

Fig. 6.16 Mutation operation

value by adding or subtracting a small value to it for real-valued chromosomes. There are a wide range of mutation operations in use in GA depending on the encoding scheme and chromosome representation. A detail description of these operators can be found in (Siddique and Adeli 2013).

A mutation probability (p_m) dictates the population of how many individuals should undergo mutation operation. The mutation probability should be kept very low as a high mutation rate will destroy fit individuals and degenerate the GA into a random walk. Much research has been done in order to determine the best setting for mutation rate especially in GA. There is no clear answer given to these problems rather different researchers provided different settings to respective problem domains. Some common settings are: $p_m = 0.001$ suggested by De Jong (1975), $p_m = 0.01$ proposed by Grefenstette (1986) and $p_m = [0.005, 0.01]$ suggested by Schaffer et al. (1989). Schaffer et al. (1989) formulated their results of the empirical study into the following expression

$$p_m \approx \frac{1.75}{N\sqrt{l}} \tag{6.11}$$

where N is the population size and l is the length of the individuals' genetic representation, i.e. length of the string. The expression is similar to the theoretical development by Hesser and Männer (1990) and given by

$$p_m(t) = \sqrt{\frac{\alpha}{\beta}} \frac{\exp(-\gamma \frac{t}{2})}{N\sqrt{l}} \tag{6.12}$$

where $\{\alpha, \beta, \gamma\}$ are constants. This expression introduces a time-dependency for the mutation rate. The time dependency eventually means the generations in an EA. The time dependency of mutation rate was also first suggested by Holland (1975) himself, although he did not give any further details on the choice of the parameter for the time-dependent increase or decrease of p_m. Fogarty (1989) used several time-dependent schedules for p_m, a measure which remarkably increased the GA performance. Both approaches use a deterministic decrease of mutation rates over time, such that $\lim_{t\to\infty} p_m(t) = 0$. In addition, the mutation rate is considered as a global parameter, i.e. single $p_m(t)$ is valid for all individuals in the population at the time instant.

McGinley et al. (2008) also proposed an adaptive mutation rate in their adaptive GA based on the measure of population diversity (PD), which divides the population into two sub-sections: an exploitation (crossover) division and an exploration (adaptive mutation) division. The adaptive mutation probability (p_m) is defined by the Eq. (6.13):

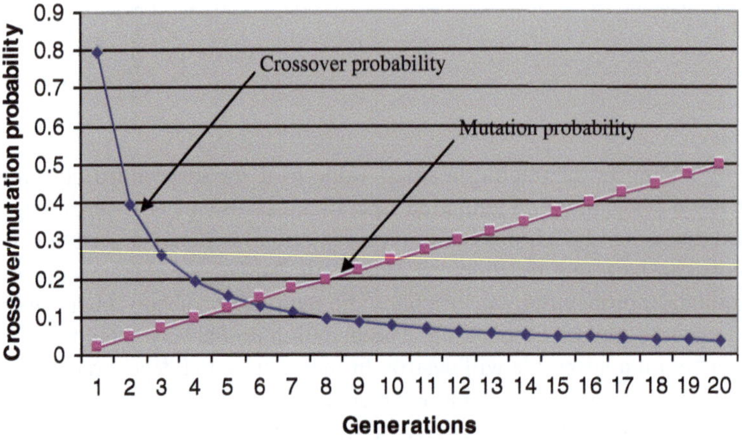

Fig. 6.17 Dynamic crossover and mutation over generation

$$p_m = \frac{-(PD - 100)}{100} .k \tag{6.13}$$

where $k = 0.2$ is the maximum mutation rate. McGinley et al. (2008) used a population size of 100 in their experiments. But for a small population size of 10 and a maximum generation of 20, this adaptive mutation rate will not improve population diversity. A faster improvement of population diversity over generation is required.

The general guidelines are drawn from empirical studies on a fixed set of test problems, and the guidelines were inadequate because the optimal use of p_c and p_m is specific to the problem under consideration. Some studies focused particularly on finding optimal crossover or mutation rates (Bäck 1992, 1993; Grefenstette 1986; Hesser and Männer 1990; Schaffer and Morishima 1987; Spears 2000). These heralded the need for self-adaptation in the crossover or mutation rates (Eiben et al. 1999).

Since the diversity in the population will fall over the generations due to decreasing rate of crossover with a small size population, addition of fresh genetic material to the population is essential. Therefore, a new dynamic mutation probability p_{dm} is proposed in this study, which is defined by

$$p_{dm} = p_m * \frac{g}{g_{\max}} \tag{6.14}$$

where $p_m = 0.2$ is the initial mutation probability and g_{max} is the maximum generation ($g_{max} = 20$).

Crossover and mutation operations are the critical operators in GA and they facilitate an efficient search and guide the search into new regions of the search space. A striking balance between the two operations should be preserved throughout the evolution. The relation between dynamic crossover probability (p_{dc}) and dynamic mutation probability (p_{dm}) over the generation is shown in Fig. 6.17. It can be seen from the figure that the mutation probability becomes higher than crossover probability after generation 6. The decreased population diversity will be compensated by the increased mutation arte at this stage of evolution.

6.5.7 Selection

The selection operator allows individual chromosome to be copied for possible inclusion in the next generation. The chance that a string will be copied is based on the individual's *fitness value*, calculated from a *fitness function*. For each generation, the selection operator chooses individuals that are placed into a *mating pool*, which is used as the basis for creating offspring for the next generation. There are many different types of selection operators. One can select the fittest and discard the worst, statistically selecting the rest of the mating pool from the remainder of the population. In general, selection is typically probabilistic, which offers better chances for individuals with high-fitness to get selected into the mating pool. Low-fit individuals are also often given a small chance. There are two important factors closely related to any GA. These are selective pressure and population diversity. A strong selective pressure essentially means to focus on best-fit individuals in the population. This in turn instigates a decrease in the population diversity, which may result in a premature convergence, whereas a weak selective pressure can make a search ineffective. It takes many generations to converge as the population increases. There are many variants of selection scheme such as random, proportional, tournament, rank-based and elitism selection. None are right or none are wrong. In fact, some will perform better than others depending on the problem domain being explored. For a detail description of these selection mechanisms, interested readers are directed to (Siddique and Adeli 2013).

Random selection is the simplest method where individuals are selected randomly with no reference to fitness at all. Each individual, good or bad, has an equal chance or probability of being selected with the probability of $\frac{1}{N}$, where N is the size of population. As a result, random selection has a low selective pressure causing a slow convergence to solution.

Elitism is the selection of a set of individuals from the current generation to survive to the next generation. The number of individuals to survive to the next generation, without mutation, is referred to as generation gap. If the generation gap is zero, the new generation will consists entirely of new individuals. For positive generation gap, say k, k individuals survive to the next generation. Elitism is generally used to prevent the loss of fittest member in a generation. Therefore, a trace is kept of the current fittest individual and always copied to the next generation.

Genetic algorithms are based on the model of biological evolutions. Thus, the chromosome string must compete with each other to survive to the next generation. Randomly selected individuals from the offspring population replace the individuals having lower fitness value than the average fitness $(\bar{f})$ of the population, i.e. replace an individual from the population with fitness f if $f < \bar{f}$. This will likely to improve the difference between the best fitness value f_{best} and average fitness value $\bar{f}$ in the population. The state of convergence can be observed by simply measuring the value of the term $(f_{best} - \bar{f})$, which is likely to be less for a population that has converged to an optimal solution (Srinivas and Patnaik 1994).

6.5.8 Initialisation

DeJong showed the best parameter setting for GA, which is now used as common settings for population size of 50–100, crossover probability of 0.6 and mutation probability of 0.001 (De Jong 1975; De Jong and Spears 1990). Grefenstette used DeJong's test suite and applied a meta-level GA to optimise the GA control parameters. He found parameter settings for best online performance as population size of 20–30, crossover probability of 0.75–0.95 and mutation rate of 0.0005–0.01 (Grefenstette 1986). Krishnakumar (1989) proposed a micro-GA with small population size for single-objective optimisation of stationary and non-stationary function optimisation. The micro-GA suggests use of 4–5 population members participating in selection, crossover, and block replacement. Lee and Takagi (1993) used a population size of 13, crossover and mutation rates of 0.9 and 0.08 respectively. Because of the evaluation constraint, a population size of 10 is considered in this study. It is reasonable to discretize and place upper and lower bounds on the solution spaces for each of these parameters of the membership functions.

6.5.9 Evaluation

The demand on computation time sometimes prohibits online application. An offline application will not serve the purpose here. The practical problem of implementation of GA online is that how to evaluate each chromosome in the population. In this case, each time the controller is applied to the real system for each individual of the population. Its performance can be evaluated by calculating the objective functions discussed in Sect. 6.5.4. Then the value is assigned to the individual's fitness. The time taken in the evaluation of genetic structures, especially in the case of fuzzy controller, imposes restriction on the size of population and also the number of generations required to run the GA to a final solution.

6.6 Some Experimental Results

The aim of the investigations carried out in this Chapter is to develop a suitable GA technique for fuzzy controller design. As mentioned earlier, a drastic improvement of the system performance is not expected rather a systematic design methodology is desired. A GA can be derived in different ways depending on the parameters chosen. The population size and the number of generations are constraints in view of evaluation involvement and time. Hence, it was reasonable to decide on a suitable population size as a starting point. The problem of a small population size is that the chromosomes created by crossover operation do not improve much over the generations. This occurs only when the crossover probability and mutation probability are fixed and generally crossover probability is chosen higher than mutation probability as found in many previous research works (Lee and Takagi 1993; Shi et al. 1999).

The proposed GA scheme was applied to the fuzzy controller for a single-link arm. The practical difficulty was how to evaluate the fitness function. The generated chromosomes were used to define fuzzy membership functions in the PD-PI-type controller and then applied to the arm system. The performance of the controller is then determined in terms of the fitness function defined earlier in Eqs. (6.7)–(6.8). The smaller the sum of absolute error, the higher is the fitness of the chromosomes. After coding each chromosome as membership functions into the controller program, it is executed on the single-link arm system for a demanded hub angle of 36 °C. The value of the hub angle position is collected from the system and the sum of absolute error is calculated. This process of evaluation is very tedious and hence the number of chromosomes in the population is restricted to only 10.

Figure 6.18 shows performance of the initial population, which was applied for a demanded hub angle of 36 °C. As can be seen from the figure some individuals did not work at all and some individuals achieved good performance with a very small steady-state error. Figure 6.19 shows the control surface of the dominant rules in the first generation.

Performance of the best individuals from initial population to generation 7 is shown in Fig. 6.20. The performance improvement in terms of rise-time, overshoot, settling time and steady state error can be seen from generation to generation shown in Table 6.3. Earlier generations show better performance in respect of rise time but have significant oscillation around set point. Later generations show better performance in respect of settling time and steady state error. The generation 5, 6 and 7 achieved an improvement in steady-state error and in total time. Overall performance of the system did not improve much after generation 4 rather the difference between performance of the best individuals and worst

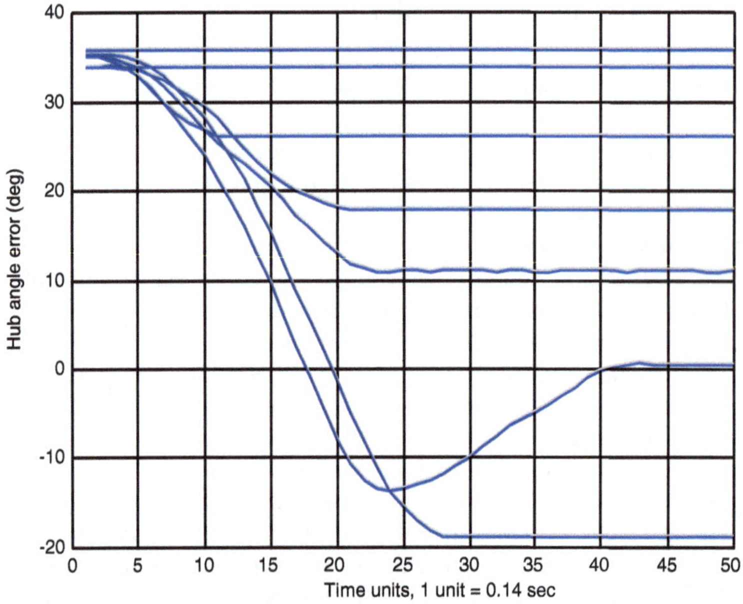

Fig. 6.18 Performance of the FLC with first population

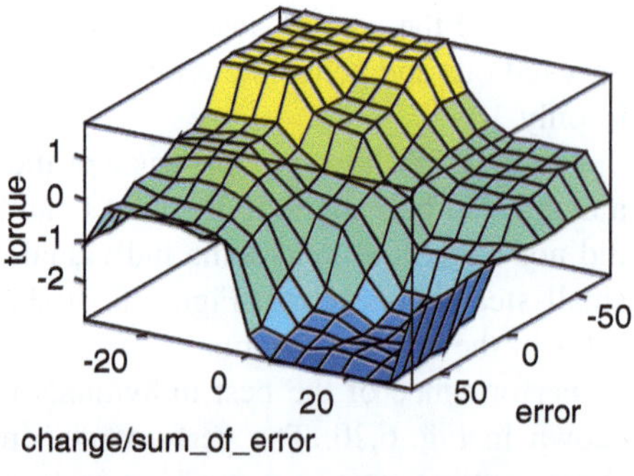

Fig. 6.19 Rule surface of the best individual showing dominant rules

individuals is minimised. This results in an improvement in the average fitness of the populations over generation. Figure 6.21a shows a difference of 15 degrees between the best and worst performance where as Fig. 6.21b shows a difference of 5 degrees between the best and worst performance in generation 3 and 7 respectively.

Table 6.3 Rise time, settling time, overshoot and steady state error for different generations

Generation	Rise time (Time units, 1 unit = 0.024 s)	Overshoot (deg)	Settling time (Time units)	Steady state error (deg)
0	85	37.01	120	−0.635
1	120	36.95	160	0.95
2	75	41.55	201	−1.32
3	120	36.84	160	0.63
4	120	37.05	175	0.42
5	128	36.90	170	0.58
6	100	35.94	115	0.52
7	120	36.58	160	0.47

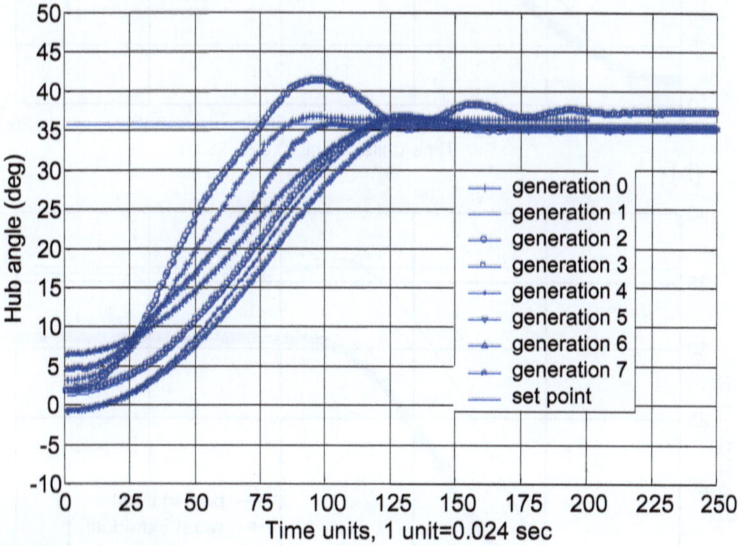

Fig. 6.20 Best individuals from 0th generation to 7th generation

Characteristics of the learning profile are given by the fitness value of the individuals of the population. Fitness is calculated as a sum of absolute error as shown in Fig. 6.22. Figure 6.23 shows the sum of squared error as a measure of performance improvement. Another way of expressing the learning profile is the mean fitness of the population. The mean fitness calculated as mean of sum of absolute error and mean of sum of squared error is shown in Figs. 6.24 and 6.25.

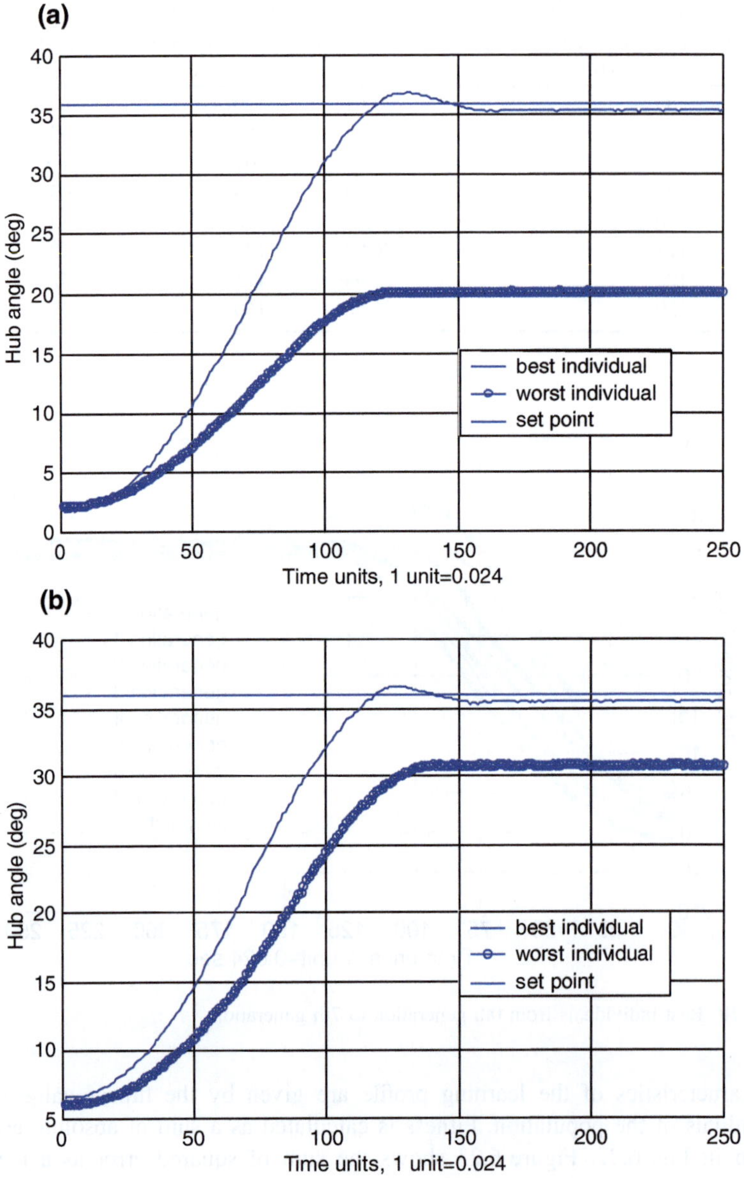

Fig. 6.21 Performance improvement from 3rd generation to 7th generation. Best and worst individuals in generation 3 (**a**); Best and worst individuals in generation 7 (**b**)

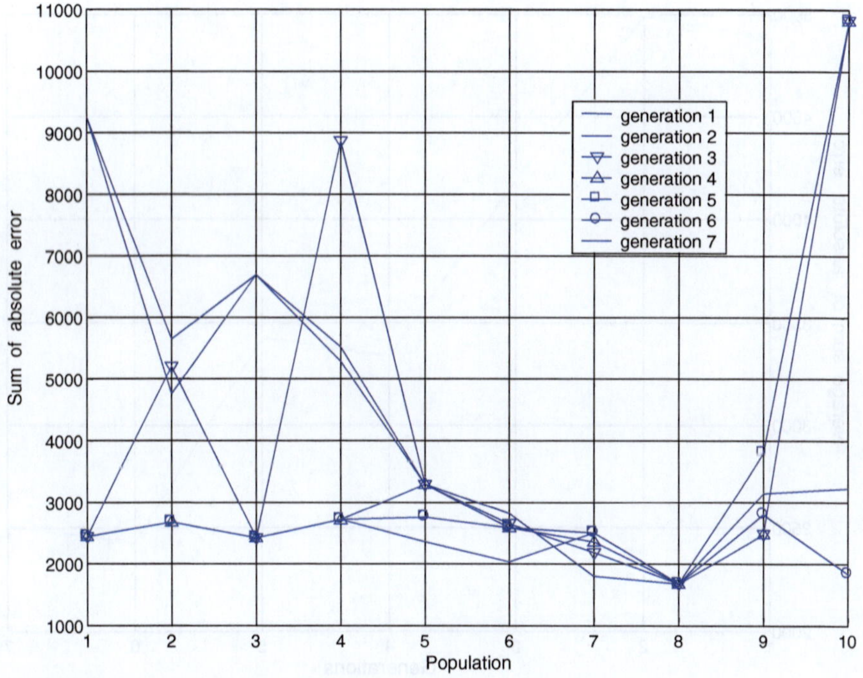

Fig. 6.22 Sum of absolute error of population from generation 1 to 7

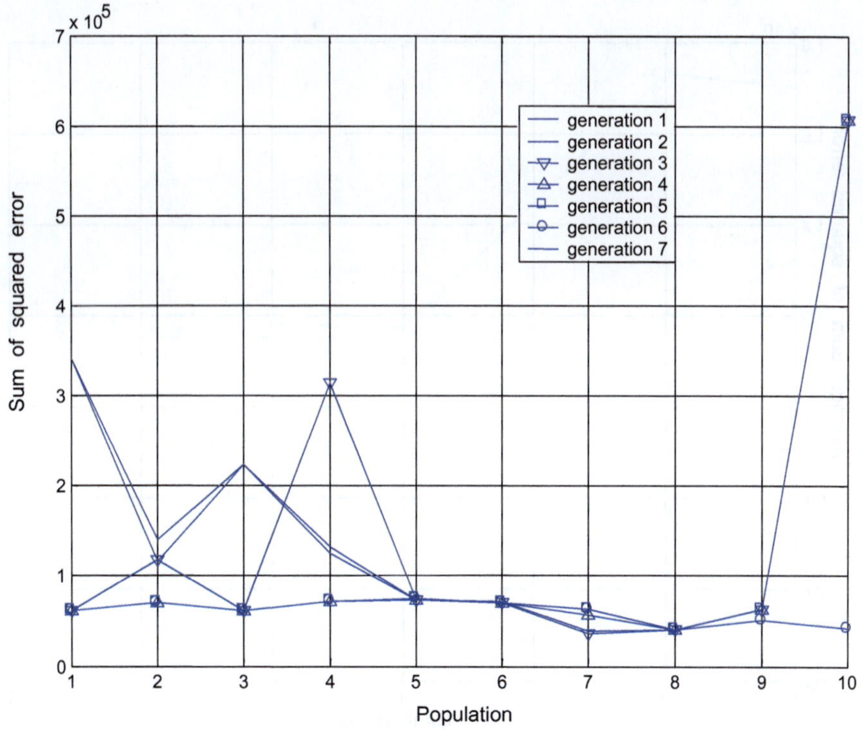

Fig. 6.23 Sum of squared error of population from generation 1 to 7

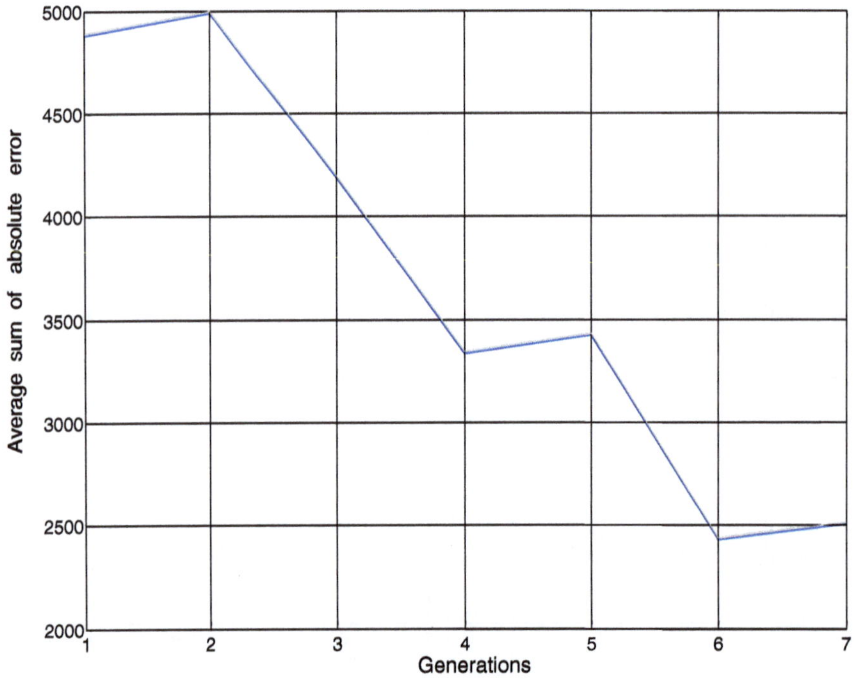

Fig. 6.24 Learning profile—Average sum of absolute error

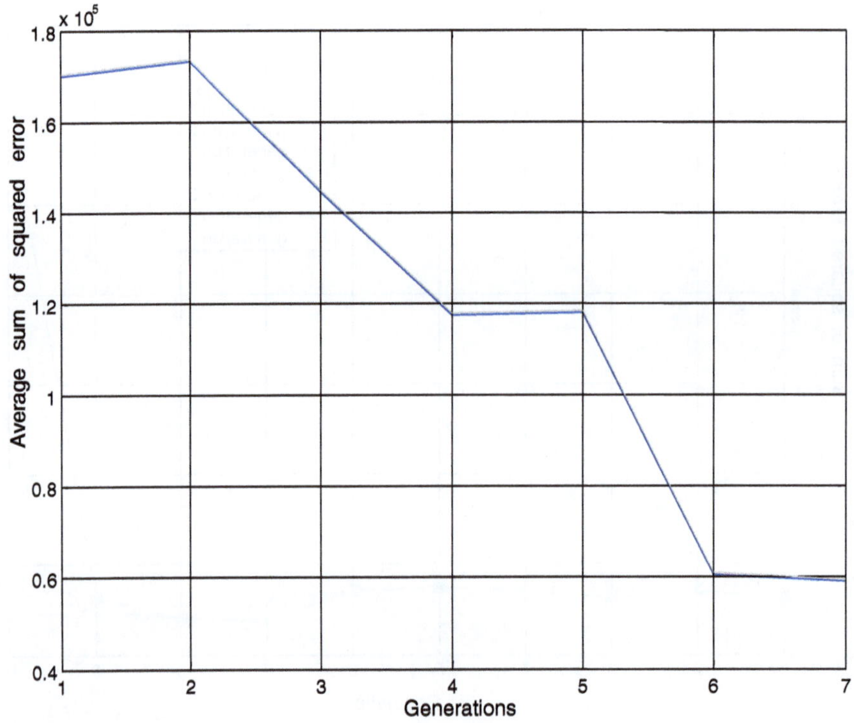

Fig. 6.25 Learning profile—Average sum of squared error

6.7 Summary

One of the most important considerations in designing fuzzy systems is construction of the membership functions as well as the rule-base. In most existing applications, the fuzzy rules are generated by an expert in the area, especially for control problems with only a few inputs. However, it is by no means trivial but plays a crucial role in the success of an application. Previously, generation of membership functions had been a task mainly done either interactively, by trial and error, or by human experts. With an increasing number of inputs and linguistic variables, the possible number of rules for the system increases exponentially, which makes it difficult for experts to define a complete set of rules and associated membership functions for a good system performance. An automated way of designing fuzzy systems might be preferable. Hence a genetic algorithm is chosen for this purpose.

This is particularly the case when a large number of parameters are needed to adjust in the same problem and higher precision is required for the final result. To overcome the problem of inefficient use of the computer memory, the underlying real-valued chromosome representation of the membership functions is adopted.

The rule-base consists of 25 parameters and will take long time for an optimisation algorithm to converge to a satisfactory level of performance. As mentioned earlier a design simplification is desired rather than a drastic improvement of the performance. Therefore, the rule-base is not optimised using GA in this study, rather it is chosen arbitrarily from the PD-part of the FLC.

Chromosomes that are created by crossover operation do not improve much over generation for a small size of population. Therefore, a dynamic crossover probability is used in this study.

Since the rate of new chromosome in the population is falling by the crossover operation with a small population, addition of fresh blood to the population is required. This is achieved by a dynamic mutation probability.

The time taken in the evaluation of genetic structures, especially in the case of fuzzy controller, imposes restriction on the size of population and also the number of generations required to run the GA to a final solution. Therefore, a small population GA is chosen and the max number of generations is limited to 10.

References

Adeli H, Hung S-L (1995) Machine learning—Neural networks, Genetic algorithms and fuzzy systems. Wiley, London

Akita S, Yoshikawa T, Tsuruoka S (2000) The 4th asian fuzzy systems symposium, May 31–June 3, Tsukuba, Japan, pp 555–560

Alacala R, Ducange P, Herrera F, Lazzerini B, Marcelloni F (2009) A Multiobjective Evolutionary Approach to Concurrently Learn Rule and Data bases of Linguistic Fuzzt-Rule-based Systems. IEEE Trans Fuzzy Syst 17(5):1106–1112

Alba E, Cotta C, Troya JM (1999) Evolutionary design of fuzzy logic controllers using strongly-typed GP. Mathware Soft Comput 6(1):109–124

Antonelli M, Ducange P, Marcelloni F (2012) Genetic Training Instanvce Selection in Multiobjective Evolutionary Fuzzy Systems: A Coevolutionary Approach. IEEE Trans Fuzzy Syst 20(2):276–290

Bäck T (1992) Self-adaptation in genetic algorithms. Proceedings of the First European conference on artificial life, pp 263–271

Bäck T (1993) Optimal mutation rates in genetic search. Proceedings of the Fifth international conference on genetic algorithms, pp 2–8

Bäck T (1996) Evolutionary algorithms in theory and practice. Oxford University Press, New York

Bäck T, Kursawe F (1994) Evolutionary algorithms for fuzzy logic: a brief overview. Proceedings of the Fifth international conference on information processing and management of uncertainty in knowledge-based systems (IPMU'94), Paris, pp 659–664

Bastian A (2000) Identifying fuzzy models utilising genetic programming. Fuzzy Sets Syst 113(3):333–350

Box GEP (1957) Evolutionary operation: a method for increasing industrial productivity. Appl Stat 6(2):81–101

Castillo L, Gonzalez A, Perez R (2001) Including a simplicity criterion in the selection of the best rule in a genetic fuzzy learning algorithm. Fuzzy Sets Syst 120(2):309–321

Celikyilmaz A, Turksen IB (2008) Uncertainty modelling of improved fuzzy functions with evolutionary systems. IEEE Trans Syst, Man and Cybern—Part B: Cybern 38(4):1098–1110

Chaiyaratana N, Zalzala AMS (2002) Time-optimal path planning and control using neural networks and a genetic algorithm. Int J Comput Intell Appl (IJCIA) 2(2):153–172

Chan ZSH, Kasabov N (2004) Evolutionary computation for on-line and off-line parameter tuning of evolving fuzzy neural networks. Int J Comput Intell Appl (IJCIA) 4(3):309–319

Chan KY, Fogarty TC, Aydin ME, Ling SH (2008) Genetic algorithms with dynamic mutation rates and their industrial applications. Int J Comput Intell Appl (IJCIA 7(2):103–128

Chen M, Linkens DA (1998) A hybrid neuro-fuzzy PID controller. Fuzzy Sets Syst 99(1998):27–36

Cheong F, Lai R (2000) Constraining the optimization of a fuzzy logic controller using an enhanced genetic algorithm. IEEE Trans Syst, Man, Cybern - Part B: Cybern 30(1):31–46

Chin TC, Qi XM (1998) Genetic algorithms for learning the rule base of fuzzy logic controller. Fuzzy Sets Syst 97:1–7

Chou CH (2006) Genetic algorithm-based optimal fuzzy controller design in the linguistic space. IEEE Trans Fuzzy Syst 14(3):372–385

Cordon O, del Jesus MJ, Herrera F, Lozano M (1999) MOGUL: A methodology to obtain genetic fuzzy rule-based systems under the iterative rule learning approach. Int J Intell Syst 14(11):1123–1153

Cordon O, Herrera F, Hoffmann F, Magdalena L (2001) Genetic fuzzy systems—evolutionary tuning and learning of fuzzy knowledge bases, World Scientific

Darwin C (1859) On the origin of species by means of natural selection, John Murray, London

De Jong KA (1975) Analysis of the behaviour of a class genetic adaptive systems, PhD Thesis, Department of computer and communications sciences, University of Michigan, Ann Arbor

De Jong KA (2006) Evolutionary computation: a unified approach. The MIT Press, Cambridge

Deb K (2008) Multi-objective Optimisation using Evolutionary Algorithms. Wiley, Chichester

De Jong KA and Spears WM (1990) An analysis of interacting roles of population size and crossover in genetic algorithms. Proceedings of 1st workshop (PPSN 1) in parallel problem solving from nature. Dortmund, West Germany, pp 38–47

Dunham B, Lewitan H, North JH (1974) Simultaneous solution of multiple problems by natural selection. IBM Tech Discl Bull 17(7): 2191–2192

Eiben A, Smith J (2007) Introduction to evolutionary computing, 2nd edn. Springer, Berlin

Eiben AE, Hinterding R, Michalewicz Z (1999) Parameter control in evolutionary algorithms. IEEE Trans Evol Comput 3(1999):124–141

Eshelman LJ (1991) The CHC adaptive search algorithm: how to have safe search when engaging in non-traditional genetic recombination. In: G.J.E Rawlin (ed) Foundations of genetic algorithms 1 (FOGA-1), San Mateo, CA: Morgan Kaufmann, pp 265–283
Eshelman LJ, Schaffer JD (1993) Real-coded genetic algorithms and interval schemata. Foundations genetic algorithms 2 (FOGA-2), pp 187–202
Fazzolari M, Alacala R, Nojima Y, Ishibuchi H, Herrera F (2013) A review of the application of multi-objective evolutionary fuzzy systems: current status and further directions. IEEE Trans Fuzzy Syst 21(1):45–65
Fogarty TC (1989) Varying the probability of mutation in genetic algorithms. In: Proceedings of the third international conference on genetic algorithms, 1989, pp 104–109
Fogel LJ (1962) Autonomous automata. Ind Res 4:14–19
Fogel DB (1991) System identification through simulated evolution: A machine learning approach to modelling, Ginn Press, Needham Heights, MA
Fogel DB (1992) Evolving artificial intelligence, Doctoral Dissertation, University of California, San Diego La Jolla, CA
Fogel DB (1995) Evolutionary computation—Toward a new philosophy of machine intelligence. IEEE Press, New York
Fogel DB (1998) Evolutionary computation—The fossil records, Edt., IEEE Press
Fraser AS (1957) Simulation of genetic systems by automatic digital computers, I. Introduction. Aust J Biol Sci 10:484–491
Friedberg RM (1958) A learning machine: Part I. IBM J. Res Dev 2(1):2–13
Friedberg RM, Dunham B, North JH (1959) A learning Machine: Part II. IBM J. Res Dev 3, pp. 282–287
Gacto MJ, Alacala R, Herrera F (2010) Integration of index to preserve the semantic interpretability in the multiobjective evolutionary rule selection and training and tuning of linguistic fuzzy systems. IEEE Trans Fuzzy Syst 18(3):515–531
Glover F (1989) Tabu Search—Part I. ORSA J Comput 1:190–206
Goldberg DE (1989) Genetic algorithms in search, Optimization and machine learning. Addison Wesley Publishing Company, Boston
Goldberg DE (1991) Real-coded genetic algorithms, virtual alphabets and blocking. Complex Syst 5(2):139–168
Goldberg, D.E. and Richardson, J. (1987) Genetic Algorithm with Sharing for Multimodal Function Optimisation, Proceedings of the 2nd International Conference on Genetic Algorithms, pp. 41-49
Goldberg DE, Korb B, Deb K (1989) Messy genetic algorithms: motivations, analysis and first results. Complex Syst 3(5):493–530
Gonzalez A, Perez R (1999) SLAVE: a genetic learning system based on an iterative approach. IEEE Trans Fuzzy Syst 7(2):176–191
Grefenstette JJ (1986) Optimisation of control parameters for genetic algorithms. IEEE Trans Syst, Man and Cybern 16(1):122–128
Hesser J, Männer R (1990) Towards on optimal mutation probability for genetic algorithms. In: Proceedings of parallel problem solving from nature conference, 1990, pp 23–32
Hoffmann F (2001) Evolutionary algorithms for fuzzy control system design. Proc IEEE 89(9):1318–1333
Hoffman E, Nelles O (2001) Genetic programming for model selection of TSK-fuzzy systems. Inf Sci 136:7–28
Holland JH (1962) Outline for a logical theory of adaptive systems. J ACM 3:297–314
Holland JH (1975) Adaptation in natural and artificial systems. University Michigan Press, Ann Arbor
Homaifar A, McCormick Ed (1995) Simultaneous design of membership functions and rule sets for fuzzy controllers using genetic algorithms. IEEE Trans Fuzzy Syst 3(2):129–139
Huang Y-P, Huang C-H (1997) Real-valued genetic algorithms for fuzzy grey prediction system. Fuzzy Sets Syst 87(3):265–276

Hwang HS (1999) Automatic design of fuzzy rule base for modelling and control using evolutionary programming. IEE Proceeding of control theory application vol 144(I), pp 9–16
Ishibuchi H, Nozaki K, Yamamoto N, Tanaka H (1995) Selecting fuzzy if-then rules for classification problems using genetic algorithms. IEEE Trans Fuzzy Syst 3:260–270
Ishibuchi H, Nakashima T, Murata T (1999) Performance evaluation of fuzzy classifier systems for multi-dimensional pattern classification problems. IEEE Trans Syst Man, Cybern B Cybern 29(5):601–618
Jang J-S, Sun CT, Mizutani E (1997) Neuro-fuzzy and soft computing, Englewood CliOEs. PTR Prentice Hall, New Jersey
Jin Y, von Seelen W, Sendhoff B (1999) On generating fc fuzzy rule systems from data using evolution strategies. IEEE Trans Syst, Man Cybern—Part B: Cybern 29(6):829–845
Juang CF (2005) Combination of on-line clustering and Q-value based GA for reinforcement fuzzy system design. IEEE Trans Fuzzy Syst 13(3):289–302
Kang S-J, Woo C-H, Hwang H-S, Woo KB (2000) Evolutionary design of fuzzy rule base for nonlinear system modelling and control. IEEE Trans Fuzzy Syst 8(1):37–45
Karr CL (1991) Design of an adaptive fuzzy logic controller using a genetic algorithm. Proceedings of the 4th international conference on genetic algorithms, Morgan Kaufmann Publishers, San Mateo, pp 450–457
Karr CL, Gentry EJ (1993) Fuzzy control of pH using genetic algorithms. IEEE Trans Fuzzy Syst 1(1):46–53
Khemliche M, Mokeddem D, Khellaf A (2002) Design of a fuzzy controller of PH by the genetic algorithms. In: Proceedings power conversion conference, vol 2, pp 912–916
Kirkpatrick S, Gelatt CD, Vecchi MP (1983) Optimization by simulated annealing. Science 220:671–680
Kovacic Z, Bogdan S (2006) Fuzzy controller design: Theory and application. CRC Press, Taylor and Francis Group
Koza John R (1992) Genetic programming: On the programming of computers by means of natural selection. The MIT Press, Cambridge
Krishnakumar K (1989) Micro-genetic algorithms for stationary and non-stationary function optimisation. Proceedings of the SPIE: intelligent control and adaptive systems, pp 289–296
Kuo Y-P, Li T-HS (1999) GA-based fuzzy PI/PD controllers for automotive active suspension system. IEEE Trans Ind Electr 46:1051–1056
Lee MA, Takagi H (1993) Dynamic control of genetic algorithms using fuzzy logic techniques. Proceedings of 5th international conference on genetic algorithms, Urbana-Campaign, IL, July 17–21, pp 76–83
Lekova A, Mikhailov L, Boyadijev D, Nabout A (1998) Redundant fuzzy rules exclusion by genetic algorithms. Fuzzy Sets Syst 100:235–243
Lin C-T, Lee CSG (1991) Neural-network-based fuzzy logic control and decision system. IEEE Trans Comput 40:1320–1336
Lin C-T and Lee CSG (1992) Real-time supervised structure-parameter learning for fuzzy neural network. Proceeding of IEEE international conference on fuzzy systems, pp 1283–1290
Lin C-T, Lee CSG (1993) Reinforced structure-parameter learning for neural-network-based fuzzy logic control systems. Proceeding of IEEE international conference on fuzzy systems, pp 88–93
Lin C-T, Lee CSG (1995) A neural fuzzy control system with structure and parameter learning. Fuzzy Sets Syst 70:183–212
Lin C-J, Chen C-H, Lin C-T (2009) A hybrid of cooperative particle swarm optimization and cultural algorithm for neural fuzzy networks and its prediction applications. IEEE Trans Syst, man Cybern—Part C: Appl Rev 39(1):55–68
Linkens DA, Nyongesa HO (1995a) Genetic algorithms for fuzzy control, Part 1: offline system development and application. IEE proceedings of control theory and application, vol 142, No. 3, pp 161–176

Linkens DA, Nyongesa HO (1995b) Genetic algorithms for fuzzy control, Part 2: online system development and application. IEE proceedings of control theory and application, vol 142, No. 3, pp 177–185

MacVicar-Whelan PJ (1976) Fuzzy sets for man-machine interaction. Int J Man-Mach Stud 8:687–697

Matyas J (1965) Random optimization. Autom Remote Control 26:244–251

McGinley B, Morgan M, O'Riordan C (2008) Maintaining diversity through adaptive selection. Crossover and Mutation, Proceedings of the GECCO'08, Atlanta, Georgia, USA, 12–16 July 2008, pp 1127–1128

Michalewicz Z (1994) Genetic algorithms+Data structures=Evolution programs, 2nd extended edn. Springer, Berlin

Nauck D, Kruse R (1992) Interpreting changes in the fuzzy sets of a self-adaptive neural fuzzy controller. Proceeding of 2nd workshop on industrial applications of fuzzy control and intelligent systems IFIS'92, College Station, Texas, Dec 2–4 1992, pp 146–152

Nauck D, Kruse R (1993) A fuzzy neural network learning fuzzy control rules and membership functions by fuzzy error back propagation. Proceeding of IEEE international conference on neural networks, pp 1022–1027

Nelder J, Mead R (1965) The downhill simplex method. Comput J 7:308–313

Park SH, Kim YH, Choi YK, Cho HC, Jeon HT (1993) Self-organization of fuzzy rule base using genetic algorithm. In: Proceedings 5th IFSA world congress, 1993, pp 881–886

Price K (1994) Genetic annealing. Dr. Dobb's J 1994:127–132

Qi XM, Chin TC (1997) Genetic algorithms based fuzzy controller for higher order systems. Fuzzy Sets Syst 91:279–284

Rechenberg I (1965) Cybernetic solution path of an experimental problem, royal aircraft establishment, library translation No. 1122, Farnborough, Hants, UK, 1965

Reynolds RG (1994) Introduction to cultural algorithms. In: Sebald AV, Fogel LJ (Eds) Proceedings of the third annual conference on evolutionary programming, World Scientific Press, Singapore, pp 131–139

Reynolds RG (1999) An overview of cultural algorithms: advances in evolutionary computation. McGraw Hill Press, USA

Schaffer JD, Morishima A (1987) An adaptive crossover distribution mechanism for genetic algorithms. In: Proceeding of the second international conference on genetic algorithms, 1987, pp 36–40

Schaffer JD, Caruana RA, Eshelman LJ, Das R (1989) A study of control parameters affecting online performance of genetic algorithms for function optimization. In: Schaffer JD (ed) Proceedings of the third international conference on genetic algorithms, San Mateo, CA: Morgan Kaufmann, pp 51–60

Schwefel H-P (1968) Projekt MHD-Strausstrhlrohr: Experimentelle Optimierung einer Zweiphasenduese, Teil I, Technischer Bericht 11.034/68, 35, AEG Forschungsinstitute, Berlin, Germany

Shi Y, Eberhart R, Chen Y (1999) Implementation of evolutionary fuzzy systems. IEEE Trans Fuzzy Syst 7(2):109–119

Siarry P, Guely F (1998) A genetic algorithm for optimizing Takagi-Sugeno fuzzy rule-bases. Fuzzy Sets Syst 99:37–47

Siddique N, Adeli H (2013) Computational intelligence: synergies of fuzzy logic, neural networks and evolutionary computing. Wiley, New York

Siddique NH, Tokhi MO (2006) GA-based neural fuzzy controller for flexible-link manipulator. J Eng Lett, Special issue on Hybrid Intelligent Systems using Neural Networks, Fuzzy Logic and Genetic Algorithms, Oscar Castillo Edt., vol 13(2), paper EL_13_2_13

Spears WM (2000) Evolutionary Algorithms: the role of mutation and recombination. Springer, Berlin

Srinivas M, Patnaik LM (1994) Adaptive probabilities of crossover and mutation in genetic algorithms. IEEE Trans Syst, Man Cybern 24(4):656–667

Stewart SH, Taylor S, Baker JM, Hoffmann F, Pfister G (1997) Evolutionary design of a fuzzy knowledge base for a mobile robot. Int J Approx Reason 17(4):447–469

Storn R (1995) Constrained optimization. Dr Dobbs J, pp 119–123

Storn R (1999) System Design by Constraint Adaptation and Differential Evolution. IEEE Trans Evol Comput 3(1):22–34

Storn R, Price K (1997) Differential evolution—a simple and efficient heuristic for global optimisation over continuous space. J Global Optim 11(4):431–459

Tarng YS, Yeh ZM, Nian CY (1996) Genetic synthesis of fuzzy logic controllers in tuning. Fuzzy Sets Syst 83:301–310

Tunsted E, Jamshidi M (1996) On Genetic programming of fuzzy rule-based systems for intelligent control. Int J Intell Autom Soft Comput 2(3):273–284

Vakula VS, Sudha KR (2012) Design of differential evolution algorithm-based robust fuzzy logic power system stabiliser using minimum rule base. IET Gener Transm Distrib 6(2):121–132

Vidyasagar M (2002) Nonlinear systems analysis, 2nd edn. Prentice Hall, Englewood Cliffs

Yun Y, Gen M (2003) Performance analysis of adaptive genetic algorithms with fuzzy logic and heuristics. Fuzzy Optim Decis Mak 2:161–175

Zajaczkowski J, Verma B (2009) A computational method using an evolutionary algorithm for finding fuzzy rules in 3-layer hierarchical fuzzy structure. Int J Comput Intell Appl (IJCIA) 8(4):467–485

Zhou Y, Joo M (2008) An Evolutionary Approach toward Dynamic Self-Generated Fuzzy Inference Systems. IEEE Trans Syst, Man Cybern—Part B: Cybern 38(4):963–969

Zhou YS, Lai LY (2000) Optimal design for fuzzy controllers by genetic algorithms. IEEE Trans Ind Appl 36(1):93–97

Chapter 7
Neuro-Fuzzy Control

7.1 Introduction

The performance of any fuzzy system or fuzzy controller mainly depends on the input–output membership functions, the If–Then rules, and tuning of both (Nauck and Kruse 1993, 1996). The choice of defuzzification method is another factor, which also influences the performance (Yager and Filev 1994). Unfortunately, there are no formal methods to define the membership functions or to construct the rule-base for fuzzy systems or controllers. The issues have been prominent in Chap. 6 and evolutionary learning[1] approaches were explored to address some of the issues. In Chap. 6, evolutionary learning is seen as an optimisation or search problem requiring a simple scalar performance index. The performance of the fuzzy system is aggregated into a scalar performance index on which basis evolutionary algorithms select outperforming rule-base, MFs or scaling parameters or their combinations. Evolutionary learning algorithms are the suitable choices where no a priori information about the MFs and the rule-base is available. There have been many successful applications of evolutionary fuzzy systems reported in the literature. Due to the nature of evolutionary algorithms, evolutionary fuzzy systems are presumably slow processes and the performance of the system inherently depends on the size of the population and the number of generations required for a solution to be robust for specific problems.

The most striking features of neural networks are their flexible structures, available learning algorithms and capability of learning from experiential data. Due to these inherent advantages, neural networks found applications in many engineering applications such as pattern recognition, signal processing, modelling and control of complex systems (Akesson and Toivonen 2006; Narendra and Parthasarathy 1990; Narendra and Mukhopadhyay 1997; Sarangapani 2006). Consequently, the combination of neural networks with fuzzy systems has been recognised as a powerful alternative approach to learning fuzzy systems. Such a combination should be able to learn linguistic rules, membership functions, or to optimise existing ones. Learning

[1] An influential paper by Hinton and Nowlan (1987) showed that learning can guide evolution and learning evolution can work synergistically together.

N. Siddique, *Intelligent Control*, Studies in Computational Intelligence 517,
DOI: 10.1007/978-3-319-02135-5_7, © Springer International Publishing Switzerland 2014

in this case means creating a rule-base or membership functions from scratch based on training data presented as a fixed or free learning problem (Nauck and Kruse 1996). The learning procedure operates on local information, and causes only local modifications in the underlying fuzzy system.

Some neuro-fuzzy systems are capable of learning and providing fuzzy rules in linguistic or explicit form. However, most of the current neuro-fuzzy approaches address parametric identification or learning only. In general, the designer chooses the shape of membership functions and the respective parameters are adjusted. As pointed out in (Jang and Sun 1995), in addition to parametric learning, structure learning problems deal with the partition of the input–output universes, the number of membership functions for each input, the number of fuzzy if–then rules, and so on. Few results on structure determination are available in the literature. For instance, in (Lin and Lu 1996) a neuro-fuzzy system was proposed with fuzzy training data and supervised learning which provides a mechanism for finding the number of rules, assuming exponential rule membership functions. The designer still has to provide the input–output space partitions. In (Figueiredo and Gomide 1999) a neuro-fuzzy network has been proposed, which provides a mechanism for obtaining rules that cover the whole input–output space as well as the membership functions including their shapes for each input variable. All these approaches are either Sugeno-type or Tsukamoto-type fuzzy systems. Mamdani-type fuzzy systems are lacking some of those approaches.

A practically suitable learning approach for designing fuzzy controller (especially the PD-PI like fuzzy controller introduced in Chap. 5 and optimised using evolutionary algorithms in Chap. 6) applying neural networks with appropriate parameter settings will be investigated further. Therefore, a brief overview of neural networks and their algorithms are presented first followed by the application of neural networks (NN) in the tuning of Mamdani-type fuzzy controllers.

7.2 Neural Networks and Architectures

A neural network consists of neurons of biological analogy. Warren McCulloch and Walter Pitts outlined the first formal model of an elementary computing neuron (McCulloch and Pitts 1943). However, the model makes use of several drastic simplifications allowing only binary states 0, 1, operates under a discrete-time assumption and synchronous operations of all neurons in a larger network. Weights and threshold are fixed in the model. Every neuron consists of a processing element with synaptic input connections and a single output. The first stage is a process where the inputs $x_1, x_2, \ldots x_n$ multiplied by their respective weights $w_1, w_2, \ldots w_n$ are summed by the neuron. The resulting summation process may be shown as:

$$net = (w_1 \cdot x_1 + w_2 \cdot x_2 + \cdots + w_n \cdot x_n) \tag{7.1}$$

It can be written in vector notation form as

$$net = \left(\sum_{i=1}^{n} w_i \cdot x_i \right) = W^T X \tag{7.2}$$

Where $\boldsymbol{w}$ is the weight vector defined as $w = [w_1, w_2, \ldots w_n]^T$ and $\boldsymbol{x}$ is the input vector defined as $x = [x_1, x_2, \ldots x_n]^T$. A threshold value b, called bias, plays an important role for some neuron models and needs to mention explicitly as separate neuron model parameter. Then the Eq. (7.2) looks like

$$net = \left(\sum_{i=1}^{n} w_i \cdot x_i \right) + b \tag{7.3}$$

In order to allow for varying input conditions and their effect on the output it is usually necessary to include a non-linear activation function *f*(.) in the neuron arrangement. This is so that adequate levels of amplification may be used where necessary for small input signals, which avoids the risk of driving the output to unacceptable limits. A number of activation functions are used with different neural networks. Detailed description on the activation functions can be found in (Haykin 2009). A perceptron neuron model is shown in Fig. 7.1. The output of the neuron is now expressed in the form

$$y = f(net) \tag{7.4}$$

Figure 7.2 shows a single layer feedforward perceptron network. The inputs to the network are the input vector x

$$x = \begin{bmatrix} x_1 \\ x_2 \\ \vdots \\ x_m \end{bmatrix} \tag{7.5}$$

The weights of the network are defined by the weight matrix W

$$W = \begin{bmatrix} w_{1,1} & w_{1,2} & \cdots & w_{1,m} \\ w_{2,1} & w_{2,2} & \cdots & \\ \vdots & \vdots & \vdots & \vdots \\ w_{n,1} & w_{n,2} & \cdots & w_{n,m} \end{bmatrix} \tag{7.6}$$

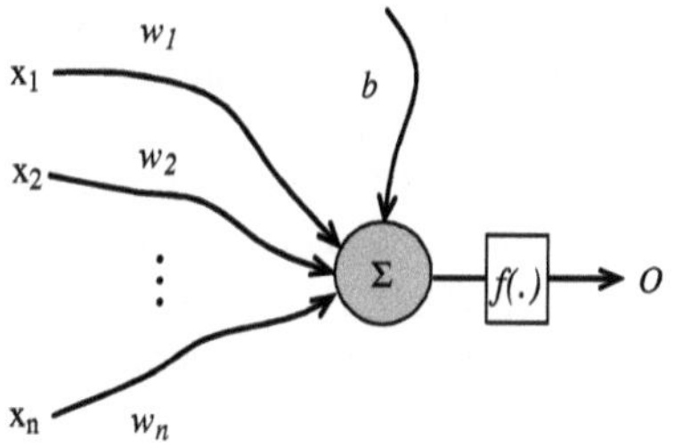

Fig. 7.1 Perceptron neuron model

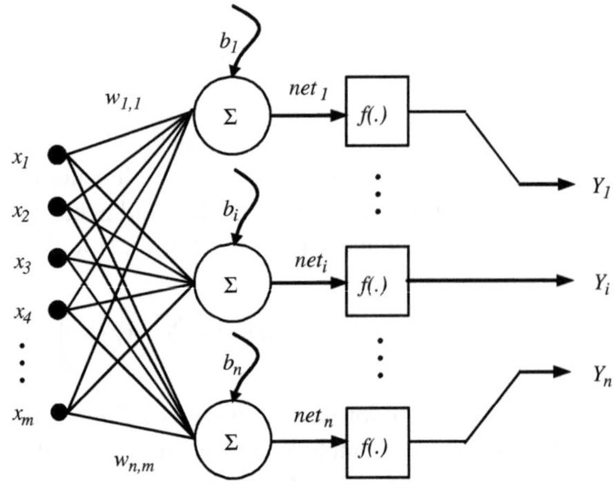

Fig. 7.2 Single layer feedforward perceptron network

and biases are defined by the bias vector b

$$b = \begin{bmatrix} b_1 \\ b_2 \\ \vdots \\ b_n \end{bmatrix} \tag{7.7}$$

The output Y of the network can be written in vector form as

$$Y = f(W \cdot x + b) \tag{7.8}$$

The notion of learning in NN is the procedure of modifying the weights. The weight update rule is formulated as the minimisation of an error function E with respect to weights w to train a network to produce a particular response to a specific input. The weight update rule is defined as

$$\Delta w = -\eta \frac{\partial E}{\partial w} \tag{7.9}$$

where η is a learning rate. The information processing ability of a neural network depends on its topology or connectivity, also called architecture (Yao 1993). The selection of network architecture is largely determined by the application domain. Usually the number of neurons, connections, and choice of activation (also called transfer) functions are fixed during the design. A variety of feedforward neural network architectures are available and widely used for modelling and control purposes in neural networks community. Among them are Multilayer Perceptron Networks (MLP), Radial Basis Function Networks (RBFN), Generalised Regression Neural Networks (GRNN), Probabilistic Neural Networks (PNN), Belief Networks (BN), Hamming Networks (HM), Stochastic Networks, Self-Organising Maps (SOM), and Learning Vector Quantisation (LVQ) and have found wide range of applications. There are many learning algorithms available for NN, which mainly depend on the architecture and availability of experiential data for training. Two different types of learning algorithms are distinguished: learning with supervision, and learning without supervision. Most of these rules are some sort of variation of the well-known and oldest learning rule, Hebb's Rule (Hebb 1949). Widrow-Hoff learning rule (Widrow and Hoff 1960) is applicable for supervised training of NN. Among the gradient descent-learning algorithms, backpropagation (BP) algorithm is most popular algorithm, which is an extension of the perceptrons to multi-layered neural network. The application of backpropagation algorithm will be demonstrated in Sects. 7.6 and 7.7. A comprehensive description of all the network architectures and learning algorithms is beyond scope of this chapter. The interested readers are directed to Haykin (2009) and Siddique and Adeli (2013).

7.3 Combinations of Neural Networks and Fuzzy Controllers

Two kinds of combinations between neural networks and fuzzy systems can be distinguished that have the goal of tuning or learning a fuzzy system. In the first approach, neural networks and fuzzy systems work independent of each other. The combination lies in the determination of certain parameters of a fuzzy system by neural networks or a neural network-learning algorithm, which can be done off-line or on-line during the execution of the fuzzy system. This kind of combination is called a cooperative neuro-fuzzy system. The second kind of combination defines a homogeneous architecture, usually similar to the structure of a neural network. This is implemented by interpreting a fuzzy system as a special kind of neural network. This kind of neuro-fuzzy system is called a hybrid neuro-fuzzy system because the resulting system can be viewed as a fuzzy system or as a neural network. There have been many hybrid architectures reported in the literature such as

- Fuzzy Adaptive Learning Control Network (FALCON) proposed by Lin and Lee (1991).
- Approximate Reasoning-based Intelligent Control (ARIC) proposed by Berenji (1992).
- Generalised Approximate Reasoning based Intelligent Control (GARIC) developed by Berenji and Khedkar (1992, 1993).
- Fuzzy Basis Function Networks (FBFN) first proposed by Wang and Mendel (1992).
- Fuzzy Net (FUN) proposed by Sulzberger et al. (1993).
- Fuzzy Inference and neural network in Fuzzy Inference Software (FINEST) proposed by Tano et al. (1994, 1996).
- Neuro-Fuzzy Controller (NEFCON) proposed by Nauck et al. (1997).
- Self Constructing Neural Fuzzy Inference Network (SONFIN) proposed by Feng and Teng (1998).

- Adaptive Neuro-Fuzzy Inferencing Systems (ANFIS) proposed by Jang (Jang 1993; Jang et al. 1997).
- MANFIS, CANFIS architecture reported by Mizutani and Jang (1995).

A very detail description of all these architectures can be found in (Siddique and Adeli 2013). All these hybrid neuro-fuzzy systems are mostly based on Sugeno-type fuzzy models discussed in Chap. 4. Sugeno-type fuzzy system has some structural advantages over the Mamdani-type fuzzy system. By replacing the consequent MFs of the rules with linear functions, Sugeno-type fuzzy system can be represented as a system of linear equations. Therefore, application of NN and learning algorithm becomes straightforward to Sugeno-type fuzzy system when a set of input–output data is available. On the contrary, Mamdani-type fuzzy system does not have this kind of structural advantage. Consequently, the combination of NN with Mamdani-type fuzzy system is required to be of cooperative in nature.

In cooperative neuro-fuzzy systems, one is the primary problem solver and the other is a supportive mechanism for pre-processing information or estimating certain parameters. In cooperative combination, both fuzzy system and neural networks work independently of each other. There can be two types of cooperative combinations:

- fuzzy-NN cooperation where neural networks play the primary role of the system and the fuzzy system as supportive for estimating system parameters.
- NN-fuzzy cooperation where fuzzy system play the primary role of the system and neural network as supportive for learning a set of parameters.

The two cooperative combinations are found very simple to apply to many applications in many ways. There are many consumer products available in the market since nineties which use both NN and fuzzy in a variety of cooperative combinations (Takagi 1992, 1995, 1997).

In cooperative fuzzy-NN systems, a fuzzy system translates linguistic statements into suitable perceptions such as feature selection, estimates learning parameters or architectural parameters for NN from available data. Takagi (1992, 1995) reported a number of cascade and developing tool type combinations of fuzzy and neural systems applied to some consumer products where fuzzy system estimates some input parameters for the neural network.

In NN-fuzzy cooperation, NN is used as a supportive technology to determine or estimate different parameters of fuzzy system or controller such as MFs, rule-base, scaling factors and rule weighting from available experiential or sensor data. Sufficient data should be available for extracting the desired parameters and training of the NN. The process of determination of the parameters can be offline or online during the operation of the fuzzy system. Besides this cooperation, NN can be used as a pre-processor or post-processor to a fuzzy system where the structure of the fuzzy system is already known. The role of NN is to assist improving the performance of the primary system. Additionally, there are other important issues to be considered such as whether the cooperation is for an existing fuzzy system to be modified or the fuzzy system has to be designed completely. A typical combination of these two techniques is the so-called neuro-fuzzy control, which is basically a fuzzy control augmented by neural networks to enhance controller's performance (Feng 2006). There are various techniques, learning algorithms and heuristic approaches reported in the literature over the last two decades (Takagi and Hayashi 1991; Takagi 1995; Yager 1994; Yea et al. 1994). The interest lies in combining a predefined fuzzy controller with an NN to provide assistance in learning parameters using available information. There can be a number of combinations possible:

- NN for correcting FLC.
- NN for learning rules.
- NN for determining MFs.
- NN for parameter tuning/learning.

7.3.1 NN for Correcting FLC

If a large number of sensor data are available for a fuzzy controller for smoother control, precision and sensitivity in fuzzy control design becomes complicated due to increased number of inputs. This also demands huge computation time too. To reduce the processing time by the fuzzy controller fewer inputs are used by FLC and the larger portion of sensor inputs is processed by an NN to provide necessary corrections to the fuzzy controller. This saves substantial processing time for the fuzzy controller. This kind of combination has been implemented by many companies for consumer products (Takagi 1992, 1995, 1997).

7.3.2 NN for Learning Rules

For a multi-input and multi-output FLC, it is difficult for an expert to formulate rules. If sufficient data are available, a neural network can be used to determine fuzzy rules from training data (Hong et al. 2004; Lin and Lin 1997; Quek and Zhou 1996; Tung and Quek 2004; Wu et al. 2001). The straightforward approach is the clustering technique and usually implemented using self-organising feature maps (SOM). SOM is trained offline and then applied to the fuzzy system. MFs of the fuzzy system are predetermined. Pedrycz and Card (1992) used SOM to extract fuzzy rules from the data. Another way to determine fuzzy rules is to use fuzzy associative memory (FAM) proposed by Kosko (1992), where fuzzy rules are interpreted as an association between antecedent and consequents. If fuzzy sets are seen as points in the unit hypercube and rules are associations, then it is possible to use neural associative memory to represent fuzzy rules. A neural associative memory is also called a bidirectional associative memory (BAM) because creating its connection matrix corresponds to the Hebbian learning rule (Kosko 1992). Kosko (1992) suggests a form of adaptive vector quantisation (AVQ) to learn

FAM from available data. AVQ is also known as learning vector quantisation (LVQ). AVQ or LVQ learning is similar to SOM and realised using competitive learning. Takagi and Hayashi (1991) proposed a comprehensive neural network based induction and tuning of fuzzy rules for a Sugeno-type fuzzy system. Chen and Linkens (1999) showed that the optimisation of the rule base corresponds to partially learning it.

7.3.3 NN for Determining MFs

Learning or optimising the membership functions is less complex than the adaptation of the rule base. Membership functions can easily be described by parameters, which can be optimised with respect to a global performance measure.

As discussed earlier in Chaps. 4 and 5 that the construction of the MFs has remained a difficult task for fuzzy control design. Poor performance of an fuzzy system is mainly caused by improper definition of MFs. Widely accepted approach is the trial and error method, which is mostly a time consuming process. Therefore, the problem of constructing MFs has been a central issue in FLC design with a number of subjective, statistical and neural approaches being proposed. The first thing for the NN-FLC systems to apply learning techniques is to parameterised membership functions of the rule antecedent and consequent parts. Then virtually any membership function can be obtained using a multi-layer perceptron network that can be trained offline separately if sufficient experiential data are available for the FLC. Using a smaller number of neurons and exploiting the possibilities of shifting, scaling and reflecting the sigmoid activation function, a satisfactory solution can be achieved without elaborating the training (Halgamuge and Glesner 1994; Halgamuge et al. 1994). NN clustering approach can also be used to extract parameter values of the MFs. In general, an NN determines the number of rules by clustering the data for designing the fuzzy system. Using this clustered data, a neural network decides on a multidimensional, nonlinear MFs, and this network is then used as a generator of the MFs. One-dimensional MFs can be constructed based on the parameters such as cluster centres and distance metric from multi-dimensional data clusters. The useful contribution of the approach is the introduction of NN into the design process of fuzzy systems. Secondly, the MFs are designed completely at one stroke, rather than separately along each input axis. The NN in the cooperative combination provides the MFs' parameters to the FLC. This simple cooperative combination can be illustrated by Fig. 7.3. Adeli and Hung (1994) proposed an algorithm for determining MFs using a topology-and-weight-change classification with two-layer NN. In this learning algorithm, the number of input nodes equals the number of patterns in each training instance and the number of output nodes equals the number of clusters. The parameters of the MFs are defined from the clusters.

When heuristic methods are applied to produce membership functions for developing FLC systems, a set of subjective membership functions are defined

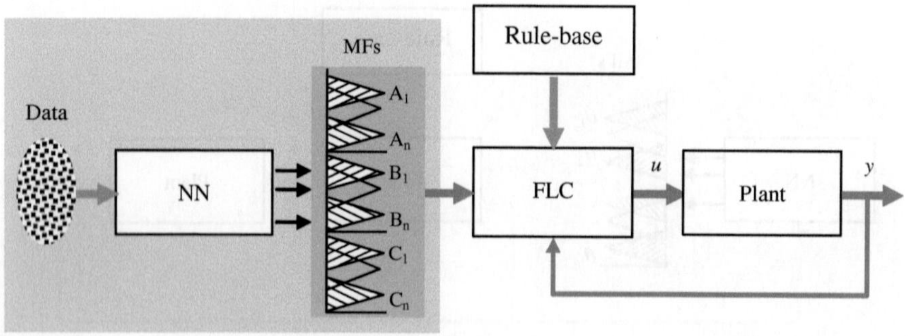

Fig. 7.3 NN learning of MFs parameters from data

within a universe of discourse, which either are too difficult to realise or cannot produce a satisfactory result. Therefore, many researchers propose to combine statistical method with fuzzy system. Using membership functions, which are generated from training data by one of the various clustering techniques, is one way to achieve this combination. A clustering algorithm can be applied to estimate the actual data distribution and the resulting clusters can be used to produce the membership functions, which will interpret the data better. Dickenson and Kosko (1993) proposed a learning technique for constructing membership functions by adaptive vector quantisation (AVQ). In this case, the NN in Fig. 7.3 is an AVQ network. To improve the performance of the fuzzy system, the chosen or generated membership functions can be further tuned by using gradient descent algorithms (Nomura et al. 1991). Ichihashi and Tokunaka (1993) proposed learning scheme based on gradient descent for adapting Gaussian MFs and Nomura et al. (1992) proposed adapting triangular MFs.

There are potential problems of determining MFs from data. The precise definition of the parameters of the MFs from data clustering is not always possible as the data distribution may not be representative of the entire input space. An alternative is to employ NN to acquire knowledge from the set of experiential data as the first step. Then multi-dimensional function is decomposed into single-dimensional functions. The error between the plant output using the designed FLC and the actual data depends on the parameters of the one-dimensional MFs. These MFs are tuned to minimise the error in a manner similar to backpropagation learning. The model in Fig. 7.4 uses an NN to optimise the parameters of the MFs by minimising the error between specification and output of the plant. This type of combination, also known as developing tool-type combination of NN-FLC, is widely used in many applications and consumer products such as washing machine, vacuum cleaners, rice cookers, dish washers and photocopiers developed by Japanese companies (Takagi 1992, 1995, 1997).

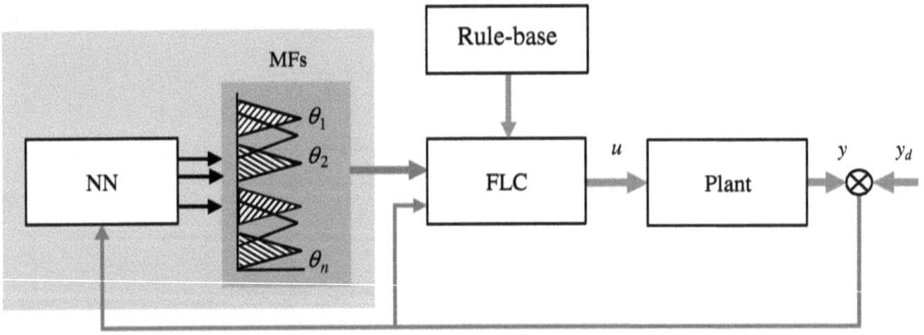

Fig. 7.4 NN learning of MFs parameters from system output

7.3.4 NN for Learning/Tuning Scaling Parameters

A neural network can be used to determine the parameters (scaling factors) of the FLC online i.e. during the use of fuzzy controller, to adapt the membership functions and it can also learn the weights of the rules online or offline. The NN can be trained off-line using the error function derived from the difference between the desired output y_d and the actual output y of the FLC in the similar way shown on Fig. 7.4. The universes of discourse of the membership functions for all the three inputs and the output have been optimised in Chap. 6. The other interpretation of this optimisation procedure is that the membership functions for change of error and sum of error have been normalised[2] within the same universe of discourse in pursuit of developing a single rule-base for the PD-PI fuzzy controller. The scaling factors will now act as the gains to fine tune or adjust the membership functions, which is analogous to the gains of a PID controller.

After designing an FLC, it is sometimes necessary to tune or adapt the MFs of the FLC with current data obtained during operation of the system. Adapting the MFs can be done using the data distribution discussed in earlier sections. The other possibility of adapting the MFs is by tuning or learning the scaling parameters of the FLC. In many cases, tuning the scaling factors or adjusting the membership functions can lead to the same result. Adjustment of membership functions requires learning of several parameters and hence scaling factor tuning is a much simpler task than adjusting the MFs parameters (Chen and Linkens 1998, 1999).

In general, increasing the scaling factors of the error and the change of error will reduce the rise-time while making the system performance sensitive around the set point. Decreasing the two scaling factors will have opposite effect. A small change of the output scaling factor will extend the rise-time resulting in sluggish response. Care should be taken to change the scaling factor for error when the actual output is within the tolerance band around the reference point as a small

[2] Normalization refers to sophisticated adjustments where the intention is to bring the entire distribution of adjusted values into alignment.

change in control action may cause the output to oscillate around the set point. When the actual output is far away from the set point, a relative larger scaling factor for error is chosen to speed up the response. The scaling factor for change of error has greater effect on the control sensitivity, i.e. transient time. A larger value of the scaling factor for change of error has a drastic impact on control action resulting in faster response, which has the possible risk of driving the system to instability. Scaling factor for the control action acts as the overall gain factor of the system and sets a trade-off between the system response and its stability. There have been many researches reported in the literature on how to determine the scaling factors without any mathematical process model (Haber et al. 2000; Hsu and Fann 1996; Linkens and Abbod 1992; Passino and Yurkovich 1998).

Figure 7.5 shows such tuning process using the cooperative combination of NN-FLC where NN determine the scaling factors for the MFs. The NN can be trained off-line using the error function derived from the difference between the desired output $y_d$and the output y of the plant. There are number of implementations reported in the literature. A method for tuning scaling factors using cross-correlation of the controller input–output was proposed by Palm (1995). Burkhardt and Bonissone (1992) defined a non-linear and discontinuous scaling function, adapting its parameters by a gradient descent method. Different nonlinear scaling functions are also proposed throughout the literature (Pedrycz et al. 1997; Gudwin et al. 1997).

7.4 Scaling Parameters of PD-PI Fuzzy Controller

In order to minimise the rule-base of the switching PD-PI fuzzy controller, two rule-bases were unified to a single rule base with $(n \times n + n \times n)$ rules and a switching mechanism was devised to ensure good controller performance such as fast rise time, minimal peak overshoot and zero steady state error. The membership functions for change of error and sum of error were redefined within the same universe of discourse and then an optimisation was carried out using genetic algorithm.

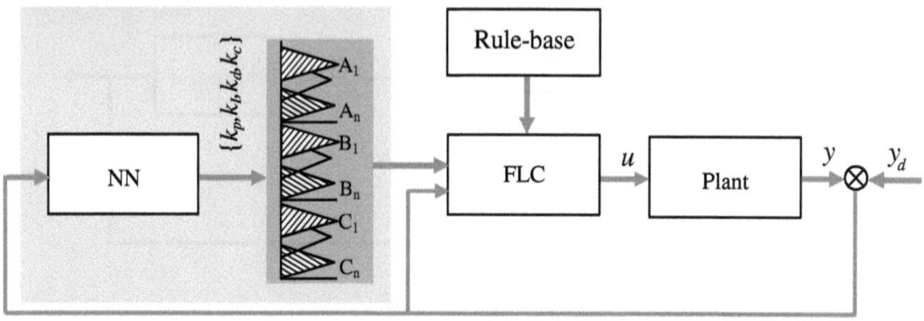

Fig. 7.5 Learning scaling parameters using NN

Though the optimisation algorithm minimises the objective criterion but the reinforced learning of the membership functions caused a deviation of the membership functions. A tuning or online adjustment of the membership functions becomes essential if such a merging procedure has been taken to reduce the number of rules and the optimisation has been carried out for the parameters of the membership functions (Chi et al. 1996). One way to achieve this is to tune the scaling parameters of the fuzzy controller. Different tuning schemes have been discussed in Sect. 7.3.

The scaling parameters of the PD- and PI-like fuzzy controller are shown with the controller's description as follows

$$k_c \cdot u = k_p \cdot e + k_d \cdot \Delta e \quad t \leq t_s \tag{7.10}$$

$$k_c \cdot u = k_P \cdot e + k_I \cdot \sum e \quad t > t_s \tag{7.11}$$

where t_s is switching time, k_p, k_d, k_I and k_c are the proportional, differential, integral and controller gain coefficients and e, Δe and $\sum e$ are the error and change of error and sum of error respectively. A block diagram of the switching PD-PI-like fuzzy controller with the scaling parameters is shown in Fig. 7.6.

The effect of scaling parameters on the performance characteristics of PD-, PI- and PID-like FLCs such as rise-time, maximum overshoot and settling time has been introduced since inception of FLC back in 1970s (Procyk and Mamdani 1979). Based on this analysis, different methods of optimal scaling parameter setting have been proposed in the literature (Zheng 1992; Daugherity et al. 1992; Hu et al. 2001; Li 1997; Lin et al. 2001).

In the following sections, a further enhancement of the performance of the switching PD-PI-like FLC is sought by tuning the scaling factors, especially the proportional, integral and derivative scaling factors by applying a neural network based learning algorithm.

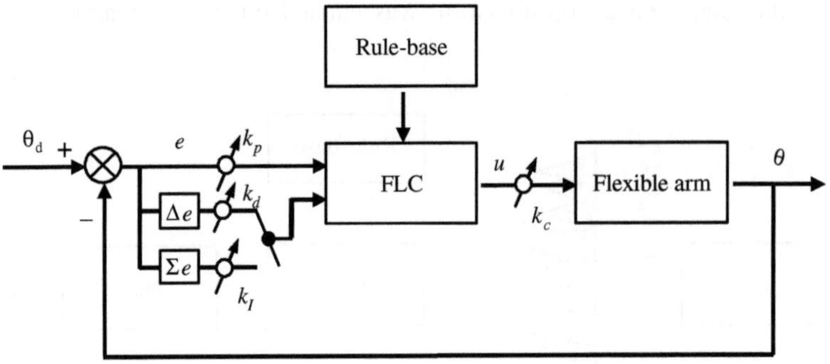

Fig. 7.6 Switching PD-PI-like FLC with 4 scaling factors

7.5 Reducing the Number of Scaling Parameters

The membership functions of the switching PD-PI-like FLC were initially defined heuristically in Chap. 5. Later on a merger of the membership functions for change of error and sum of error and the rule bases for the PD-like and PI-like FLCs were carried out while keeping the membership functions for error unchanged. An optimisation procedure using genetic algorithm was carried out to adjust the membership functions and adapt to the reduced single rule-base. It has been suggested by Chi et al. (1996) that a tuning or online adjustment of the membership functions becomes essential at this stage. It is now suggested to re-adjust the membership functions for change of error and sum of error by tuning the scaling factor k_d and k_I using a neural network. There is no need to re-adjust or tune the membership functions for error, since it remained the same in PD- and PI-like FLCs and moreover it is already optimised by genetic algorithm. Therefore, for simplicity the scaling factor k_p is not tuned further. This suggests elimination of the scaling factor k_p from Eqs. (7.10) and (7.11). Dividing both sides of Eqs. (7.10) and (7.11) by k_p yields

$$k_c' \cdot u = e + k_d' \cdot \Delta e \tag{7.12}$$

$$k_c' \cdot u = e + k_I' \cdot \sum e \tag{7.13}$$

Where the modified scaling factors k_d', k_I' and k_c' are defined as

$$k_d' = \frac{k_d}{k_p} \tag{7.14}$$

$$k_I' = \frac{k_I}{k_p} \tag{7.15}$$

$$k_c' = \frac{k_c}{k_p} \tag{7.16}$$

Thus the resulting switching PD-PI-like fuzzy controller with three modified scaling factors becomes as shown in Fig. 7.7.

The scaling factors k_d' and k_I' are learned by a neural network while the scaling factor k_c' is chosen by some heuristic rules in the same way as k_c. The elimination of k_p from the set of scaling factors has further impact on the tuning procedure as the size of the neural network will be reduced to some extent demanding less data and time to train and consequently the processing time will also be shortened.

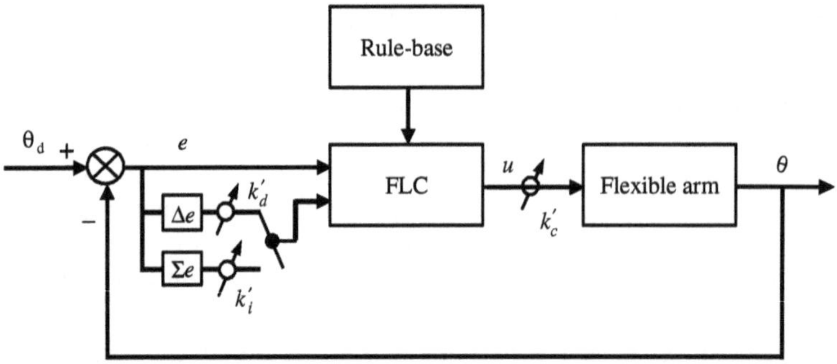

Fig. 7.7 Switching PD-PI-like FLC with modified 3 scaling factors

7.6 Neural Network for Tuning Scaling Factors

Simulation results in Zheng (1992) showed that the tuning of membership functions can achieve a marginal improvement in transient response of a second-order linear system. In this specific case, tuning has resulted in asymmetric triangular membership functions with unequal base for error e, i.e., specifically the width of membership functions increased around $e = 0$. Such membership functions contradict the usual practice (Driankov et al. 1993; Harris et al. 1993) where the membership functions get narrower and move closer to the origin to provide increased sensitivity at steady state. Such tuning process of the membership functions cannot guarantee improved performance under disturbance, which is an important criterion for performance evaluation of fuzzy control system. Moreover, a training scheme such as backpropagation algorithm is bounded by its input–output data set though it is minimising the objective function, where the objective function is a measure of minimisation of error function defined by the distance between the actual output and desired output. The same data set is used for training epochs to minimise objective function. Such a training procedure does not guarantee any improved performance of the controller (Choi and Choi 1992; Stevenson et al. 1990; Yeung and Sun 2002).

Though a multilayer neural network with sufficient number of neurons in the hidden layer can approximate the non-linearity better but the use of multilayer perceptron could simply exhaust the system by calculating exponential terms in the network, causing very slow response of the system and consequently resulting in degraded performance of the system. A neural network simpler in its structure and smaller in size will meet the demand for the computation of the scaling factors. Therefore, a single neuron network with non-linear activation function is proposed for this implementation that can better represent the system's nonlinearity and the nonlinearity can also be controlled by optimising the shape of the sigmoidal function.

Considering the above analysis as a design criterion, a single-neuron network is recommended, which can be employed for the PD-PI fuzzy controller. The self-learning task of a multilayer perceptron is replaced with a single unbiased neuron with activation function. Thus the architecture of the neural network becomes very simple as shown in Fig. 7.8. The network has two inputs: one input is for the error and the other input is for change of error before switching time or sum of error after switching time. The bias is set to zero. Two cases for the activation function are investigated: linear and nonlinear activation function. The neural network can be trained using the backpropagation algorithm with a set of input and output data. In this case, the training procedure will employ segmented data set for the backpropagation algorithm, which is termed as multi-resolution training. This issue will be discussed under Sect. 7.7.

The block diagram of the cooperative neuro-fuzzy controller is shown in Fig. 7.9 along with the single-neuron network based tuning unit.

7.6.1 Backpropagation Learning with Linear Activation Function

The purpose here is to construct an adaptive mechanism, which is able to adjust the control parameters to minimise the error function defined as

$$E = \frac{1}{2} e^2 \tag{7.17}$$

where $e = y_d - y$.

The backpropagation learning algorithm is used for parameter tuning. The weight update rule with a momentum term is defined as

$$\Delta w_i(t) = -\eta \frac{\partial E}{\partial w_i} + \alpha \, \Delta w_i(t-1) \tag{7.18}$$

where $\eta \in [0, 1]$ is the learning rate and $\alpha \in [0, 1]$ is the momentum.

Using the chain rule of derivative, $\partial E / \partial w_i$ can be written as

$$\frac{\partial E}{\partial w_i} = \frac{\partial E}{\partial y} \frac{\partial y}{\partial u} \frac{\partial u}{\partial k} \frac{\partial k}{\partial w_i} \tag{7.19}$$

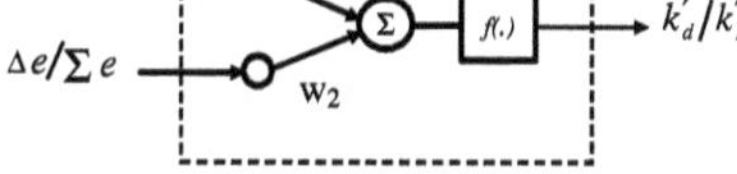

Fig. 7.8 Single-neuron network

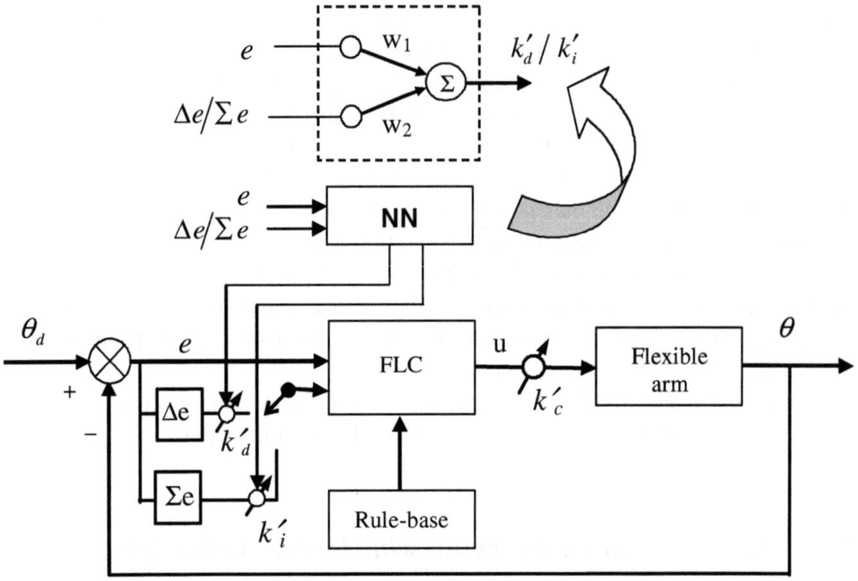

Fig. 7.9 Block diagram of the cooperative neuro-fuzzy controller

where

$$\frac{\partial E}{\partial y} = -e \tag{7.20}$$

For linear activation function

$$\frac{\partial k}{\partial w_i} = \frac{\partial f(net)}{\partial w_i} = \frac{\partial f(\sum w_i o_i + b)}{\partial w_i} = o_i \tag{7.21}$$

$$\frac{\partial y}{\partial u}\frac{\partial u}{\partial k} = \frac{\partial y}{\partial k} \tag{7.22}$$

and the bias $b = 0$ for unbiased neuron.

It is practically difficult to calculate the term $\partial y/\partial k$ because quantitative knowledge of the process is not generally available. One way is to use a numerical approximation such that $\partial y/\partial k$ can be replaced by $\Delta y/\Delta k$ at each iteration. Thus, Eq. (7.19) is rewritten as

$$\frac{\partial E}{\partial w_i} = -e\frac{\Delta y}{\Delta k}o_i \tag{7.23}$$

where $\Delta y = y(t) - y(t-1)$ and $\Delta k = k(t) - k(t-1)$.

Gradient descent search does not necessarily demand quantitative information of the process, since a search direction is enough to converge to the optimal point.

The sign of $\Delta y/\Delta k$ is enough for the calculation of the weight update equation, thus giving the simplified back-propagation algorithm as

$$\Delta w_i(t) = \eta\, e\, \mathrm{sgn}\left(\frac{\Delta y}{\Delta k}\right) o_i + \alpha\, \Delta w_i(t-1) \tag{7.24}$$

Thus, k'_d and k'_I are updated at every iteration as

$$k'_{d/I} = w_i(t) \cdot o_i \tag{7.25}$$

where o_i and w_i are defined as

$$o_i = [e \quad \Delta e]^T \text{ for } k = k'_d \tag{7.26}$$

$$o_i = \left[e \quad \sum e\right]^T \text{ for } k = k'_I \tag{7.27}$$

$$w_i = [w_1 \quad w_2] \tag{7.28}$$

The network is trained offline with the input–output data available. Once the weights of the neural network are learned, the parameters, i.e., scaling factors k'_d and k'_I are updated at each iteration (Fig. 7.10).

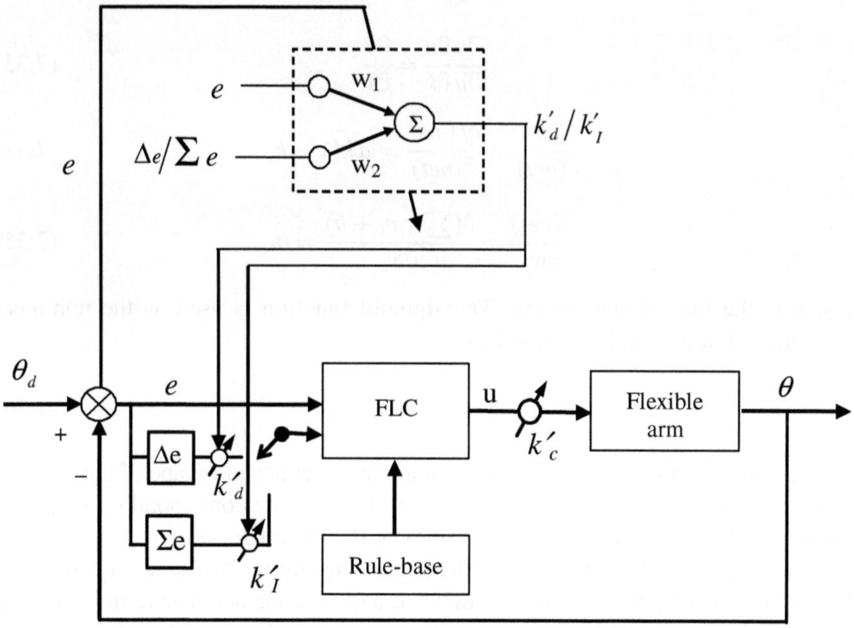

Fig. 7.10 Backpropagation learning of NN with linear activation function

7.6.2 Learning with Non-Linear Activation Function

The purpose here is to model the unseen nonlinearity in the adaptive mechanism discussed in sect. 7.6.1 by incorporating a nonlinear activation function into the neural network, which is able to adjust the control parameters so as to minimise the error function defined as

$$E = \frac{1}{2} e(t)^2 \tag{7.29}$$

where $e(t) = y_d - y(t)$. The backpropagation learning algorithm is used for parameter tuning. The weight update rule with a momentum term is defined as

$$\Delta w_i(t) = -\eta \frac{\partial E}{\partial w_i} + \alpha \, \Delta w_i(t-1) \tag{7.30}$$

where $\eta \in [0, 1]$ is the learning rate and $\alpha \in [0, 1]$ is the momentum. Using the chain rule of derivative, $\partial E / \partial w_i$ can be written as

$$\frac{\partial E}{\partial w_i} = \frac{\partial E}{\partial y} \frac{\partial y}{\partial u} \frac{\partial u}{\partial k} \frac{\partial k}{\partial (net)} \frac{\partial (net)}{\partial w_i} \tag{7.31}$$

where

$$\frac{\partial E}{\partial y} = -e \tag{7.32}$$

$$\frac{\partial y}{\partial u} \frac{\partial u}{\partial k} = \frac{\partial y}{\partial k} \tag{7.33}$$

$$\frac{\partial k}{\partial (net)} = \frac{\partial f(net)}{\partial (net)} = a\,k(1-k) \tag{7.34}$$

$$\frac{\partial (net)}{\partial w_i} = \frac{\partial (\sum w_i o_i + b)}{\partial w_i} = o_i \tag{7.35}$$

and b is the bias of the neuron. The sigmoid function is used as the nonlinear activation function and is defined as

$$f(x) = \frac{1 - e^{-ax}}{1 + e^{-ax}} \tag{7.36}$$

where x is the network output and the parameter a defines the shape of the sigmoid function chosen by trial and error. Figure 7.11 shows the corresponding shapes of the sigmoidal function for different values of the parameter a.

Practical difficulty arises in calculating the term $\partial y / \partial k$, as quantitative knowledge of the process is not available at any time instant. The term $\partial y / \partial k$ can be approximated numerically as

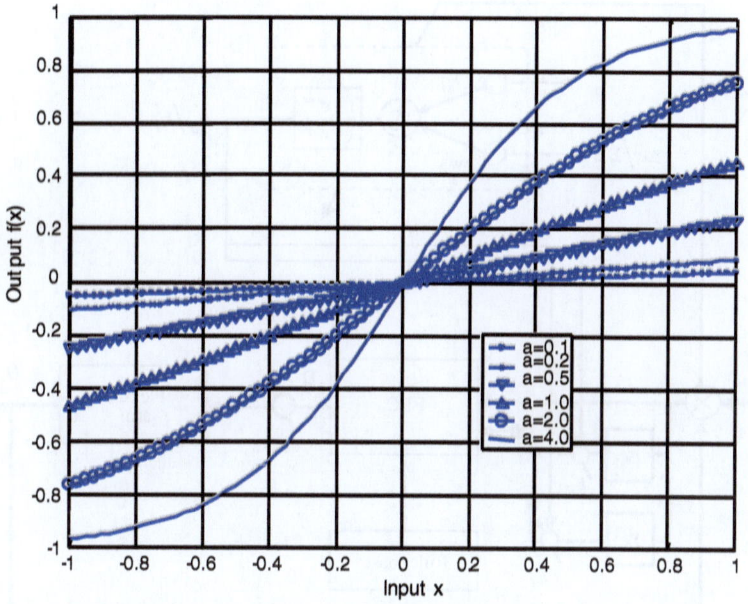

Fig. 7.11 Shape of sigmoidal function for different values of a

$$\frac{\partial y}{\partial k} \approx \frac{\Delta y}{\Delta k} \tag{7.37}$$

Thus, Eq. (7.31) is rewritten as

$$\frac{\partial E}{\partial w_i} = -e\,a\,k(1-k)\frac{\Delta y}{\Delta k}o_i \tag{7.38}$$

where $\Delta y = y(t) - y(t-1)$ and $\Delta k = k(t) - k(t-1)$. Gradient descent search does not necessarily demand quantitative information of the process, since a search direction is enough to converge to the optimal point. The sign of $\Delta y/\Delta k$ is accordingly enough for the calculation of the weight update equation. This results in the simplified back-propagation algorithm as

$$\Delta w_i(t) = \eta\, a\, e\, k(1-k)\, sgn\left(\frac{\Delta y}{\Delta k}\right) o_i + \alpha\,\Delta w_i(t-1) \tag{7.39}$$

where

$$o_i = [e \quad \Delta e]^T \text{ for } k = k'_d \tag{7.40}$$

$$o_i = \left[e \quad \sum e\right]^T \text{ for } k = k'_I \tag{7.41}$$

$$w_i = [w_1 \quad w_2] \tag{7.42}$$

This is thus used to update k'_d and k'_I at every iteration (Fig. 7.12).

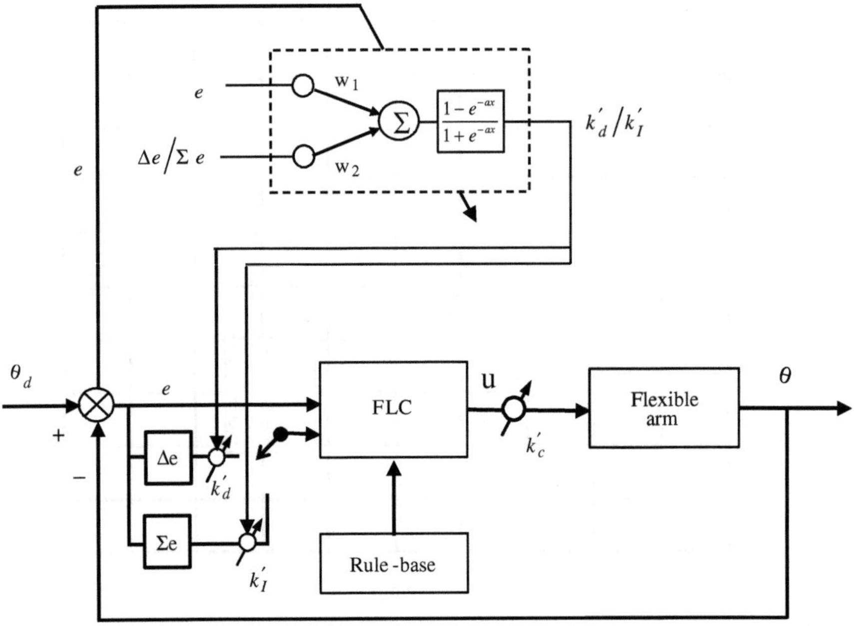

Fig. 7.12 Backpropagation learning of NN with nonlinear activation function

7.7 Multi-Resolution Learning

A general problem with the training neural networks using backpropagation algorithm is the convergence of learning, which mainly depends on the initialisation of the weights, experiential data set representative of the input–output space, learning parameters, the predefined architecture of the NN and the objective function for performance metric. This is actually an exhaustive trial and error process to find the correct set of combination for all these architectural, learning and training parameters. The choice of architecture and the performance metric is mainly application dependent. The initial weights have the most decisive influence on the convergence speed of the learning algorithm. Some initial weights lead to very slow convergence to a solution or, in the worst case, to divergence. Nguyen and Widrow (1990) showed that the initial weights and biases generated with certain constraints result in a faster learning speed for an NN than randomly generated initial weights. In Ngugen and Widrow's (1990) approach, NN weights are randomly initialised within the range of the data set and the weights are updated epoch by epoch during training. Also known weight training procedures for NNs is biased towards the data and parameters sets used for a particular training scheme. The performance of NNs can be improved and the convergence can be accelerated if appropriate data sets, learning parameters and initial weights are found.

To help avoid such divergence, some researchers proposed that random weight re-initialization be used whenever convergence becomes slow. To determine when convergence to a solution is slow, the performance metric, usually the sum of squared error defined as $sse = \sum e^2$, is checked after some N number of epochs to determine the speed of the convergence. Let $sse(k)$ denote the sse value after k epochs and $sse(k+N)$ denote the sse value after $k+N$ epochs. If $0 < sse(k) - sse(k+N) < T$ where T is some pre-selected threshold, then the convergence is considered slow and a re-initialization of the weights should be instigated.

The relationship between the resolution of training data and the steepness of activation function of neurons has been explored and investigated further. This section investigates how to maximize the effectiveness of multi-resolution learning by adapting neurons' sigmoid activation functions during the learning process. It is a new concept and the method is to adapt the slope of the sigmoid activation function to the training data at different resolution during multi-resolution learning process. The experimentation will demonstrate how this simple approach can further improve the generalization ability and robustness of the constructed neural networks and maximize the effectiveness of learning paradigm.

Multi-resolution learning is based on multi-resolution analysis (Liang 2000; Liang and Page 1997, 1998) in wavelet theory. The multi-resolution analysis framework is employed for decomposing the original signal and approximating it at different levels of detail. Unlike traditional neural network learning which employs a single data representation for the entire training process, multi-resolution learning exploits the approximation sequence representation-by-representation, from the coarsest version to finest version during the neural network training process. In this way, the original data can be segmented from coarse to the finest resolution in the approximation sequence and will be used in the learning process.

Assume that a given sampled data s^m is to be learned with $0 < M < m$ and $M \in Z$. Let a learning activity $A_j(r_j)$ denote a specific training phase conducted on the representation r_j of the training data (this of course includes some form of pre-processing) with a given learning algorithm. The learning dependency operator is denoted as "$\rightarrow$". $A_j \rightarrow A_i$ means that the learning activity A_j should be conducted before the learning activity A_i. Multi-resolution learning then can be defined as a sequence of learning activities $\{A_J(r_j)\}_{j \in Z \wedge j \geq M}$ associated with the sequence of approximation subspaces $\{V_j\}$ in multi-resolution analysis such that the following requirements are satisfied.

1. The representation r_j is associated with the approximation $s^j$of the original signal s^m in the approximation subspace $\{V_j\}$.
2. From the definition of $A_j(r_j) + A_{j+1}(r_{j+1})$, it can be seen that the multi-resolution learning paradigm generates an ordered sequence of learning activities such as

$$A_M(r_M) \rightarrow A_{M+1}(r_{M+1}) \rightarrow \cdots \rightarrow A_m(r_m) \quad (7.43)$$

where the parameter M indicates the approximation level of the original data s^m used to initiate the learning process. The first learning activity $A_M(r_M)$ starts with randomly generated initial network weights, and each subsequent learning activity $A_j(r_j)_{j>M}$ starts with the connection weights resulting from the previous learning activity.

The approximation $s^j (j<m)$ will contain fewer data samples than the original signal s^m. However, the training vectors in each learning activity A_j should occupy the full dimension of the neural network inputs in V_m to guarantee the smooth transition between subsequent learning activities. Therefore, a method is needed for constructing the representation r_j of training data for A_j based on s^j. This is achieved by setting the data details $d_k(k>j)$ to zero and reconstructing s^j in V_m.

7.7.1 Adaptive Neural Activation Functions

Since training data at different resolution level can potentially expose different inherent characteristics and correlated structure, it would be desirable to employ different neural processing capability of neurons which will be more appropriate and effective for each different resolution training data. The approach is to adapt the sigmoid activation functions of neurons (in feed-forward neural networks) to the training data at different resolutions by means of adjusting the slope (or steepness) of the activation functions in the region. It will show how this novel approach can improve the generalization ability and robustness of the constructed neural networks, and therefore maximize the effectiveness of multi-resolution learning paradigm. The effect of the activation function and steepness parameter on the performance metric (cost function) has been investigated as shown in Fig. 7.24. This will help choosing the appropriate activation function for the neural network.

The traditional trial and error approach uses the entire data set for training the network with an initial random weights and parameter setting. When the learning convergence is not to the satisfactory level defined by the performance metric, training is re-instigated with new set of weights and parameter settings. Rather than training the network with the entire data set, the idea here is to decompose the entire training data s^m into segments defined as

$$S^m = S^{m-2} + d^{m-2} + d^{m-1} \quad (7.44)$$

The Haar wavelet basis can be used for the decomposition. From this decomposition, two approximation versions at coarser resolutions of training data S^{m-2} and $S^{m-1} = S^{m-2} + d^{m-2}$ are obtained. The corresponding multi-resolution learning process for the data series then will contain $A_{m-2}(r_{m-2})$, $A_{m-1}(r_{m-1})$ and $A_m(r_m)$ where

$$r_j = \begin{cases} S^j & j = m \\ S^j + \sum_{k=j}^{m-1} d^k & d^k = 0; j = m-2, m-1 \end{cases} \tag{7.45}$$

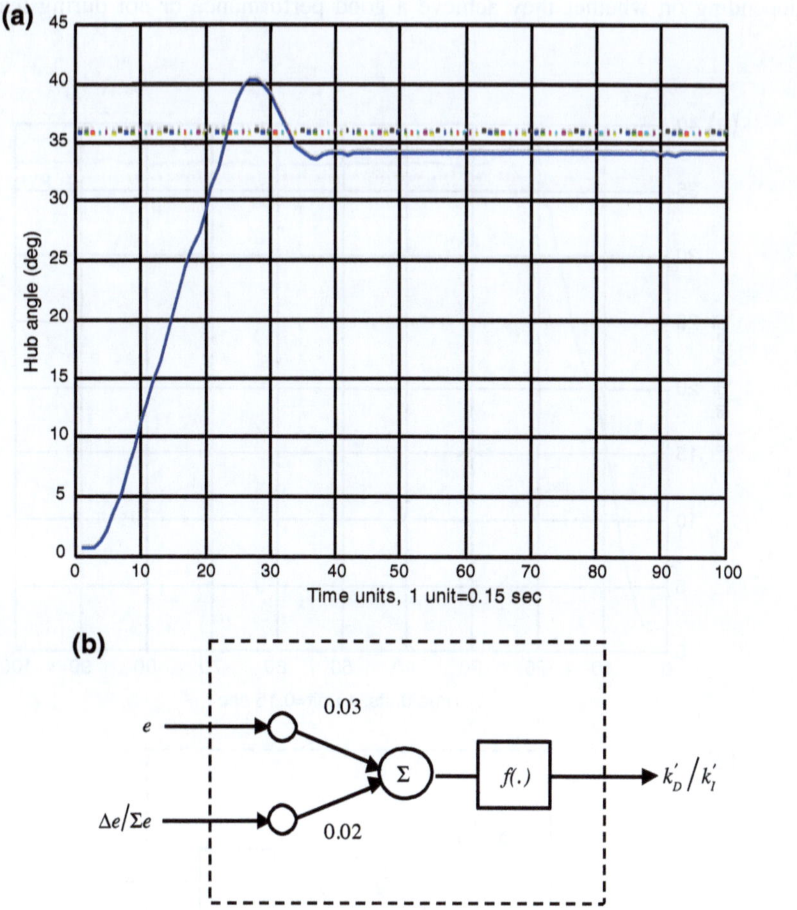

Fig. 7.13 Performance of the neuro-fuzzy controller after first 100 epochs training. Response of the flexible arm (**a**); Neural network with initial weights (**b**)

7.8 Some Experimental Results

Multilayered neural networks and the backpropagation algorithm were originally developed for pattern classification problems. In pattern classification applications, the training patterns are static, the training procedure and error function are straightforward, and real-time learning is not necessary. In control applications, training patterns for the neural network change with time, the backpropagation algorithm needs to be simplified, the error function needs to be defined in the context of applications, and real-time learning is a practical need. Generally, the training of neural networks for control can be performed on-line or off-line, depending on whether they achieve a good performance or not during learning.

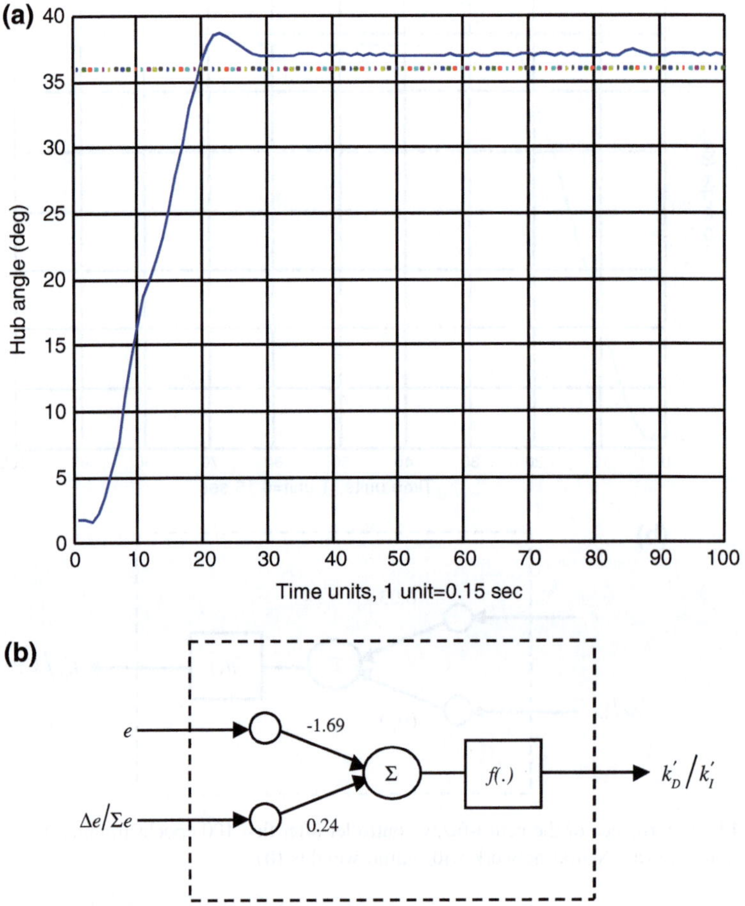

Fig. 7.14 Performance of the neuro-fuzzy controller after second 100 epochs of training. Response of the flexible arm (**a**); Neural network with weights learnt in first training (**b**)

Although off-line training is usually straightforward, conditions for assuring good generalisation of the neural network through the control space are difficult to attain, which makes on-line training always necessary in control applications. In fact, training should ideally occur exclusively on-line with neural networks learning from any initial set of weights.

The standard backpropagation algorithm with a linear and a non-linear activation function and the error functions are simplified for the purpose of real-time training in Sects. 7.6 and 7.7. To implement an on-line training, the learning rate, momentum and initial weights are set. A higher learning rate may need fewer epochs to train the network. On the other hand, a lower learning rate needs a longer time and a higher number of epochs to train the network. Now the on-line training of the network has two goals: firstly, it has to attain the control space within a

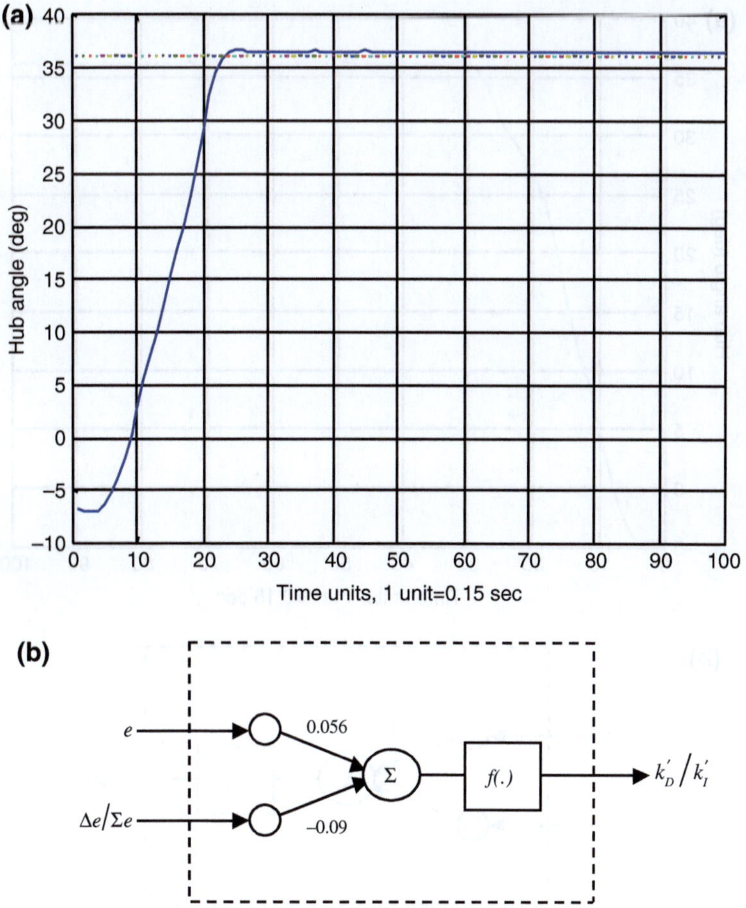

Fig. 7.15 Performance of the neuro-fuzzy controller after third 100 epochs of training. Response of the flexible arm (**a**); Neural network with weights learnt in second training (**b**)

limited number of epochs and secondly it has to achieve the desired error goal, which is not possible within the limited epochs. To resolve these conflicting goals, the following approach is adopted: train the network for a limited number of epochs within which it can attain a control space and then repeat until error goal is achieved. In other words, the error goal is defined as the difference between the output of the system and the set point. Now the objective of the training is to force the error goal asymptotically to zero (theoretically) or to a predefined tolerant region ε(practically) within a time interval, by repeatedly operating the system. Specifically, it requires that $|e_k(t)| \to 0$ or $|e_k(t)| < \varepsilon$ uniformly in $t \in [0, T]$ as $k \to \infty$ where k denotes the iteration number.

According to the approach mentioned above, the training of the network was performed in the following way: the learning rate and momentum were fixed at 0.02 and 0.75 respectively. The single-link flexible arm was operated for 100

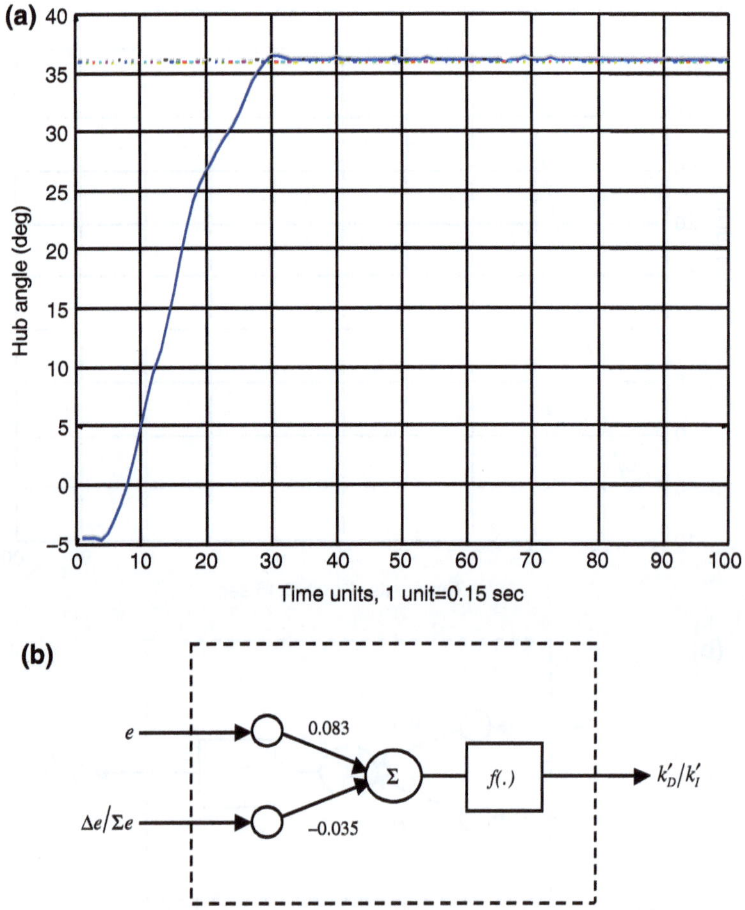

Fig. 7.16 Performance of the neuro-fuzzy controller after fourth 100 epochs of training. Response of the flexible arm (**a**); Neural network with weights learnt in third training (**b**)

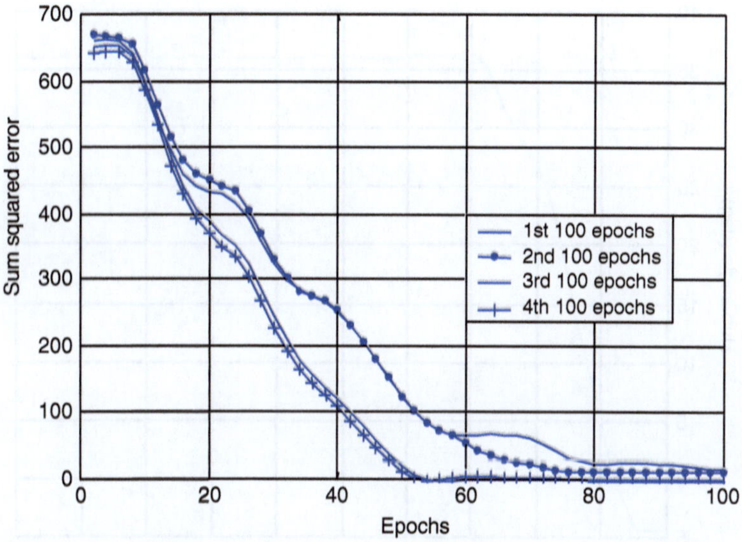

Fig. 7.17 Learning profile of the network in first, second, third and fourth learning

iterations with initial weights of $w_1 = 0.03$ and $w_2 = 0.02$. Figure 7.13 shows the system response of the neuro-fuzzy controller for a demanded hub angle of 36° with linear activation function. PD-PI-like fuzzy controller parameters k'_d and k'_I are updated at every iteration and a value of 76 was chosen heuristically for k'_c whereas the switching point was kept fixed at 25 time units. After the first 100 epochs of training the network weights were $w_1 = -0.016$ and $w_2 = 0.08$. A hub angle of 34.72° was reached after the first training.

In the second 100 epochs of training, the network was initialised with weights $w_1 = -0.016$ and $w_2 = 0.08$ obtained from the first training, and the learning rate and momentum were the same as before. The performance of the system for the same demanded hub angle, value of k'_c and switching point is shown in Fig. 7.14 along with the network. This time, a hub angle of 37.05° was achieved and the learnt weights were $w_1 = -1.69$ and $w_2 = 0.24$. In a similar way, the weights learnt from second training were set for training in the third round and the weights learnt from the third round training were set for training in the fourth 100 epochs of training keeping the all other parameters same as before. Hub angles of 36.52° and 36.26° were achieved in the third and fourth round training respectively. Figures 7.15 and 7.16 show the system performance with the corresponding network and their weights. The learning profiles of the four 100 epochs of training are shown in Fig. 7.17. As can be seen from the figure, the third and fourth round of training are close to each other and showed improvement over the first and second 100 epochs of training. After the fourth training, an error goal of 0.012 was achieved. An error goal of less than 0.012 depends on the precision of the interface card and may require costly high precision equipment. Figure 7.18 shows the

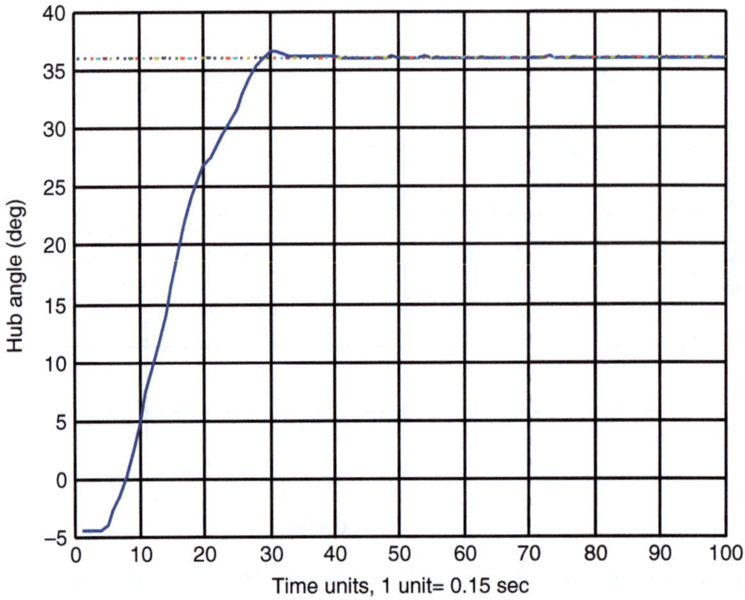

Fig. 7.18 Response of the flexible arm after training of the network

performance of the system after training of the network, that is, after the fourth 100 epochs of training. There is a significant improvement in the performance of the neuro-fuzzy controller in respect of rise time, maximum overshoot, settling time and steady state error as shown in Table 7.1.

Introduction of a non-linear sigmoid activation function to the single neuron network was another investigation in this chapter. The aim of this investigation is to figure out the possible effect of the sigmoid function shape on the performance of the network. To investigate these effects, training was performed in the same way as before. This time weights and bias were learnt with different sigmoidal function shape. It was found that for different sigmoidal function shape the learnt weights and bias were different and hence affected the performance of the controller.

Table 7.1 Performance of the Neuro-fuzzy controller with linear activation function

Training epochs	Rise time (time units, 1 unit = 0.12 s)	Overshoot	Settling time	Steady state error
First 100 epochs	23	40.50	39	1.747
Second 100 epochs	20	38.75	28	−1.058
Third 100 epochs	23	36.79	27	−0.529
Fourth 100 epochs	28	36.68	33	−0.264
After training	28	36.68	35	-0.164

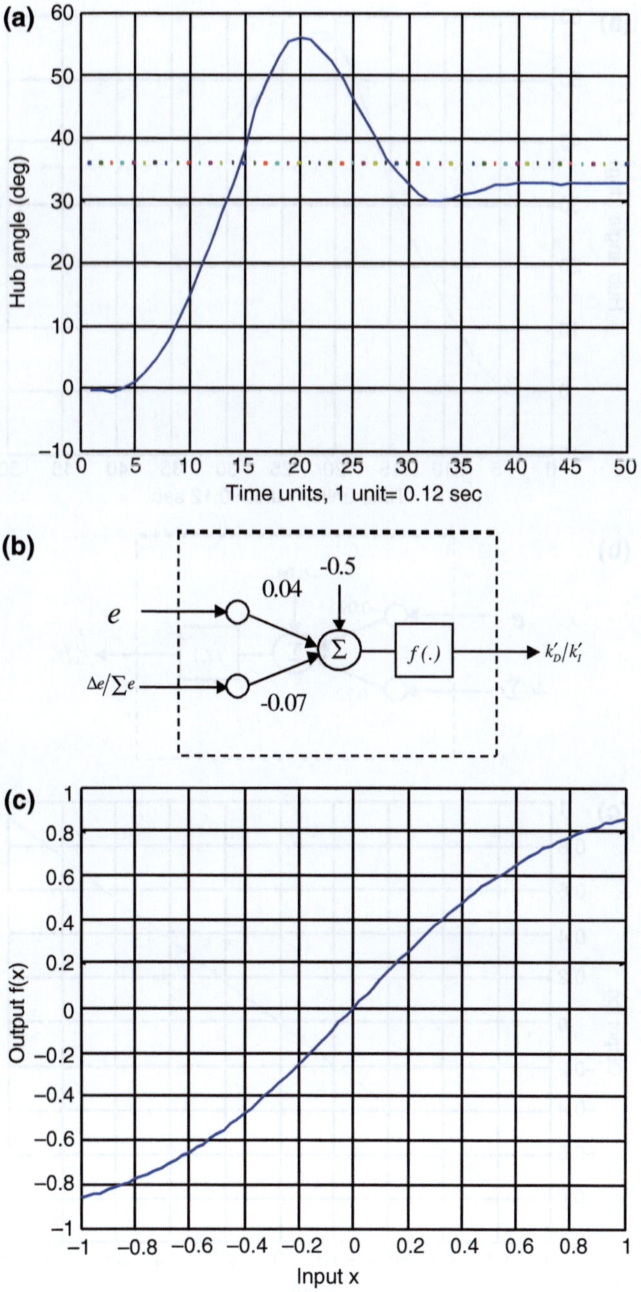

Fig. 7.19 Performance of the flexible arm with neuro-fuzzy controller. Response of the flexible arm (**a**); Neural network with weights and bias (**b**); Sigmoidal function shape with $a = 2.6$ (**c**)

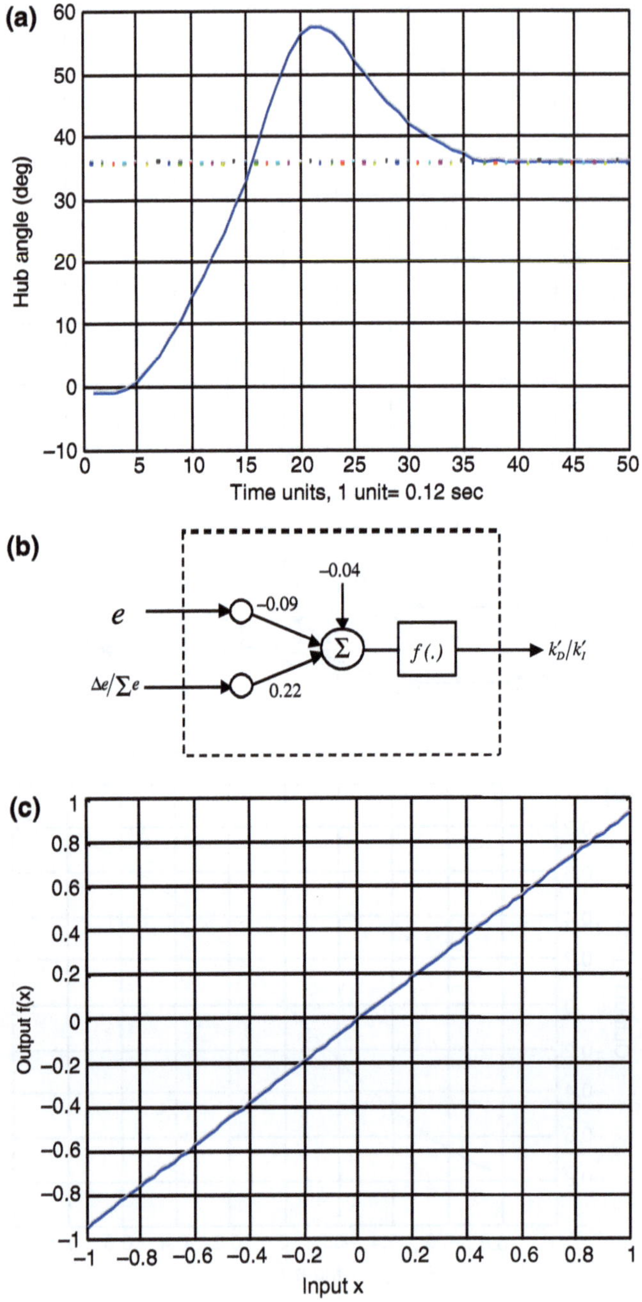

Fig. 7.20 Performance of the flexible arm with neuro-fuzzy controller. Response of the flexible-link (**a**); Neural network with weights and bias (**b**); Sigmoidal function shape with $a = 1.19$ (**c**)

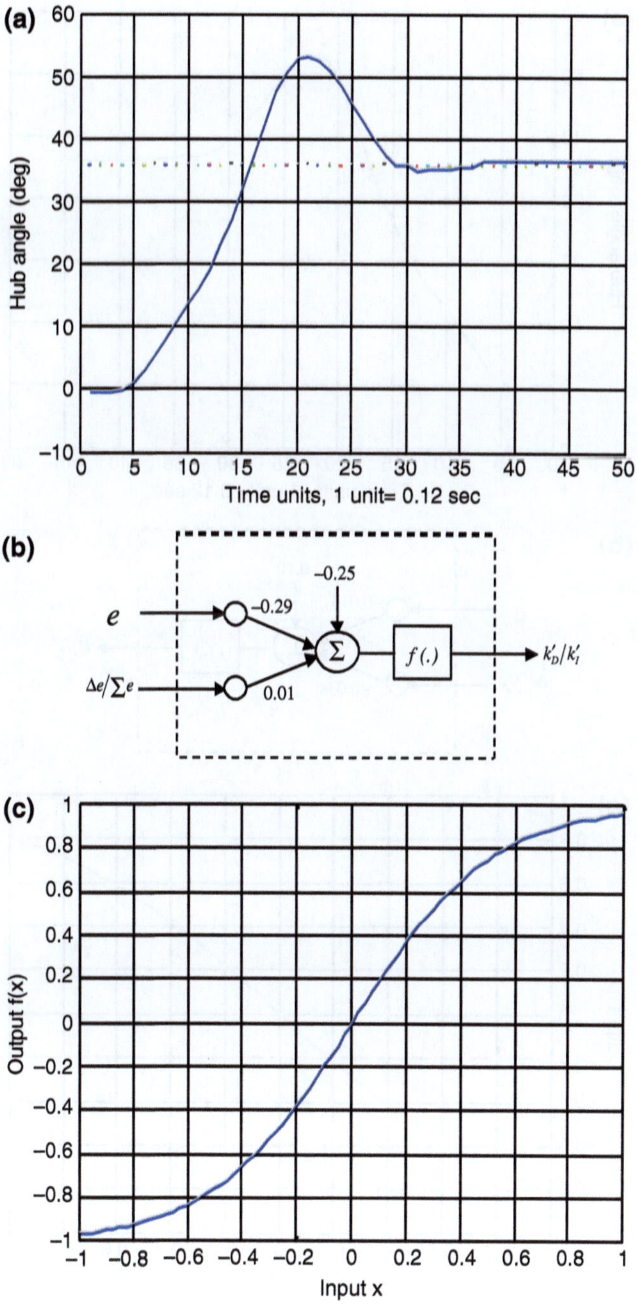

Fig. 7.21 Performance of the flexible arm with neuro-fuzzy controller. Response of the flexible-link (**a**); Neural network with weights and bias (**b**); Sigmoidal function shape with $a = 3.94$ (**c**)

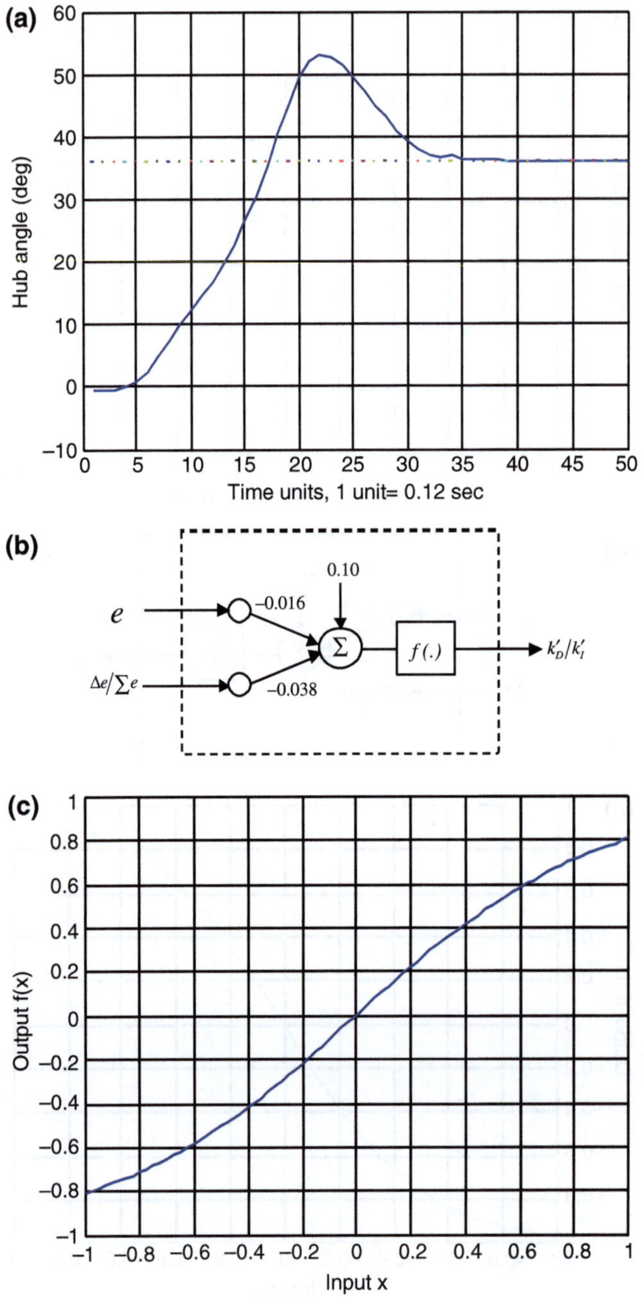

Fig. 7.22 Performance of the flexible arm with neuro-fuzzy controller. Response of the flexible arm (**a**); Neural network with weights and bias (**b**); Sigmoidal function shape with $a = 2.2$ (**c**)

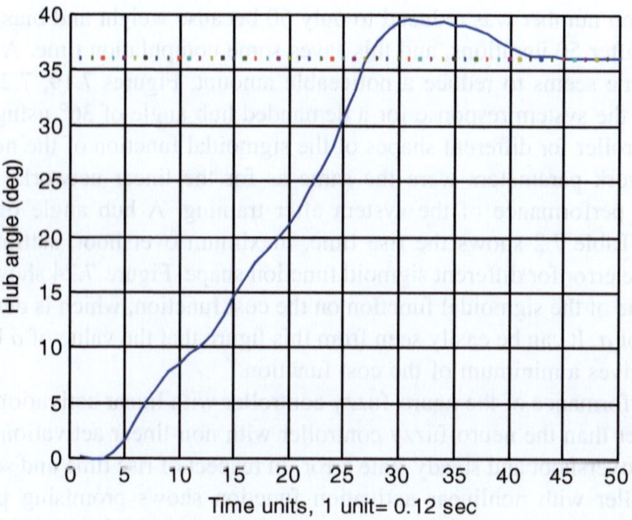

Fig. 7.23 Performance of the flexible arm with neuro-fuzzy controller after training

Table 7.2 Performance of the controller with nonlinear activation function

Sigmoidal function shape	Rise time (time units, 1 unit = 0.12 s)	Overshoot (deg)	Settling time	Steady state error (deg)
$a = 1.19$	16	57.24	38	−0.13
$a = 2.2$	17	51.92	39	−0.03
$a = 2.6$	14	56.67	45	2.1
$a = 3.94$	16	52.65	37	−0.89
After training	27	39.32	45	−0.19

Fig. 7.24 Effect of a on cost function

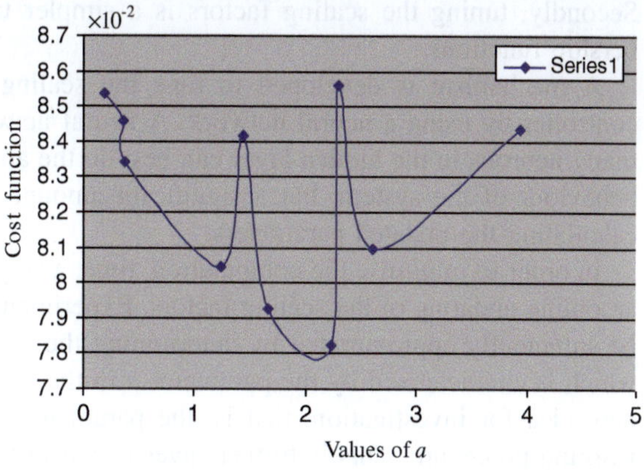

The iteration number was reduced to only 50 because weight and bias changes do not occur after 50 iterations, and this saves some computation time. Also the total training time seems to reduce a noticeable amount. Figures 7.19, 7.20, 7.21 and 7.22 show the system response for a demanded hub angle of 36° using the neuro-fuzzy controller for different shapes of the sigmoidal function of the network. The other network parameters were the same as for the linear network. Figure 7.23 shows the performance of the system after training. A hub angle of 36.19 was achieved. Table 7.2 shows the rise time, maximum overshoot, settling time and steady state error for different sigmoid function shape. Figure 7.24 shows the effect of the shape of the sigmoidal function on the cost function, which is determined by the value of a. It can be easily seen from this figure that the value of a between 2.0 and 2.23 gives a minimum of the cost function.

The performance of the neuro-fuzzy controller with linear activation function is much better than the neuro-fuzzy controller with non-linear activation function in respect of overshoot and steady state error. In respect of rise time and settling time, the controller with nonlinear activation function shows promising performance than the linear activation function. Shape of the sigmoid function is playing an important role in determining the performance of the controller, which needs further investigations.

7.9 Summary

Redefinition of the MFs for change of error and sum of error within a common universe of discourse can significantly influence the performance of the PD-PI fuzzy controller. A readjustment of the MFs is required at this stage. Neuro-fuzzy approaches are mostly used in such readjustment of membership functions, and this involves several parameters to be adjusted. In many cases, tuning the scaling factors gives the same performance as with membership function adjustment. Secondly, tuning the scaling factors is a simpler task than adjusting the membership functions.

A mechanism is developed to tune the scaling factors of the PD-PI fuzzy controller by using a neural network. A neural network with multiple layers and many neurons in the hidden layer can best do the approximation of the non-linear behaviour of the system, but a significant amount of time will be consumed in calculating the updated parameters.

In order to minimise the computation, time, a single neuron network is used for an online updating of the scaling factors. Experiments show that non-linearity can be sufficiently approximated by determining the shape of the sigmoidal function which is characterised by the parameter a in the activation function. This gives a new idea for investigation, that is, the parameter a can also be included in the learning procedure. This is further investigated in the next chapter.

References

Adeli H, Hung S-L (1994) An Adaptive Conjugate Gradient Learning Algorithm for Effective Training of Multilayer Neural Networks. Appl Math Comput 62(1):81–102

Akesson BM, Toivonen HT (2006) A Neural Network Model Predictive Controller, J Process Control 16:937–946

Berenji HR (1992) A reinforcement learning-based architecture for fuzzy logic control. Int J Approximate Reasoning 6:267–292

Berenji HR, Khedkar P (1992) Learning and tuning fuzzy logic controllers through reinforcements. IEEE Trans Neural Networks 3(5):724–740

Berenji HR, Khedkar P (1993) Clustering for product space for fuzzy inference. In: IEEE international conference on neural networks, San Francisco, pp 1402–1407

Burkhardt D, Bonissone PP (1992) Automated fuzzy knowledge-base generation and tuning. In: Proceedings of the first IEEE international conference on fuzzy systems (FUZZ-IEEE'92), San Diego, pp 179–188

Chen M-Y, Linkens DA (1998) A hybrid neuro-fuzzy PID controller. Fuzzy Sets Syst 99:27–36

Chen M-Y, Linkens DA (1999) A hybrid neural network architecture for self-construction of fuzzy rule-based systems. In: Proceedings of the 14th world congress of IFAC, Beijing, China, pp 109–114

Chi Z, Yan H, Phan T (1996) Fuzzy algorithms: with applications to image processing and pattern recognition. World Scientific, Singapore

Choi JY, Choi C-H (1992) Sensitivity analysis of multilayer perceptron with differentiable activation functions. IEEE Trans Neural Networks 3:101–107

Daugherity WC, Rathakrisnan B, anf Yen J (1992) Performance evaluation of a self-tuning fuzzy controller. In: Proceedings of the first IEEE international conference on fuzzy systems (FUZZ-IEEE'92), San Diego, USA, pp 389–397

Dickenson J, Kosko B (1993) Fuzzy function learning with covariance ellipsoids. In: Proceedings of IEEE international conference on neural networks, San Francisco, USA, vol 3, pp 1162–1167

Drainkov D, Hellendorn H, Reinfrank M (1993) An introduction to fuzzy control. Springer, New York

Feng G (2006) A survey on analysis and design of model-based fuzzy control systems. IEEE Trans Fuzzy Syst 14(5):676–697

Feng JC, Teng LC (1998) An online self constructing neural fuzzy inference network and its applications. IEEE Trans Fuzzy Syst 6(1):12–32

Figueiredo M, Gomide FAC (1999) Design of fuzzy systems using neurofuzzy networks. IEEE Trans Neural Networks 10(4):815–827

Figueiredo M, Gomide F, Rocha A, Yager R (1993) Comparison of Yager's level set method for fuzzy logic control with Mamdani's and Larsen's method. IEEE Trans Fuzzy Syst 1(2):156–159

Gudwin R, Gomide F, Pedrycz W (1998) Context adaptation in fuzzy processing and genetic algorithms. Int J Intell Syst 13(10–11):929–948

Haber RE, Haber RH, Alique A, Ros S (2000) Hierarchical fuzzy control of the milling process with self-tuning algorithm. In: Proceedings of the 2000 IEEE international conference on intelligent control, Rio Patras, Greece, pp 115–120

Halgamuge SK, Glesner M (1994) Neural networks in designing fuzzy systems for real world applications. Fuzzy Sets Syst 65:1–12

Halgamuge SK, Mari A, Glesner M (1994) Fast perceptron learning by fuzzy controlled dynamic adaptation of network parameters. In: Kruse R, Gebhardt J, Palm R (eds) Fuzzy systems in computer science. Vieweg, Braunschweig

Harris CJ, Moore CG, Brown M (1993) Intelligent control: aspects of fuzzy logic and neural nets. World Scientific, Singapore (World scientific series vol. 6)

Haykin S (2009) Neural networks and learning machines, 3rd edn. Pearson Education Inc, NJ
Hebb DO (1949) The organization of behavior: a neuropsychological theory. Wiley, New York
Hinton GE, Nowlan SJ (1987) How learning can guide evolution. Complex Syst 1(3):495–502
Hong X, Harris CJ, Chen S (2004) Robust neuro-fuzzy rule base knowledge extraction and estimation using subspace decomposition combined with regularization and D-optimality. IEEE Trans Syst Man Cybern Part B Cybern 34(1):598–608
Hsu P-L, Fann W-R (1996) Fuzzy adaptive control of machining process with a self-learning algorithm. ASME J Manuf Sci Eng 118:522–530
Hu B-G, Mann GKI, Gosine RG (2001) A systematic study of fuzzy PID controllers—function-based evaluation approach. IEEE Trans Fuzzy Syst 9(5):699–712
Ichihashi H, Tokunaka M (1993) Neuro-fuzzy optimal control backing up a trailer truck. In: Proceedings of the international conference on neural networks, San Francisco, pp 306–311
Jang J-SR (1993) ANFIS: adaptive-network-based fuzzy inference system. IEEE Trans Syst Man Cybern 23(3):665–685
Jang J, Sun C (1995) Neuro-fuzzy modelling and control. Proc IEEE 83:378–406
Jang J-SR, Sun C-T, Mizutani E (1997) Neuro-fuzzy and soft computing. Prentice Hall, NJ
Kosko B (1992) Neural networks and fuzzy systems: a dynamical systems approach to machine learning. Prentice Hall, Englewood Cliffs
Leng G, McGinnity TM, Prasad G (2005) An approach for on-line extraction of fuzzy rules using a self-organising fuzzy neural network. Fuzzy Sets Syst 150(2):211–243
Li H-X (1997) A comparative design and tuning for conventional fuzzy control. IEEE Trans Syst Man Cybern Part B Cybern 27(5):884–889
Liang Y (2000) Adaptive neural activation functions in multi-resolution learning. In: Proceedings of the IEEE conference on system, man and cybernetics, pp 2601–2606
Liang Y, Page EW (1997) Multi-resolution learning paradigm and signal prediction. IEEE Trans Signal Process 45(11):2858–2864
Liang Y, Page EW (1998) Applications of multi-resolution learning in neural network robustness studies. In: Proceedings of international joint conference on information sciences (JCIS), vol 2, pp 163–166
Lin TC, Lee CS (1991) Neural network based fuzzy logic control and decision system. IEEE Trans Comput 40(12):1320–1336
Lin C, Lu Y (1996) A neural fuzzy system with fuzzy supervised learning. IEEE Trans Syst Man Cybern Part B 26:744–763
Lin CJ, Lin CT (1997) An ART-based fuzzy adaptive learning control network. IEEE Trans Fuzzy Systs 5(4):477–496
Lin W-S, Tsai C-H, Liu J-S (2001) Robust neuro-fuzzy control of multivariable systems by tuning consequent membership functions. Fuzzy Sets Syst 124:181–195
Linkens DA, Abbod MF (1992) Self-organising fuzzy logic control and the selection of its scaling factors. Trans Inst Meas Control 14(3):114–125
McCulloch WS, Pitts WH (1943) A Logical Calculus of the Ideas Imminent in Nervous Activity, Bull Math Biophys 5:115–133
Mizutani E, Jang J-SR (1995) Coactive neural fuzzy modelling. In: Proceedings of the international conference on neural networks, Nov 1995, pp 760–765
Narendra KS, Parthasarathy K (1990) Identification and Control of Dynamic Systems using Neural Networks, IEEE Trans Neural Networks 1(1):4–27
Narendra KS, Mukhopadhyay S (1997) Adaptive Control using Neural Networks and Approximate Models, IEEE Trans Neural Networks 8:475–485
Nauck D, Kruse R (1993) A fuzzy neural network learning fuzzy control rules and membership functions by fuzzy error backpropagation. In: Proceeding of IEEE international conference on neural networks, pp 1022–1027
Nauck D, Kruse R (1996) Designing neuro-fuzzy systems through backpropagation. In: Pedrycz W (ed) Fuzzy modelling: paradigms and practice. Kluwer, Boston, pp 203–228
Nauck D, Klawoon F, Kruse R (1997) Foundations of neuro-fuzzy systems. Wiley, Chichester

Nguyen D, Widrow B (1990) Improving the learning speed of 2-layer neural networks by choosing initial values of adaptive weights. Int Joint Conf Neural Networks 3:21–26
Nomura H, Hayashi I, Wakami N (1991) A self-tuning method of fuzzy control by descent method. In: Proceedings of IFSA'91, Brussels, pp 155–158
Nomura H, Hayashi H, Wakami N (1992) A learning method of fuzzy inference rules by descent method. In: Proceedings of the first IEEE international conference on fuzzy systems, San Diego, USA, pp 203–210
Palm R (1995) Scaling of fuzzy controllers using cross-correlation. IEEE Trans Fuzzy Syst 3(1):116–123
Passino KM, Yurkovich S (1998) Fuzzy control. Addison-Wesley, Reading
Pedrycz W, Card HC (1992) Linguistic interpretation of self-organising maps. In: Proceedings of the IEEE international conference on fuzzy systems, San Diego, CA, pp 371–378
Pedrycz W, Gudwin R, Gomide F (1997) Nonlinear context adaptation in the calibration of fuzzy sets. Fuzzy Sets Syst 88(1):91–97
Procyk TJ, Mamdani EH (1979) A linguistic self-organising process controller. Automatica 15(1):15–30
Quek C, Zhou RW (1996) POPFNN-TVR: A Pseudo Outer-Product based Neural Fuzzy Network. Neural Netw 9(9):1569–1581
Sarangapani J (2006) Neural Network Control of Nonlinear Discrete-Time Systems. CRC Press, Taylor & Francis, London, New York
Siddique N, Adeli H (2013) Computational Intelligence: Synergies of Fuzzy Logic. Neural Networks Evol Comput John Wiley and Sons
Stevenson M, Winter R, Widrow B (1990) Sensitivity of feedforward neural networks to weight errors. IEEE Trans Neural Networks 1:71–80
Sulzberger SM, Tschicholg-Gurman NN, Vestli SJ (1993) FUN: optimization of fuzzy rule based systems using neural networks. In: Proceedings of the IEEE conference on neural networks, San Francisco, pp 312–316
Takagi H (1992) Applications of neural networks and fuzzy logic to consumer products. In: First international workshop on industrial applications of fuzzy control and intelligent systems, Nov 21–22, 1991, Texas, pp 1629–1633
Takagi H (1995) Applications of neural networks and fuzzy logic to consumer products. In: Yen J, Langari R, Zadeh L (eds) Industrial applications of fuzzy control and intelligent systems. IEEE Press, Piscataway, NJ, pp 93–106 (Chap. 5)
Takagi H (1997) Introduction to fuzzy systems, neural networks and genetic algorithms. In: Ruan D (ed) Intelligent hybrid systems. Kluwer Academic Publishers, Boston, pp 3–33 (Chap. 1)
Takagi H, Hayashi I (1991) NN-driven fuzzy reasoning. Int J Approximate Reasoning 5(3):191–212
Tano S, Oyama T, Arnould T, Bastian A (1994) Definition and Tuning of Unit-based Fuzzy Systems in FINEST. FUZZ-IEEE 94:436–441
Tano S, Oyama T, Arnould T (1996) Deep Combination of Fuzzy Inference and Neural Network in Fuzzy Inference. Fuzzy Sets Syst 82(2):151–160
Tung WL, Quek C (2004) Falcon: Neural Fuzzy Control and Decision Systems Using FKP and PFKP Clustering Algorithms. IEEE Trans Syst, Man and Cybernetics - Part B: Cybernetics, 34(1):686–695
Wang LX, Mendel JM (1992) Fuzzy Basis Functions, Universal Approximation and Orthogonal Least-squares Learning. IEEE Trans Neural Networks 3(5):807–814
Widrow B, Hoff ME Jr (1960) Adaptive switching circuits. IRE Western Electric Show and Convention Record, part 4, pp 96–104
Wu S, Joo-Er M, Gao Y (2001) A Fast Approach for Automatic Generation of Fuzzy Rules by Generalized Dynamic Fuzzy Neural Networks. IEEE Trans Neural Networks 9(4):578–594
Yager RR (1994) Modelling and formulating fuzzy knowledge bases using neural networks. Neural Networks 7(8):1273–1283
Yager RR, Filev DP (1994) Essentials of fuzzy modelling and control. Wiley, NY

Yao X (1993) Evolutionary Artificial Neural Networks. Int J Neural Syst 4(3):203–222
Yea B, Konishi R, Osaki T, Sugahara K (1994) Discrimination of many kinds of odor species using fuzzy reasoning and neural networks. Sens Actuators, A Phys 45(2):159–165
Yeung DS, Sun X (2002) Using function approximation to analyze the sensitivity of MLP with anti-symmetric squashing activation function. IEEE Trans Neural Networks 13(1):34–44
Zheng L (1992) A practical guide to tune proportional and integral (PI) like fuzzy controllers. In: Proceedings of the first IEEE international conference on fuzzy systems (FUZZ-IEEE'92), San Diego, USA, pp 633–640

Chapter 8
Evolutionary-Neuro-Fuzzy Control

8.1 Introduction

Fuzzy systems (FS) have been shown to be able to approximate or model any complex and nonlinear system, capable of modelling vagueness, handling uncertainty and supporting human-type reasoning. Linguistic analysis and modelling of complex systems has become one of the most popular methods since its inception by Zadeh (1973). A linguistic model is a knowledge-based representation of a system, its rules and input–output variables are described by linguistic variables represented by membership functions (MF) and hedges. Fuzzy logic systems are capable of modelling ambiguity, supervising uncertainty and also to support manual interpretation. The only drawback of a fuzzy system is that it does not have any mechanism for acquiring knowledge and mainly depends on experts in the domain who relies on some heuristic rules. Even such a priori information is not readily available always. Efforts have been made to automate knowledge acquisition for fuzzy systems by means of learning using neural-networks (NN). NNs are capable of learning without any prior intervention when provided with sufficient data which are available or measurable. A number of learning algorithms are available to train NNs. NNs, similarly, are widely used for system modelling, control and classification because of their ability to approximate complex non-linear systems using experiential data. The problem with the existing learning algorithms for NNs is that the experiential data may not always be available or available data may not be representative of the system's entire input–output space and the convergence of learning depends on many other structural and learning parameters. Fuzzy-logic and neural systems, however, have very contrasting application requirements and their integration can offer a facility to bridge linguistic knowledge processing and connectionist learning. The significance of the integration becomes more apparent by considering their disparities. NNs do not provide a suitable mechanism for knowledge representation, while fuzzy systems do not possess any learning capabilities. In most of the cases, the inferencing mechanism and the rule-base are replaced by a neural network so that

N. Siddique, *Intelligent Control*, Studies in Computational Intelligence 517,
DOI: 10.1007/978-3-319-02135-5_8, © Springer International Publishing Switzerland 2014

suitable neural learning algorithms can be applied to the combined system. In that sense, the neural networks and fuzzy logic systems are dynamic parallel processing systems which can estimate any input–output functions to a reasonable accuracy. The neuro-fuzzy system is the combination in terms of the number of practical real time algorithms. In neuro-fuzzy systems, the fuzzy system is the main focal point of the combination procedure and the neural network includes the learning capability to the inference engine.

The main drawback of the neuro-fuzzy systems is that the learning technique is based on the gradient descent optimization technique (Jang et al. 1997). Backpropagation algorithm is the most popular and widely used learning algorithm also applied to neuro-fuzzy systems. The problem of backpropagation learning is that the convergence of the algorithm depends on many other parameters and the tuning of the fuzzy system's parameters through neural learning is not guaranteed. The algorithm is very often trapped in local minima. The solution obtained using back propagation algorithm will lead to suboptimal performance of the neuro-fuzzy systems. Another difficulty in the existing neuro-fuzzy systems is that they fail to quantify the rules or the MFs. Moreover, the architecture of the neural network is also a trial and error process and the known learning algorithms do not guarantee a robust optimal solution due to variations in architectures and learning parameter settings.

On the other hand another learning method has emerged out of this complexities to meet the demand of global optimality, robustness and convergence as an alternative to inductive techniques used with neural networks, namely, evolutionary learning. Evolutionary Algorithms (EA) are derivative-free population based iterative probabilistic algorithms, which are used to real-life problems to find global optimal solutions that helps learning of such hybrid systems. Different variants of EAs, representation schemes and genetic operators have been discussed in Chap. 6.

Though the problem of convergence and assurance of global optimal solution is met, yet another issue of convergence property of the evolutionary learning algorithms depends on several parameters such as population size, recombination, mutation and selection. EA itself has no internal mechanism of controlling these parameters rather chosen by the user arbitrarily. A fuzzy logic controller or a neural network can aid the parameter adjusting process and help speeding up the convergence based on a suitable performance measure of the evolutionary algorithm. Besides these synergisms, fuzzy, neural and evolutionary systems can work independently on a single problem, where one of them is the primary problem solver and the other two are the supporting systems. Researchers are making efforts to represent an adaptive system applying the synergism of the three technologies, where a complex system is described by simple linguistic expressions and rules, the system is capable of learning from the experience within an environment and the system evolves to adjust certain features (expressed by parameters) to the changing environment.

The three paradigms have their own advantages and disadvantages. The advantage of a fuzzy system is that it can work using approximate information and can represent knowledge in a linguistic form, which resemble human-like reasoning process. It is not easy for human to understand the way knowledge is extracted through learning. The NN has the advantage of learning from experience provided that a set of input–output data is available and some structural flexibility is amenable. Whereas an EA can optimise or learn through adaptation of a system's structure or parameters by evolving a random population while exploring the search space in several directions ensuring a global and robust solution. The fusion of the three paradigms will certainly fabricate a system with certain degree of autonomy and enhanced performance.

This chapter will briefly present the synergism and different approaches of integration of the three paradigms and then focus on the evolutionary optimisation of the parameters of the neuro-fuzzy system developed throughout the chapters from five to seven.

8.2 Integration of Fuzzy Systems, Neural Networks and Evolutionary Algorithms

In the previous two chapters, different approaches to cooperative combinations of evolutionary fuzzy and neural fuzzy systems are discussed. A fuzzy system (FS) is described by a fuzzy inference system *FIS*, the predetermined membership functions *MF* and a rule-base *R* that produces a set of crisp values *Y* when an input *X* is presented and expressed by

$$Y = \Phi[X, MF, FIS, R] \tag{8.1}$$

$\Phi[X, MF, FIS, R]$ also represents the fuzzy inference system (FIS) which can be of Mamdani, Sugeno or Tsukamoto-type discussed in detail in Chap. 4.

A neural network (NN) is described by a set of inputs X, a predetermined neural network architecture *NNA* and a learning algorithm Λ where Γ is a mapping $\Gamma : X \rightarrow D$ defined by

$$D = \Gamma[X, W, NNA, \Lambda] \tag{8.2}$$

The learning updates the weights W. $\Gamma[X, W, NNA, \Lambda]$ produces the decisions D and *NNA* can be of any type of network architecture. The learning algorithm Λ can be of any learning algorithm, which uses suitable error function derived from the output.

An evolutionary algorithm (EA) can be described by the difference Eq. (8.3) where a population $\Pi(t)$ at time t evolves by the random variation $v(.)$ and selection $\Psi(.)$ and produces a new generation of population $\Pi(t+1)$ at time $t+1$.

$$\Pi(t+1) = \Psi[v(\Pi(t))] \tag{8.3}$$

$\Pi(t)$ is a population generated randomly for a set of parameters that describe a system. Evolutionary algorithm can help finding an optimal set of parameters of the system based on a predefined optimality criterion. For example, it is the *MF* and rule-base *R* for the FS and architecture *NNA* and learning parameters of Λ for the *NN* to be optimised or learnt by the EA. The detail description of different FS, NNs and EAs can be found in (Siddique and Adeli 2013).

There can be of various forms of synergistic combination of the three technologies FS, NNs and EAs. Firstly, two way integration of the three technologies FS, NN and EA, as shown in Fig. 8.1, may exist in the real world which is beyond explanation and our comprehension. Such a system may be too complicated to realise using the traditional and existing computational tools. A one-way integration of the three technologies is very much possible as shown in Fig. 8.2a–b. An FS provides a suitable input *X* to an NN. The NN estimates the parameters Ω of the EA and EA optimises the membership functions *MF* or the rule-base *R* or the scaling parameters *K* of the FS. This type of one-way integration of FS, NN and EA is shown in Fig. 8.2a. The integration can be in the other way round where an NN can tune or estimate the *MF*, the rule-base *R* or the scaling parameters K of an FS. The FS then controls the parameters Ω of the EA and EA optimises the architecture or the weights of the NN. This type of integration of NN, FS and EA is shown in Fig. 8.2b.

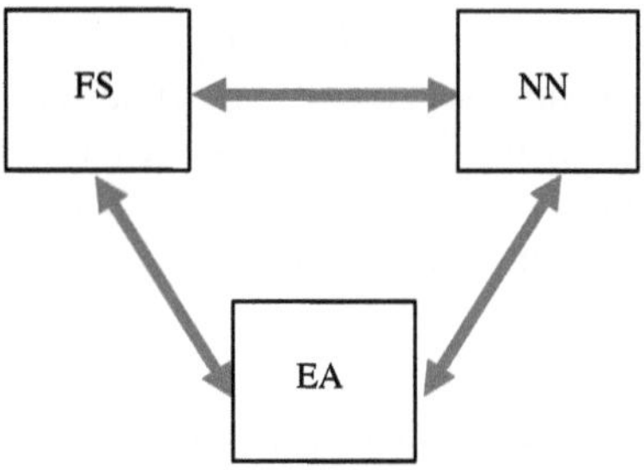

Fig. 8.1 Two-way integration of FS, NN and EA

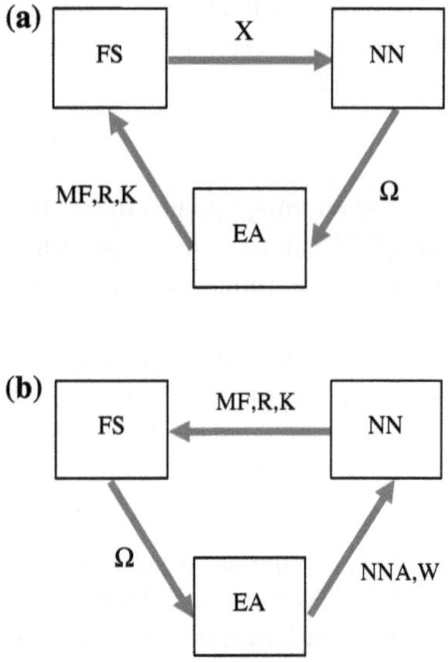

Fig. 8.2 One-way integration of FS, NN and EA

Mingling of the three technologies FS, NN and EA as shown in Figs. 8.1 and 8.2 is rather theoretical and has little application in solving practical problems. It would be practical if one of the technologies FS or NN is used in developing a model of a real world system as the technologies allow modelling complex systems without much a priori information and apply the other to control or tune the system. Finally, EA is to provide the appropriate structure and/or parameters of the controller or the tuner (FS or NN) based on a performance index derived from the plant output. This kind of combinations is shown in Fig. 8.3a, b. In Fig. 8.3a, the system model is developed using NN and a fuzzy controller is used to provide the control input. An EA is used to optimise the parameters of MFs, rule-base or scaling parameters based on the performance metric calculated from the output of the system. In Fig. 8.3b, the system model is developed using FS and an NN is used to provide the control input to the system. An EA is used to optimise the parameters of the NN controller such as weights, architecture, or activation functions based on the performance metric calculated from the output of the system. Model development using NN or FS, controller development using FS or NN and optimising the controllers' parameters $\{MF, R, K\}$ or $\{W, NNA, f(.)\}$ is not a trivial task and is a huge undertaking. This will never be a realistic approach for integration of the three technologies and there have been very few researches reported in the literature hitherto.

The general practice is that a plant or a plant model is used to be controlled by a neuro-fuzzy controller and the controller is optimised by an EA as shown in Fig. 8.4. There can be of different combinations between NN and FLC such as

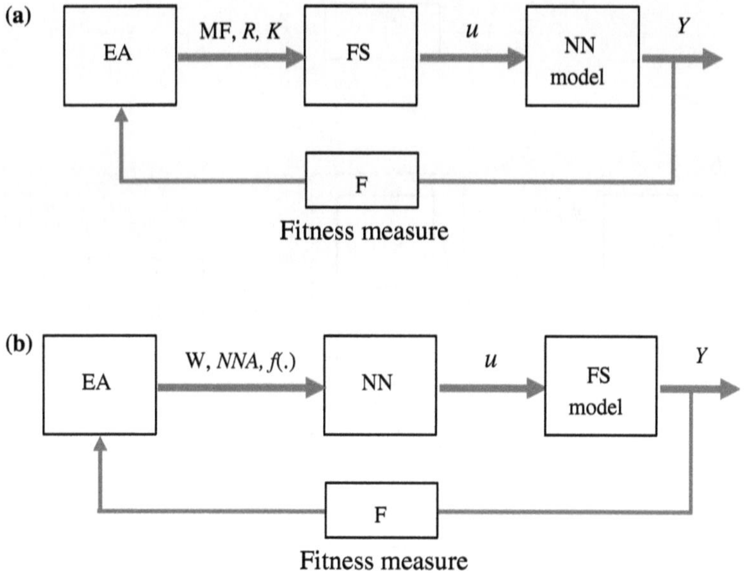

Fig. 8.3 One-way integration of FS, NN and EA. EA-based FS control of NN-model (**a**); EA-based NN control of FS-model (**b**)

cooperative, collaborative or hybrid. In the FLC-NN cooperative system shown in Fig. 8.4a, FLC parameters are evolved by EA and the FLC controls the parameters of NN. In the NN-FLC cooperative system shown in Fig. 8.4b, NN parameters such as $\{W, NNA, f(.)\}$ are evolved by EA where NN is defined by $\Gamma[X, NNA, \Lambda]$. The NN learns the parameters of the FLC such as $\{MF, R, K\}$ where the FLC is defined by $\Phi[FIS, MF, R]$ with a predefined fixed inference mechanism. $\Phi[FIS, MF, R]$ represents a fuzzy system which can be of Mamdani, Sugeno or Tsukamoto-type.

In most of the applications of the cooperative neuro-fuzzy systems shown in Fig. 8.4a, b, the structure and the parameters of one of the subsystems (FLC or NN) are predefined and fixed. They are designed based on the local information available independently. Optimality may not be reached by only learning or optimising

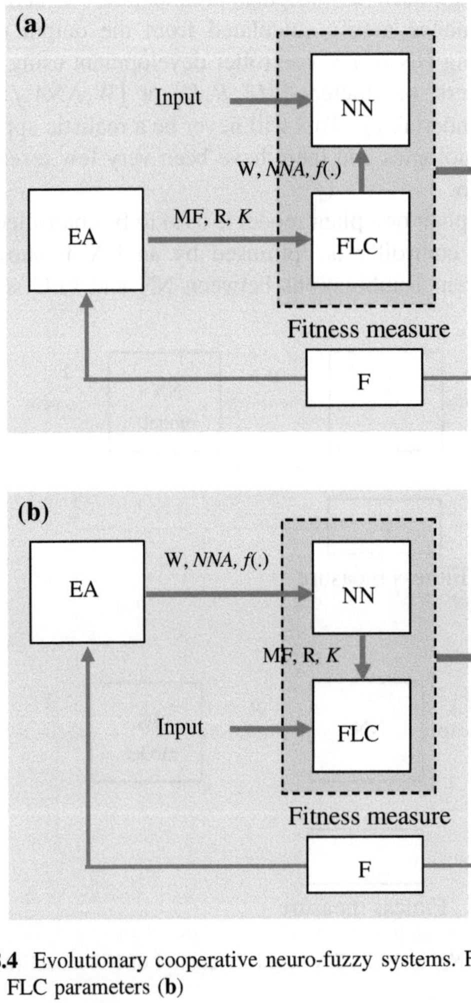

Fig. 8.4 Evolutionary cooperative neuro-fuzzy systems. FLC controls NN parameters (**a**); NN learns FLC parameters (**b**)

the parameters of one subsystem as it may be correlated to structural elements of the other subsystem. Therefore, an optimisation or adaptation of the structure and parameters in an operating environment may be necessary for a desired optimal performance. Therefore, EA can also be applied for structure optimisation and parameter learning of the FLC and NN at the same time. All the structural parameters of the FLC and architectural parameters of the NN are coded into the chromosome. For example, parameters of the antecedent and consequent MFs, rule-base of the FLC, parameters of the NN architecture defined by the number of hidden layers and number of neurons per layer, activation functions, and learning rules are coded into the chromosome representation. This would be a huge undertaking for an optimisation procedure, which will simply increase the number of parameters in the chromosome representation resulting in a long chromosome for the evolutionary algorithm and involving time consuming computation. An alternative would be to hybridise the FLC and NN to reduce the number of parameters.

A hybrid NN-FLC combination is shown in Fig. 8.5 where the hybrid system is defined by the mapping $\Theta[FIS, MF, P, NNA, \Lambda]$. In the hybrid system, the rule-base R is replaced by an NN whose architecture NNA has to be optimised by a suitable mechanism. A predefined fuzzy inference system FIS is normally used with adapting membership functions MF for the inputs, rule-base is defined by a neural system with a flexible architecture NNA, the consequent MFs are replaced by suitable parameterised linear functions represented by a set of parameters P and the learning algorithm Λ. The parameters of the hybrid system are the parameters of the MF, structure and parameters of the NNA, parameters of the Λ and the consequent parameters P. This results in a huge set of parameters which need to be learnt or optimised. Another problem of the neuro-fuzzy (NN-FLC) system is the difficulty of determining the appropriate number of rules and the number of MFs for each input and output.

There are number of different implementations of hybrid architectures of NN-FLC available in the literature such as FALCON (Lin and Lee 1991), GARIC (Berenji and Khedkar 1992), ANFIS (Jang 1993), FUN (Sulzberger et al. 1993), FINEST (Tano et al. 1996), NEFCON (Nauck et al. 1997), MANFIS, CANFIS (Jang et al. 1997), SONFIN (Feng and Teng 1998) and NFN (Figueiredo and

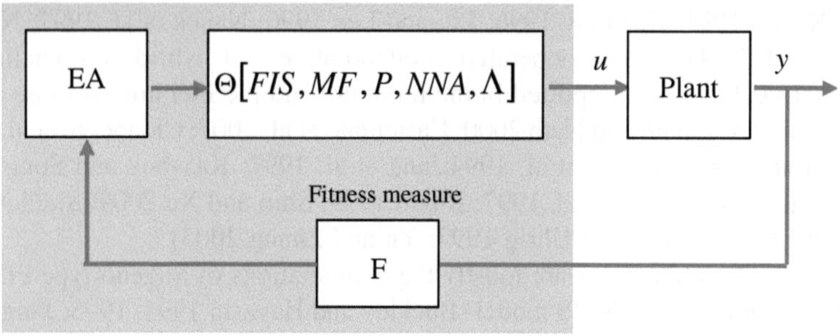

Fig. 8.5 Evolutionary hybrid neuro-fuzzy system

Gomide 1999). Detail discussion of these architectures is beyond scope of this Chapter. The interested readers are directed to (Siddique and Adeli 2013).

In all of these hybrid architectures (or models), the rule-base is replaced with a neural network and consequent MFs are replaced with suitable linear functions while keeping the antecedent MFs unchanged. The idea behind the hybridisation was to apply suitable learning algorithm for estimating the antecedent and consequent parameters. Mostly the learning algorithms were a combination of least square and backpropagation (BP) methods, which again require a set of input–output data. A set of input–output data may not be available always. It is particularly useful in the cases where complex interaction among independent variables necessitates training for all system parameters. Since BP training uses the gradient descent optimization technique to minimize the error function, it is very often stuck at the local minima. When global minima are hidden among the local minima, BP algorithm can end up bouncing between local minima without much overall improvement, which leads to very slow training. BP is a method requiring the computation of the gradient of error with respect to weights, which again needs differentiability. As a result, BP cannot handle discontinuous optimality criteria or discontinuous node transfer functions. BP's speed and robustness are sensitive to parameters such as learning rate, momentum and acceleration constant and the best parameters to use seem to vary from problem to problem. Besides, BP training performance depends on the initial values of the system parameters, and for different network topologies one has to derive new mathematical expressions for each network layer (Cantú-Paz and Kamath 2005; Yao and Liu 1997, 1998; Yao 1999). The intended tuning of the antecedent MFs' and consequent parameters through backpropagation-based learning of the NN is also not guaranteed.

Considering the disadvantages mentioned above one may end up with suboptimal performance even for a suitable fuzzy neural network topology. Hence, techniques capable of training the system parameters and finding the global solution while optimizing the overall structure are needed. In this respect, EA appears to be better candidates and several EA based approaches have appeared in the literature. Several methodologies have also been proposed to develop a form of evolutionary-neural-fuzzy (EA-NN-FLC) hybridisation. In EA-NN-FLC, the learning algorithm Λ is to be replaced with an EA as shown in Fig. 8.5. Hybrid combinations are the most useful and widely used systems in intelligent control paradigm (Jang et al. 1997; Kosko 1991; Kazabov 1996; Lin and Lee 1996; Nauck et al. 1997; Nie and Linkens 1995). Different cooperative, collaborative and hybrid combinations of EA-NN-FLC have been reported in the literature and the literature is quite rich in this domain (Abraham and Nath 2000; Castellano et al. 2007; Chiaberge et al. 1995; Farag et al. 1998; Fukuda et al. 1994;Jang et al. 1997; Kasabov and Song 1999; Loila et al. 2000; Nauck et al. 1997; Russo 1999; Shin and Xu 2009; Siddique and Adeli 2013; Tsoukalas and Uhrig 1997; Yu and Zhang 2005).

The hybrid architectures are mostly the combinations of Sugeno-type FLC and NN. For example, the ANFIS model (Buckley and Hayashi 1994, 1995; Jang 1993; Jang et al. 1997) implements a Sugeno-type fuzzy system in a network structure, and applies a mixture of backpropagation algorithms and least squares procedure

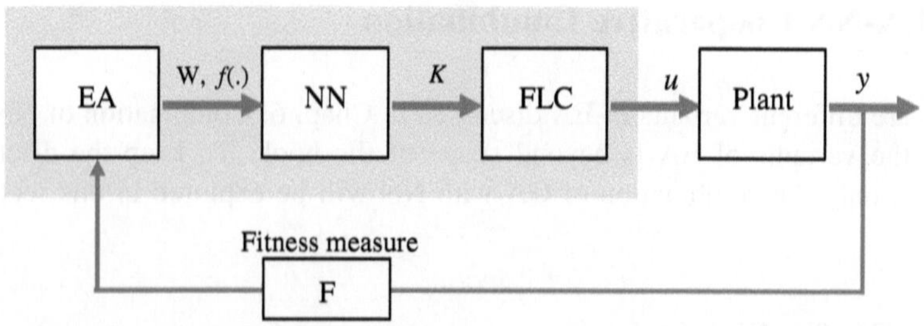

Fig. 8.6 Cascade combination of EA-NN-FLC

to train the system. The problem associated with these types of neuro-fuzzy models is that they sometimes are not as easy to interpret for Mamdani-type fuzzy systems (Nauck et al. 1997). Very few attempts have been made to hybridise Mamdani-type FLC with NN. It appears that Mamdani-type FLC is better suited for cooperative combination with NN. Some of the combinations of NN and FLC such as NN for correcting FLC, NN for learning fuzzy rules, NN for determining MFs, and NN for parameter tuning or learning of FLC have been discussed in Sect. 7.3 in Chap. 7. The intent is now to demonstrate the utility of the third technology EA in further enhancing the performance of the cooperative NN-FLC system developed in Chap. 7.

An appropriate application of EA mainly depends on the combination of the two technologies, i.e. NN and FLC. In the simplest approach, an EA is used to learn or tune the parameters of NN, which assists a predefined Mamdani-type FLC. Based on the analysis on the combinations shown in Figs. 8.3b and 8.4b, a cascade combination of cooperative EA-NN-FLC is presented in Fig. 8.6. There have been many of such cooperative combinations reported in the literature (Chen et al. 2009; Lin et al. 2008, 2011; Rahmoun and Berrani 2001; Seng et al. 1999; Wang and Li 2003). EA encode all the parameters of the NN model. The number of parameters depends on the architecture of the NN, activation functions and weights of the connectivity. The cooperative combination of NN with a Mamdani-type PD-PI FLC has been investigated in Chap. 7. A multilayer NN could simply exhaust the system by calculating exponential terms in the network, causing very slow response of the system. A single neuron network with non-linear activation function was used assuming that it can better represent the system's non-linearity avoiding huge processing by a multilayer NN and can be further enhanced by optimising the shape of the sigmoidal function. Performance of BP-based training of the NN was also investigated. Therefore, the rationale is now to combine an EA with the NN-FLC towards the development of an integrated[1] EA-NN-FLC to further investigate the cooperative relationship between the three technologies with the pursuit of improving the performance of the PD-PI FLC.

---

[1] The term integrated is better suited here as the three technologies are in cooperative combination rather than hybrid as the term hybrid indicates some kind of amalgamated system in general.

8.3 EA-NN Cooperative Combination

There are different variants of EA discussed in Chap. 6. Combination of NN with all of the variants of EA is beyond scope of the book. To keep the discussion concise, only the combination of GA with NN will be explored in this section.

8.3.1 *EA for Weight Learning*

Supervised or unsupervised learning in NN has mostly been formulated as weight training process in which efforts are made to find an optimal set of connection weights according to some optimality criteria. To overcome the shortcomings in gradient-descent learning algorithms such as backpropagation, global search procedure like EA can be used effectively in the training process as an evolution of connection weights towards an optimal set defined by a fitness function. The other advantages of EA is that it can handle large, complex, multimodal and non-differentiable functions.

The EA approach to weight training in NN consists of three phases: chromosome representation of connection weights, definition of a fitness function, and definition of genetic operators in conjunction with the representation scheme. Different representations and applicable genetic operators can lead to different training performance. A typical weight training process for an NN applied to a plant using an EA is shown in Fig. 8.7.

Chromosome Representation for NN

Different chromosome representations can lead the EA to quite different training performance in terms of training time and accuracy. The most convenient representation of connection weights is with binary strings. In such a representation scheme each connection weight is represented by some value (binary bits of

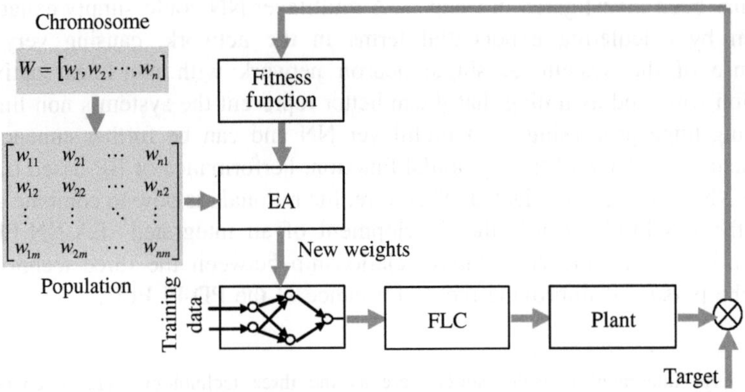

Fig. 8.7 Weight training of an NN using EA

Fig. 8.8 Generic chromosome represented in string form

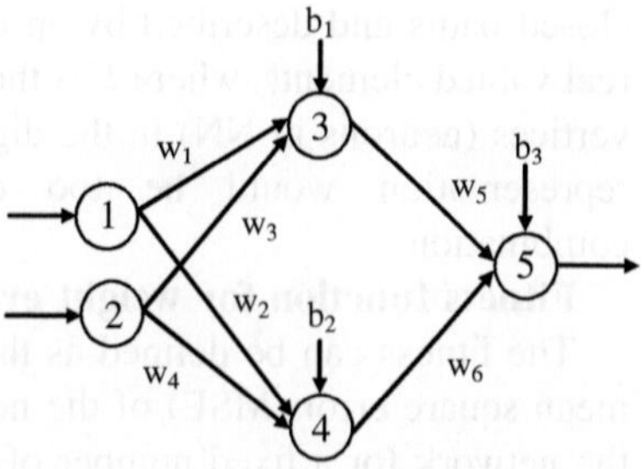

$$\{w_1, w_2, w_3, w_4, w_5, w_6, b_1, b_2, b_3\}$$

certain length or real). The most convenient and straightforward chromosome representation of connection weights and biases is in string form. Each connection weight and bias is represented by some value $\{w, b\} \in \Re$ where $\Re$ is a real number or binary bit string. An example of such string representation scheme for a feed forward NN with 5 neurons is shown in Fig. 8.8.

A limitation of binary representation is the required precision of connection weights. If too few bits are used to represent weights, training may take an extremely long time or even fail. On the other hand, if too many bits are used, chromosome string for large NN become very long, which will prolong the evolution and make the evolution impractical.

To overcome the inherent problems encountered in binary representation scheme, real values are proposed $\{w, b\} \in \Re$ i.e. $\Re$ is a real number per connection weight or bias. Chromosome is then represented by concatenating these numbers as a string. For example, a real number representation of the chromosome for the NN is shown in Fig. 8.9.

The advantages of real coding are many-fold such as shorter string length with increased precision and easy application of genetic operators. Standard mutation operation used in binary coding cannot be applied directly to the real valued representation. In such circumstances, an important task is to carefully design a set of genetic operators suitable to real encoding scheme. For example, mutation in real number chromosome representation can be as follows

$$w_i(t) = w_i(t-1) \pm random(0, 1) \tag{8.4}$$

Montana and Davis (1989) defined a large number of domain-specific genetic operators incorporating many heuristics about training NN.

There are other representation schemes for NN such as matrix representation, where a feed-forward NN is thought of as a weighted digraph $G = \{E, V\}$ with no

Chromosome representation: $\{w_1, w_2, w_3, w_4, w_5, w_6, b_1, b_2, b_3\}$

Chromosome in real-valuedcoding: $\{1.91 | 2.55 | 1.9 | 3.12 | 0.88 | 0.91 | 1.1 | 1.9 | 0.98\}$

Fig. 8.9 Real valued chromosome representation

closed paths and described by an upper or lower diagonal adjacency matrix with real valued elements, where E is the set of all edges of the graph and V is the set of vertices (neurons in NN) in the digraph (Siddique and Tokhi 2001). This kind of representation would be too exhaustive for the proposed EA-NN-FLC combination.

Fitness function for weight evolution

The fitness can be defined as the minimum of the sum squared error (SSE) or mean square error (MSE) of the network over a set of training data after training the network for a fixed number of iterations as follows:

$$f(NN) = \sum_{P} e^2 \tag{8.5}$$

$$f(NN) = \frac{1}{P}\sum_{P} e^2 \tag{8.6}$$

where NN is neural network with predefined and fixed architecture, P is the number of patters used for training and e is the difference between the desired output and the actual output of the system.

Some researchers used a fitness function based on the sample counter changing method (Gao 2003). As to network individual training of each generation, not the whole set of training sample is used, i.e., a part of training sample set (say about the 80 % of the sample) is randomly chosen to train the individual of each generation. So, the used training sample set (denoted as x^a in the equation) for neural network of each generation is changed, and then the fitness of individual whose generalization capacity is poor will be smaller while the fitness of individual whose generalization capacity is strong will become large. Consequently, the performance of the whole NN model is improved through selection. The error function of neural network is expressed as follows.

$$E = \frac{1}{2}\sum_{a=1}^{N}\sum_{k=1}^{M}\left[y_k(w_k; x^a) - t_k^a\right]^2 \tag{8.7}$$

where y_k is the network output and t_k^a is the target output for the sample set x^a. The individual fitness of the neural network is expressed by the following transformation of error function of neural network.

$$F = \frac{1}{1+E} \tag{8.8}$$

Some researchers defined the fitness as the number of correctly labelled instances returned by NN among inputs (Tong and Mintram 2010). This fitness function may be better suited for feature selection than weight training.

8.3.2 EA for Weights and Activation Functions Learning

In neural networks, each neuron's output is transferred through an activation function, which is chosen arbitrarily by an expert and assumed to be fixed in neural network architecture. A variety of activations are commonly used in neural networks (Haykin 1999). It is common to use the same activation function for all neurons in the same layer and different activation functions for different layers. The choice of activation function is important in designing NN as part of the architecture, which is to be very sensitive to the performance of the networks (Stork et al. 1990; DasGupta and Schniter 1992; Tong and Mintram 2010). Most neural network applications for supervised learning use sigmoid or radial basis functions as the gradient information for those activation functions are easy to obtain. Some applications may require more complex kind of neuron with nonlinear activation functions. How such nonlinearity can be incorporated within a neuron is a research issue in neural computing domain.

There is another issue with sigmoidal activation function is its shape, which is assumed fixed throughput the network. However, the parameter such as the optimum shape of the sigmoid function is determined by trial and error or heuristically in most of the cases. There have been few studies on the optimum shape of the sigmoid function. Yamada and Yabuta (1992) proposed an auto-tuning method for the sigmoid function shape in order to apply it to a servo control system. Their method is based on the gradient descent method and confirmed the characteristics and practicality of the method with simulation results. The usual tan-sigmoid function *f*(*x*) is defined as

$$f(x) = \frac{1 - e^{-ax}}{1 + e^{-ax}} \tag{8.9}$$

where x is the network output and a defines the shape of the activation function. The shape of sigmoid function for varying a is shown in Fig. 8.10. The activation function is defined in Eq. (8.9) and the parameter a defines the shape of the sigmoidal function. The use of different shapes of sigmoidal function can lead to different weights and biases during learning with the backpropagation algorithm, which is experienced in many applications (Yamada and Yabuta 1992). That is, the shape of sigmoidal functions should be fixed during execution of the backpropagation algorithm. This type of activation function is characterised by its gain (slope) and seriously affects the control characteristics. If this gain tuning is used in control applications, the network output may become unstable in certain cases. When the usual sigmoid function is used only in the hidden layer, sigmoid function shape tuning is the same as weight tuning. A mathematical proof is given in (Yamada and Yabuta 1992). Therefore, sigmoid function shape learning in NN can achiev the intended nonlinearity and consequently contribute significantly to the improvement of the system performance.

EA based training of the neural networks with non-logistic-function neurons is a viable alternative. Stork et al. (1990) were the first to apply EA to the selection of

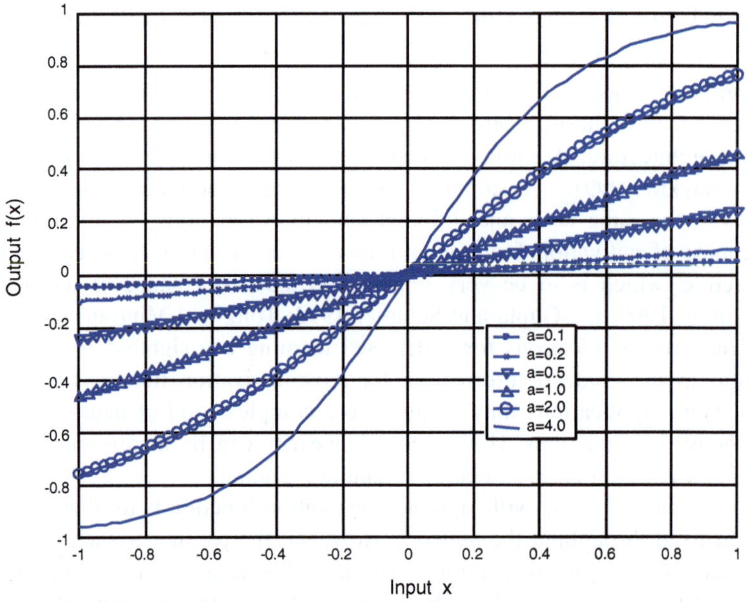

Fig. 8.10 Shape of sigmoidal function for different values of a

node transfer function in a neural network. Activation function used in this application was more complex than usual sigmoid function and was specified in the genotype representation of the chromosome. Liu and Yao (1996) applied EP for the selection of sigmoidal or Gaussian nodes in a neural network, where the EP allowed growth or shrinking of the neural network by adding or deleting a sigmoidal or Gaussian node. Experimentation on a set of benchmark problems demonstrated good performance.

Consider a neural network with three neurons, two inputs and single output shown in Fig. 8.11. The two activations functions $f_1(x)$ and $f_2(x)$ at the hidden layer are tansigmoidal functions with shaping parameters a_1 and a_2 respectively and the activation function $f_3(x)$ at the output layer is a linear function with scaling parameter a_3 defined in (8.10).

$$f_1(x) = \frac{1 - e^{-a_1 x}}{1 + e^{-a_1 x}}, f_2(x) = \frac{1 - e^{-a_2 x}}{1 + e^{-a_2 x}}, f_3(x) = a_3 x \tag{8.10}$$

The activation functions of the three neurons in the network can be different due to the shaping parameters $\{a_1, a_2, a_3\}$, which will impose restrictions on using the backpropagation algorithm for training of the network. Moreover, training of the weights $\{w_1, w_2, \cdots, w_6\}$ and biases $\{b_1, b_2, b_3\}$ of the network along with the sigmoidal function shape parameters $\{a_1, a_2, a_3\}$ using backpropagation learning algorithm would be computationally intensive and cumbersome. Evolutionary learning can automatically decide the optimal shape of the sigmoid function as well as optimise the weights and biases.

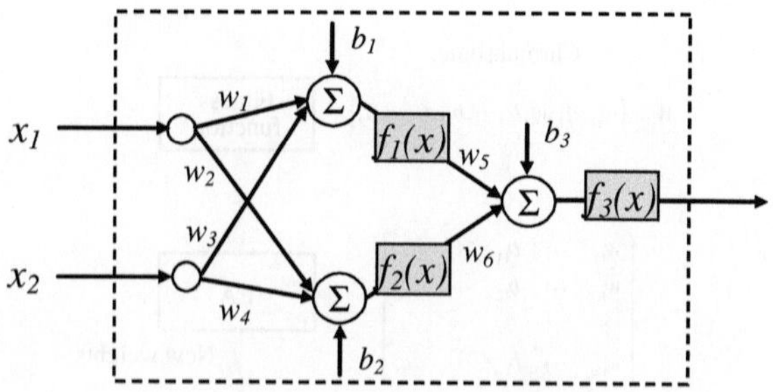

Fig. 8.11 Neural network with non-linear activation function

The decision on how to encode the activation functions in the chromosome representation depends on the a priori information and computational time allowed for the training of the network. In general, nodes in the same layer tend to have the same type of activation function with possible different in the parameter set, e.g. parameters $\{a_1, a_2\}$ of $\{f_1(x), f_2(x)\}$ in layer two in Fig. 8.11. Nodes in different layers can have different activation functions, e.g. $\{f_1(x), f_2(x)\}$ in layer two are tansigmoidal and $f_3(x)$ in layer three is linear. The training is possible in two ways: firstly, using EA for optimising activation functions and weight training (weights and bias) using backpropagation algorithm and secondly, using EA for both the function and weight training. Use of EA and backpropagation would be slow and computationally exhaustive. Furthermore, it may not guarantee optimal network performance due to the nonlinear relationship between the two parameter spaces. Evolution of both transfer functions and weights at the same time would be advantageous since they constitute a complete architecture. Encoding weights and parameters of the activation functions into the same chromosome would be easier to find the optimal performance of the network by exploring the two sets of parameter space. The chromosome representation is very straightforward as shown in Fig. 8.12, the w_i are the weights, b_i are the biases and a_i are the activation function shape parameters.

The different evolutionary algorithms discussed in Chap. 6 can be applied for the weights, biases and activation function parameters. A simple example is shown in Fig. 8.13.

The fitness function for the evolutionary NN in Fig. 8.13 could be any of the fitness function used for weight learning such as SSE or MSE defined in (8.5–8.8). The connectivity of the network is assumed to be fixed during learning of the weights and parameters in the evolutionary learning shown in Fig. 8.13.

$$\{w_1, w_2, \cdots, w_6\}, \{b_1, b_2, b_3\}, \{a_1, a_2, a_3\}$$

Fig. 8.12 Chromosome for weights, bias and function shape parameters

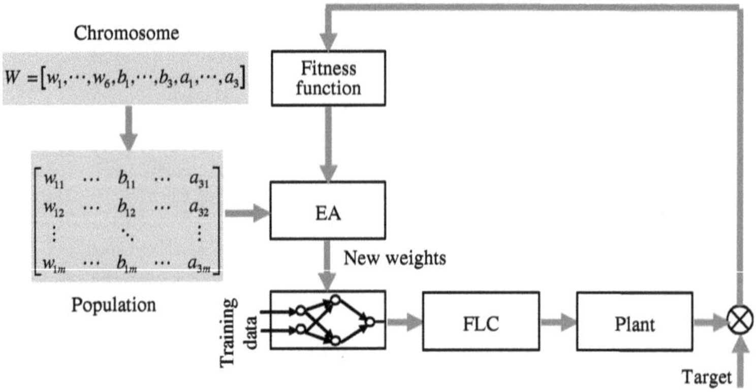

Fig. 8.13 Evolution of weights and activation function shape parameters

8.4 Optimal Sigmoid Function Shape Learning

Introduction of a non-linear activation function to the single neuron network and its performance over a linear activation function was investigated in Chap. 7. The aim of this investigation was to figure out the possible effect of the nonlinear activation function on the performance of the network as well as the performance of the neuro-fuzzy controller for the flexible arm. A tan-sigmoidal function has been used where the shape of the sigmoid function represents the nonlinearity of the activation function. Backpropagation algorithm has been applied to train the single neuron network and the shape of the sigmoidal function was chosen arbitrarily. The results in Chap. 7 show that the shape parameter, denoted as a in Chap. 7, has significant effect on the cost function. It was also found that for different sigmoidal function shape, the learnt weights and bias were different and hence affected the performance of the controller. In respect of rise time and settling time, the controller with nonlinear activation function shows promising performance than the linear activation function. The shape of the sigmoid function is playing an important role influencing the performance of the controller, which needs further investigations.

A criterion is required for the selection of an optimal neural network to represent the non-linearity of the system. However, the parameter of the non-linear activation function, such as the optimal shape of a sigmoid function, is determined by trial and error. In this Chapter, a genetic algorithm-based technique is used to optimise the shape of the activation function together with the weights and bias of the single-neuron network.

Nonlinearity can be represented with suffient number of hidden layers with fixed activation functions. However, many parameters such as the optimum shape of the sigmoid function are determined by trial and error in most of the cases. This, as noted in the previous Chapter, limits the application of the network. There have been few studies on the optimum shape of the sigmoid function. Yamada and Yabuta (1992) proposed an auto-tuning method for the sigmoid function shape in order to apply it to a servo control system. Their method is based on the gradient descent method and confirms the characteristics and practicality of the method with simulation results.

The usual tan-sigmoid function $f(x)$ is defined as

$$f(x) = \frac{1 - e^{-ax}}{1 + e^{-ax}} \tag{8.11}$$

where x is the network output and a defines the shape of the activation function. The shape of sigmoid function is shown in Fig. 8.10 with varying shape parameter a.

This type of activation function is characterised by its gain (slope), which seriously affects the control characteristics. If this gain tuning is used in control applications, the plant output may become unstable in certain cases. When the usual sigmoid function is used only in the hidden layer, sigmoid function shape tuning is the same as weight tuning. A mathematical proof is given in (Yamada and Yabuta 1992). Therefore, sigmoid function shape tuning in the single neuron network can contribute more in improving performance of the controller. In the following section, genetic algorithm based learning of the shape of sigmoid function as well as the weights and bias of the network is provided.

8.5 Evolutionary-Neuro-Fuzzy PD-PI-like Controller

The interest in training neural networks using genetic algorithms has been growing rapidly in recent years (Caudell and Dolan 1989; Montana and Davis 1989; Whiteley et al. 1990; Yam and Chow 1997). The interest in this study is to explore possible benefits arising from the interactions between neural networks and evolutionary search procedures. Accordingly the most popular models of NN and evolutionary search procedures are considered, such as the feed forward network and GA. Supervised learning has mostly been formulated as a weight training process, in which effort is made to find an optimal set of connection weights for a neural network according to some optimality criteria. The most popular training algorithms for feed forward NNs is the backpropagation (BP) algorithm as discussed in Chap. 7. This is a gradient descent search algorithm based on minimization of the total mean squared error between the actual output and the desired output. However, the BP algorithm suffers from a number of problems. It is very often stuck in local minima and is very inefficient in searching for a global minimum of the search space. BP's speed and robustness are sensitive to several parameters of the algorithm and the best parameters to use appear to vary from problem to problem (Caudell and Dolan 1989).

8.5.1 GA-Based Neuro-Fuzzy Controller

A block diagram of the GA-based neuro-fuzzy control system is shown in Fig. 8.14, which incorporates a single neuron network shown in Fig. 8.15.

The activation function is defined in Eq. (8.11) and the parameter a defines the shape of the sigmoidal function. The use of different shapes of sigmoidal function can lead to different weights and biases during learning with the backpropagation algorithm, which is experienced in the previous Chapter. That is, the shape of sigmoidal functions should be fixed during execution of the backpropagation algorithm.

A mechanism is sought to learn the weights, bias and the parameter a of the network. Two approaches present themselves instantly for this purpose: firstly, backpropagation algorithm learning of weights and bias and trial and error method for parameter a and secondly, genetic algorithm based learning of the weights, biases and the parameter a simultaneously. The first approach was investigated in Chap. 7. It seemed somewhat tedious and slow because of the computation involved in updating the weights and bias for each parameter a, which prolonged the computation in each learning epoch. Moreover, the system response was not very promising in comparison with linear activation function. This chapter aims to investigate the possible benefit of learning the shape of the sigmoid function together with the weights and bias, which will reduce the computing time greatly and can exploit the non-linearity involved in the system. Genetic algorithm can best serve such a learning objective.

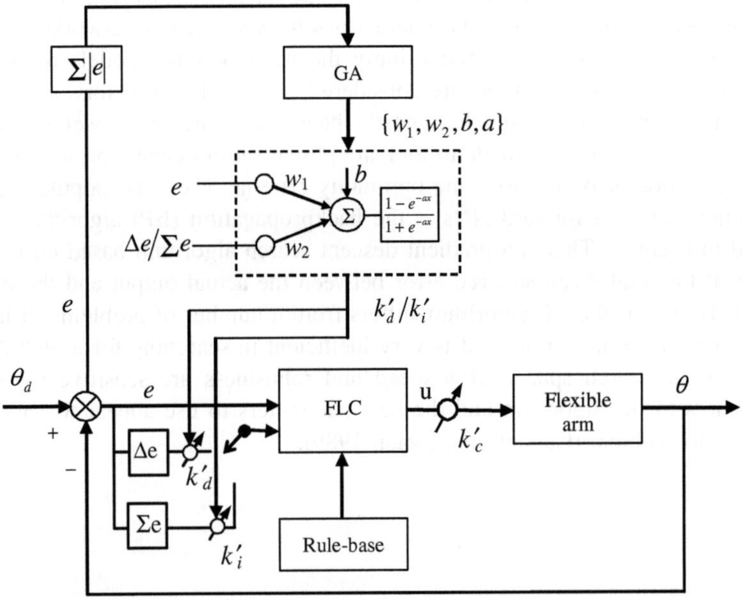

Fig. 8.14 Block diagram of the GA-based neuro-fuzzy control system

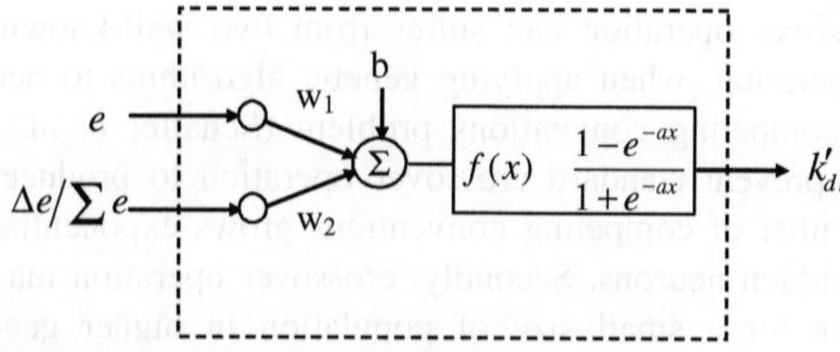

Fig. 8.15 Single neuron network with non-linear activation function

w_1, w_2, b, a

Fig. 8.16 String representation of chromosome of the neural network

In this section, the weights (w_1, w_2), bias b and the parameter a of the neural network, shown in Fig. 8.15, are learnt by genetic algorithm. The chromosome representation is straightforward and it is shown in Fig. 8.16.

The sum of absolute error is chosen as the objective function defined as

$$J = \sum_{k=1}^{N} |e(k)| \tag{8.12}$$

where $e(k)$ is hub-angle error of the flexible arm and N is some reasonable number of time units by which the system can be assumed to have settled close to steady state. The evaluation of the objective function is performed by applying the controller on the experimental flexible arm.

Experience from the experiments in the previous Chapter says that the values of the weights-bias and sigmoid function shape parameter are within the ranges $[-0.5, +0.22]$ and $[2.0, 2.6]$ respectively. The use of this a priori information about the interval of the weights-bias and parameter of the sigmoid function is very useful: firstly the region of the search space is known and secondly the initial population will be well distributed over the search space. This will help the evolutionary process to converge in a reasonable time (i.e. in generations) otherwise it would have been taken a long time. Considering this a priori information, a population of 10 chromosomes is initialised within the ranges of values.

Elitists single point crossover operation is used. Elitism is an optional characteristic of genetic algorithm. When used, it makes sure that the fittest chromosome of a population is passed on to the next generation unchanged. In this investigation, an extended form of elitism is used where best m ($m = 8$ in this study) chromosomes are retained from N chromosomes, N is the population size ($N = 10$ in this study). In other words, the worst two chromosomes are replaced by two offsprings created by crossing two best chromosomes in the population.

The crossover operation can suffer from two well-known problems: firstly, crossover operation, when applying genetic algorithms to neural networks, can result in a competing conventions problem (Schaffer et al. 1992). Competing conventions prevent standard crossover operation to produce useful offsprings. Also the number of competing conventions grows exponentially with respect to number of hidden neurons. Secondly, crossover operation may not produce new chromosomes for a small size of population in higher generations. Mutation operation can thus strike a balance to these problems encountered by crossover. Montana and Davis used three different types of mutation operators (Montana and Davis 1989) to overcome such problems. In this study, a mutation operation with a higher mutation rate is applied to genetic algorithm based learning of the neural network. A randomly chosen value from the offspring is mutated with a mutation rate of 0.5. This mutation rate will ensure changes of at least two values in the offspring chromosome.

8.6 Some Experimental Results

Considering the results in Chap. 7, a population of 10 chromosomes is initialised within the range of $[-0.5, +0.5]$ and $[0, 4]$ for weights and bias and for the parameter a respectively. The practical constraint of applying the GA-based Neuro-Fuzzy controller to the flexible arm involved is how to evaluate the objective function. The easiest way is to operate the Neuro-Fuzzy controller repeatedly and evaluate its performance by calculating the absolute sum of error. The population is tested up to the 13th generation. Figure 8.17 shows the system responses of the best 4 individuals in the 1st generation. In earlier generations, some of the chromosomes needed a longer time to settle, and the chromosomes were required to evaluate for 250 iterations in the program loop. This has caused some ties of the fitness values. To help resolve the ties, only 50 iterations were evaluated in the later generations.

Figure 8.18 shows the system response for a target hub angle of 36 degrees of the best individual in generations 5, 7 and 9. In these generations positive and negative overshoots are big and only generation 9 achieved the desired hub angle at 43 time units with a very big overshoot of 65°. Figure 8.19 shows the system response of the best individual in generations 11, 12 and 13 for the same demanded hub angle. This shows significant improvement of the performance in respect of rise time, maximum overshoot, settling time and steady state error. A numerical comparison of these response parameters is shown in Table 8.1. Figure 8.20 shows fitness convergence of GA over generations. The weights, bias and the parameter a after learning were found to be $w_1 = -0.029$, $w_2 = 0.01$, $b = 0.23$ and $a = 2.18$.

The performance of a three-neuron network with linear activation function and with non-linear activation function was also verified. The performance of the

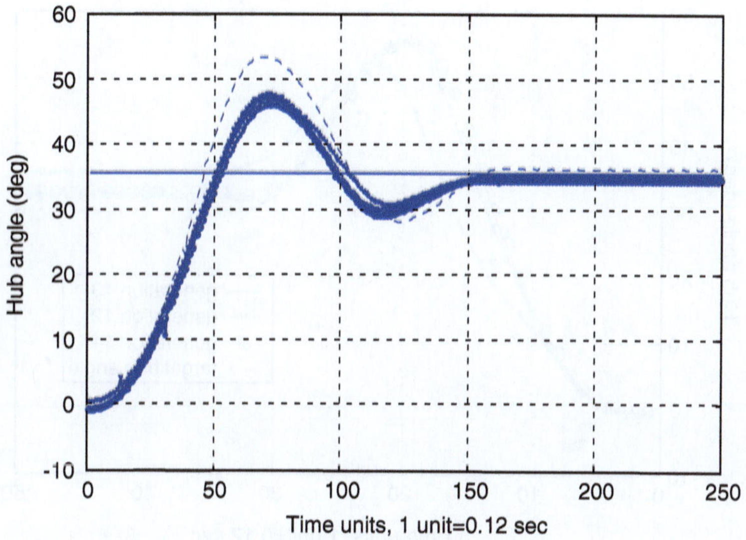

Fig. 8.17 System response using best 4 individuals in generation 1

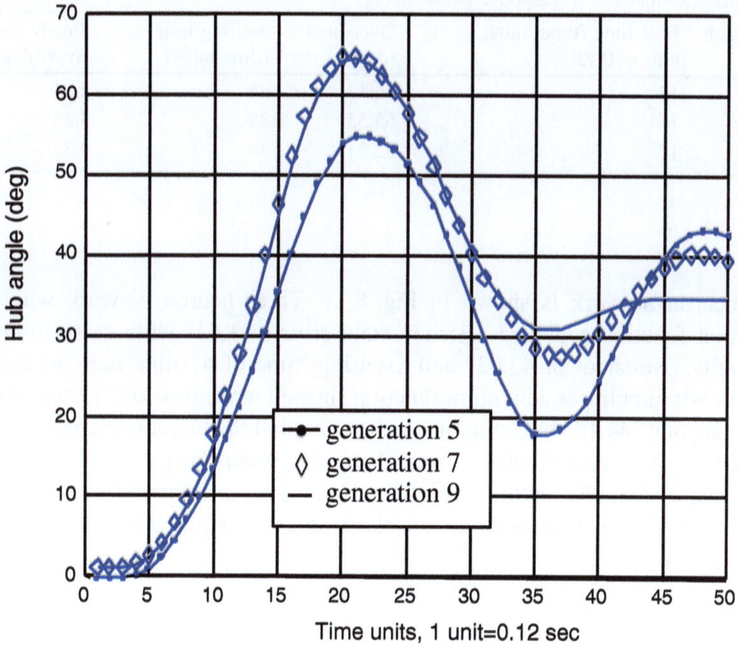

Fig. 8.18 System response of best individuals in generations 5, 7 and 9

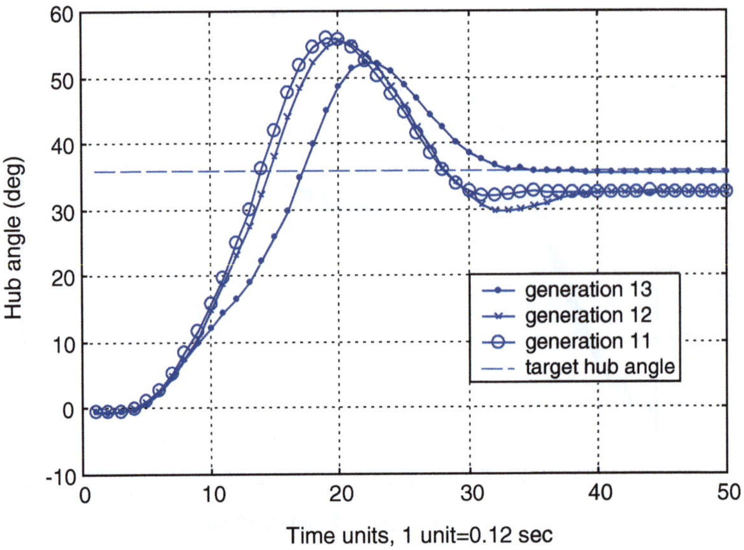

Fig. 8.19 System response with best individuals in generations 11, 12 and 13

Table 8.1 Comparison of response parameters

Generations	Rise time (time units, 1 unit = 0.12 s)	Overshoot (deg)	Settling time (time units)	Steady state error (deg)
11	13	56.13	35	3.24
12	14	55.55	39	3.34
13	17	52.51	34	0.33

three-neuron network is shown in Fig. 8.21. Three-neuron network with linear activation function achieved a steady state error of 1.61° with a rise time of 22 time units, overshoot of 43.03° and a settling time of 47 time units whereas the network with nonlinear activation function showed a rise time of 21 time units and an overshoot of 44.18° but could not settle at around the target hub angle within 50 time units. The network was trained using the backpropagation algorithm. The performance degradation of the system is obvious and possibly caused by the excessive calculation of weights and biases updates required in backpropagation algorithms.

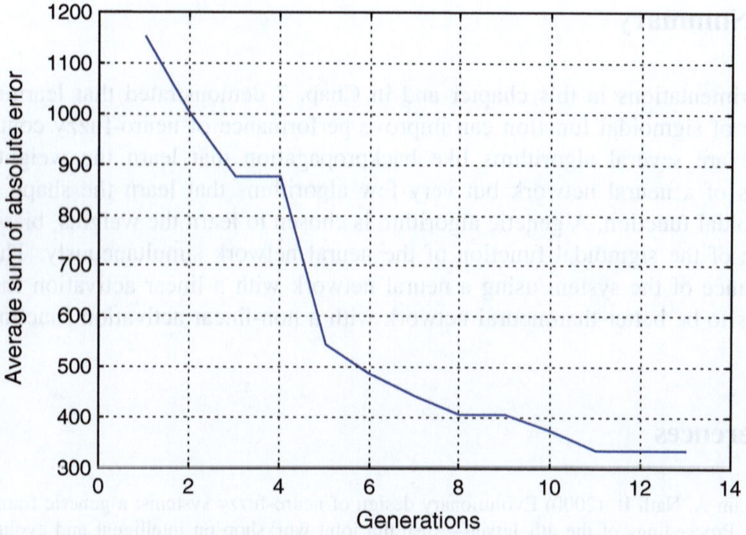

Fig. 8.20 Convergence of the fitness

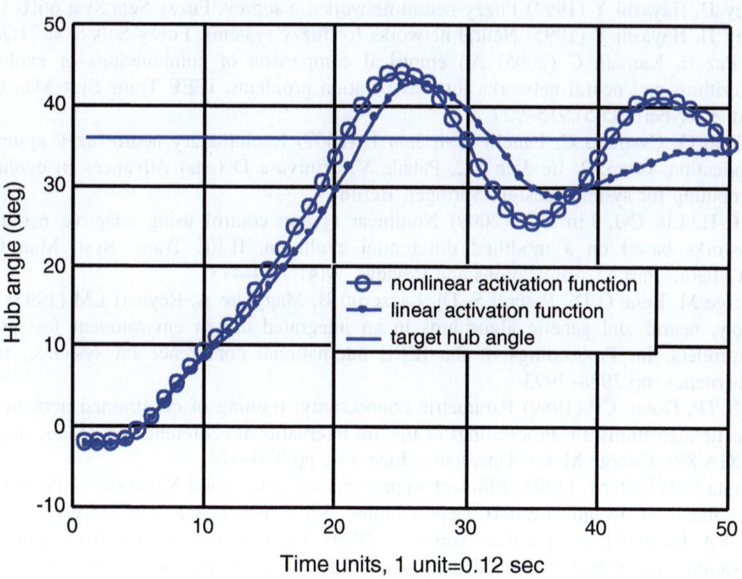

Fig. 8.21 System response using Neuro-fuzzy controller with 3-neuron network

8.7 Summary

Experimentations in this chapter and in Chap. 7 demonstrated that learning the shape of sigmoidal function can improve performance of neuro-fuzzy controller. There are several algorithms like backpropagation that learn the weights and biases of a neural network but very few algorithms that learn the shape of the sigmoidal function. A genetic algorithm is chosen to learn the weights, biases and shape of the sigmoidal function of the neural network simultaneously. The performance of the system using a neural network with a linear activation function seems to be better than neural network with a non-linear activation function.

References

Abraham A, Nath B. (2000) Evolutionary design of neuro-fuzzy systems: a generic framework. In: Proceedings of the 4th Japan—Australia joint workshop on intelligent and evolutionary systems, Japan, November 2000

Berenji HR, Khedkar P (1992) Learning and tuning fuzzy logic controllers through reinforcements. IEEE Trans Neural Netw 3:724–740

Buckley JJ, Hayashi Y (1994) Fuzzy neural networks: a survey. Fuzzy Sets Syst 66:1–13

Buckley JJ, Hayashi Y (1995) Neural networks for fuzzy systems. Fuzzy Sets Syst 71:265–276

Cantú-Paz E, Kamath C (2005) An empirical comparison of combinations of evolutionary algorithms and neural networks for classification problems. IEEE Trans Syst Man Cybern, Part B: Cybern 35(5):915–927

Castellano G, Castiello C, Fanelli AM, Jain L (2007) Evolutionary neuro-fuzzy systems and application, Chap. 2. In: Jain LC, Palade V, Srinivasa D (eds) Advances in evolutionary computing for systems design, Springer, Berlin

Chen C-H, Lin C-J, Lin C-T (2009) Nonlinear system control using adaptive neural fuzzy networks based on a modified differential evolution, IEEE Trans. Syst. Man Cybern. B Cybern.- Part C: Applications and Review, 39(4): 459–473

Chiaberge M, Bene G Di, Pascoli S. Di, Lazzerini B, Maggiore A, Reyneri LM (1995) Mixing fuzzy, neural and genetic algorithms in an integrated design environment for intelligent controllers. In: Proceedings of the IEEE international conference on systems, man and cybernetics, pp 2988–2993

Caudell TP, Dolan CP (1989) Parametric connectivity: training of constrained networks using genetic algorithms. In: Proceedings of the 3rd international conference on genetic algorithms (ICGA'89), George Mason University, June 4–7, pp 370–374

DasGupta B, Schniter G (1992) Efficient Approximation with neural Networks: A comparison of gate functions, technical report, Dept. Comput. Sci. Pennsylvania State University.

Farag WA, Quintana VH, Lambert-Torres G (1998) A genetic-based neuro-fuzzy approach for modeling and control of dynamical systems. IEEE Trans Neural Netw 9(5):756–767

Feng JC, Teng LC (1998) An online self constructing neural fuzzy inference network and its applications, IEEE Trans. Fuzzy Syst. 6(1):12–32

Figueiredo M, Gomide FAC (1999) Design of fuzzy systems using neurofuzzy networks. IEEE Trans. Neural Networks 10(4):815-827

Fukuda T, Shimojima K, Shibata T (1994) Fuzzy, neural networks and genetic algorithm based control systems. In: Proceedings of the IEEE international conference on industrial electronics, control and instrumentation, pp 1220–1225

Gao W (2003) Study on new evolutionary neural network. In: Proceedings of the second international conference on machine learning and cybernetics, Wan, 2–5 November 2003, pp 1287–1292

Haykin S (1999) Neural Networks - A Comprehensive Foundation, Prentice Hall, Upper Saddle River, NJ

Jang J-SR (1993) ANFIS: adaptive-network-based fuzzy inference system. IEEE Trans Syst Man Cybern 23(3):665–685

Jang JSR, Sun CT, Mizutani E (1997) Neuro-fuzzy and Soft Computing. Prentice Hall, Upper Saddle River

Kazabov N (1996) Foundations of neural networks, fuzzy systems and knowledge engineering. MIT Press, Cambridge, MA

Kasabov N, Song Q (1999) Dynamic evolving fuzzy neural networks with 'm-out-of-n' activation nodes for on-line adaptive systems, Technical Report TR99/04, Department of Information Science, University of Otago

Kosko B (1991) Neural networks and fuzzy systems. Prentice-Hall, Upper Saddle River

Lin TC, Lee CS (1991) Neural network based fuzzy logic control and decision system. IEEE Trans Comput 40(12):1320–1336

Lin CT, Lee CSG (1996) Neural fuzzy systems: a neuro-fuzzy synergism to intelligent systems. Prentice-Hall, Upper Saddle River

Lin C-J, Chen C-H, Lin C-T (2008) Efficient self-evolving evolutionary learning for neuro-fuzzy inference systems. IEEE Trans Fuzzy Syst 16(6):1476–1490

Lin CT, Han MF, Lin YY, Liao SH, Chang JY (2011) Neuro-fuzzy system design using differential evolution with local information. In: 2011 IEEE International conference on fuzzy systems, June 27–30, Taipei, Taiwan, pp 1003–1006

Liu Y, Yao X (1996) Evolutionary design of artificial neural networks with different nodes. In: Proceedings of the 1996 IEEE international conference on evolutionary computation (ICEC'96), Nagoya, Japan, pp. 670–675

Loila V, Sessa S, Staiano A, Tagliaferri R (2000) Merging fuzzy logic, neural networks and genetic computation in the design of a decision support system. Int J Intell Syst 15:575–594

Montana DJ, Davis L (1989). Training feedforward neural network using genetic algorithms. In: Proceedings of 11th international joint conference on artificial intelligence, San Mateo, CA, Morgan Kaufmann, pp 762–767

Nauck D, Klawonn F, Kruse R (1997) Foundations of neuro-fuzzy systems. Wiley, Chichester, NY, Weinheim, Brisbane, Singapore, Toronto

Nie J, Linkens D (1995) Fuzzy-neural control: principles, algorithms and applications. Prentice-Hall International (UK) Ltd, Upper Saddle River

Rahmoun A, Berrani S (2001) A Genetic-based Neuro-Fuzzy Generator: NEF-GEN, ACS/IEEE Int. Conf. on Comput. Syst. and Appl. pp 18–23

Russo F (1999) Evolutionary neural fuzzy systems for noise cancellation in image data. IEEE Trans Instrum Measurement 48(5):915–920

Schaffer JD, Whitley D, Eshelman LJ (1992) Combinations of genetic algorithms and neural networks: a survey of the state of art. IEEE International workshop on combinations of genetic algorithms and neural networks, Baltimore, pp 1–37

Seng TL, Khalid MB, Yusof R (1999) Tuning of a Neuro-Fuzzy Controller by Genetic Algorithm, IEEE Trans. Syst. Man Cybern. B Cybern.- Part B: Cybernetics, 29(2):226–236.

Shin YC, Xu C (2009) Intelligent systems: modelling, optimisation and control. CRC Press, Taylor and Francis Group, Boca Raton, Fl

Siddique NH, Tokhi MO (2001) Training neural networks: backpropagation vs genetic algorithms. In: IEEE international joint conference on neural network, Washington DC, 14–19 July, pp 2673–2678

Siddique N, Adeli H (2013) Computational Intelligence: Synergies of fuzzy logic, neural networks and evolutionary computing, John Wiley and Sons

Sulzberger SM, Tschicholg-Gurman NN, Vestli SJ (1993) FUN: Optimization of fuzzy rule based systems using neural networks, In: Proceedings of the IEEE Conference on Neural Networks, San Francisco, pp 312–316

Stork DG, Walker S, Burns M, Jackson B (1990) Pre-adaptation in neural circuits. In: Proceedings of the international joint conference on neural networks, vol I. Washington DC, pp 002–205

Tano S, Oyama T, Arnould T (1996) Deep combination of fuzzy inference and neural network in fuzzy Inference, Fuzzy Sets Syst. 82(2):151–160

Tong DL, Mintram R (2010) Genetic algorithm- neural network (GANN): a study of neural network activation functions and depth of genetic algorithm search to feature selection. Int J Mach Learning Cybern 1:75–87

Tsoukalas LH, Uhrig RE (1997) Fuzzy and neural approaches in engineering. Wiley, Hoboken

Wang WY, Li YH (2003) Evolutionary learning of BMF fuzzy-neural networks using a reduced-form genetic algorithm. IEEE Trans Syst Man Cybern Part B Cybern 33(2003):966–976

Whiteley D, Starkweather T, Bogart C (1990) Genetic algorithms and neural networks: optimizing connections and connectivity. Parallel Comput 14:347–361

Yamada T, Yabuta T (1992) Neural network controller using auto tuning method for nonlinear functions. IEEE Trans Neural Netw 3(4):595–601

Yam JYF, Chow TWS (1997) Extended least squares based algorithm for training feedforward networks. IEEE Trans Neural Netw 8(3):806–810

Yao X, Liu Y (1997) A new evolutionary system for evolving artificial neural networks. IEEE Trans Neural Netw 8(3):694–713

Yao X, Liu Y (1998) Making use of population information in evolutionary artificial neural networks. IEEE Trans Syst Man Cybern Part B: Cybern 28(3):417–425

Yao X (1999) Evolving artificial neural networks. Proc IEEE 87(9):1423–1447

Yu L, Zhang Y-Q (2005) Evolutionary fuzzy neural networks for hybrid financial prediction. IEEE Trans Syst Man Cybern -Part C: Appl Rev 35(2):244–249

Zadeh LA (1973) Outline of a new approach to analysis of complex systems and decision process. IEEE Trans Syst Man Cybern SMC-3:28–44

Chapter 9
Stability Analysis of Intelligent Controllers

9.1 Introduction

Stability theory plays an important role in analyzing dynamical systems and especially designing and analyzing control systems. When roughly speaking, stability means that the system outputs and its internal signals are bounded within admissible limits, i.e. bounded-input/bounded-output. Strictly speaking, the system outputs tend to an equilibrium state of interest, i.e. asymptotic stability (Khalil 2002). The basic notion of stability emerged from the study of equilibrium of rigid body mechanical system under gravitational force by E. Torricelli around 1644. G. Lagrange formulated the classical stability theorem of conservative mechanical system in 1788. The theorem states that if the potential energy of a conservative system has a minimum, then this equilibrium point of the system is stable (Merkin 1997). Lyapunov published his general theory of stability of motion in 1892, a fundamental theory that dominated systems and control engineering over a century and still modern control engineering depends on the principles of Lyapunov's stability theory (Lyapunov 1992). Over the century, there have been many theories developed on stability and analysis of dynamic systems and control design (Martynnyuk 2003).

Since fuzzy controllers are essentially nonlinear and the dynamic behaviour of the system is not well known (as because a precise mathematical model is not required for fuzzy control), the closed-loop behaviour is complex. Therefore, it is very important that the stability of the fuzzy controllers should be investigated before deployment as to make sure that the control system is safe for operation and the desired performance is guaranteed in the presence of variations, disturbance and uncertainties during operation.

N. Siddique, *Intelligent Control*, Studies in Computational Intelligence 517,
DOI: 10.1007/978-3-319-02135-5_9,

9.2 Mathematical Preliminaries

Approaches to stability of dynamic systems can be categorised into two broad classes of stability theory:

(i) Qualitative theory and
(ii) General stability theory

The state space approach is formalised with the definition of stability and robustness indices based on concepts from qualitative theory of dynamical systems. In qualitative theory, two classes of indices are used. One is related to the relative stability and gives a measure of the degree of stability of the equilibrium point at the origin. The other one is related to the global stability of the system and measures how far the system is from a bifurcation giving rise to an attractor-repulsor pair. Aracil et al. (1989, 1988) used these indices for the analysis and design of fuzzy control systems. A detail description of the stability and robustness indices can be found in Driankov et al. (1993).

In the framework of the general stability theory, there are two directions:

(i) Stability in the Lyapunov sense, which means that the state vector tends towards zero,
(ii) Input–output stability, which means bounded input and bounded output stability.

Roughly speaking, the Lyapunov stability of a system with respect to its equilibrium of interest is about the behaviour of the system outputs towards the equilibrium state, wondering nearby or around the equilibrium state (also called stability in the sense of Lyapunov), or gradually approaching towards the equilibrium state (also called asymptotic stability). Consider the general non-autonomous system

$$\dot{x} = f(x, t; u), t \in [t_0, \infty) \tag{9.1}$$

$$u(t) = h(x(t), t) \tag{9.2}$$

where $x = x(t)$ is the state of the system, $u = u(t)$ is the control input, $f(.)$ is a locally Lipschitz map or continuously differentiable function, so that the system has a unique solution for each admissible input and suitable initial condition $x(t_0) = x_0 \in R^n$ with initial time $t_0 \geq 0$, R^n is the entire state space to which the system states belong to and $h(.)$ is a mapping function that specifies the control input. It is to be noted that a special case of the system (9.1), with or without control, is said to be autonomous if the variable t does not appear independently from the state vector in the system function f. An equilibrium of the system (9.1) at the origin of R^n, if exists, is a solution x^* of the algebraic equation $f(x^*) = 0$ and $\dot{x}^* = 0$, which means that an equilibrium of a system must be a constant state.

The system (9.1) is said to be stable in the sense of Lyapunov at the equilibrium $\dot{x}^* = 0$ if for each $\varepsilon > 0$, there exists a constant $\delta = \delta(\varepsilon, t_0) > 0$ such that

$$\|x_0\| < \delta \Rightarrow \|x(t)\| < \varepsilon, \ \forall t \geq t_0 \tag{9.3}$$

where $\|.\|$ defines a norm of the vector. The system (9.1) is said to be asymptotically stable, if it is stable at the equilibrium $\dot{x}^* = 0$ and $\delta = \delta(t_0) > 0$ can be chosen such that

$$\|x_0\| < \delta \Rightarrow \underset{t \to \infty}{Lim} \|x(t)\| = 0 \tag{9.4}$$

An equilibrium point $\dot{x}^* = 0$ that is Lyapunov stable but not asymptotically stable is called marginally stable.

Lyapunov stability theory has been extensively applied to control systems design. There have been many applications of Lyapunov stability criteria to fuzzy control design reported in the literature (Feng 2003, 2004; Ohtake et al. 2006, 2007; Shin and Xu 2009; Wang et al. 1995; Wang 1997; Zhang et al. 2006; Zhang and Feng 2008). A brief survey reveals that most of the applications of Lyapunov stability theory are applied to design and analysis of Takagi–Sugeno type fuzzy controllers. It is mainly due to the fact that Takagi–Sugeno type fuzzy controllers comprise more mathematical rigor in describing the models compared to Mamdani-type fuzzy controllers. Tanaka and Sugeno (1992) modelled a plant using hybrid fuzzy/linear system and provided a sufficient condition for the system's asymptotic stability. A good discussion on Takagi–Sugeno fuzzy systems' stability analysis using Lyapunov techniques is provided in Shin and Xu (2009).

Input–output stability theory is a powerful tool for analysing the stability and robustness of controllers. Especially it is applicable to fuzzy controllers as it requires minimal assumptions to make about the process to be controlled, i.e., the functional gain. A fuzzy controller is described explicitly as an input–output map, where the input–output theory is directly applicable. This provides a "bounded-input/bounded-output" stability test for a pre-synthesized static controller, i.e. relative small output with respect to small inputs. Let us consider a feedback system shown in Fig. 9.1 for further explanation of the input-output stability.

In Fig. 9.1, typically, G represents a plant, H represents a controller, y is the control variable and u is the control input to the plant. The input–output relation is represented by

$$y = Gu \tag{9.5}$$

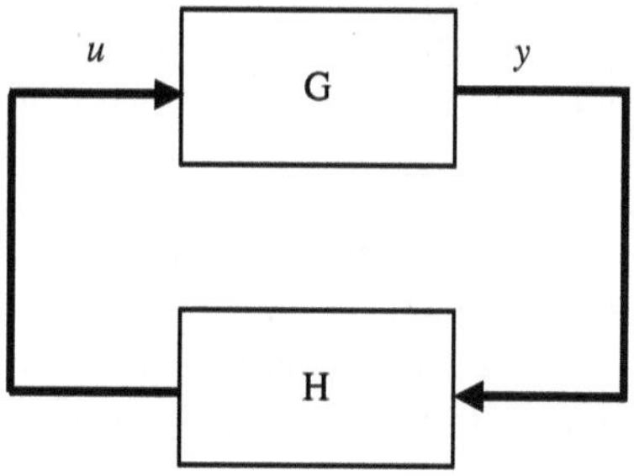

Fig. 9.1 Closed-loop feedback system

where G is some mapping or operator that specifies y in terms of u. The input u belongs to a space that map the time interval $[0, \infty)$ into the Euclidean space R^m, i.e. $u : [0, \infty) \rightarrow R^m$. For the space of piecewise continuous bounded functions, the norm is defined as

$$\|u\|_{L_\infty} = \sup_{t \geq 0} \|u(t)\| < \infty \tag{9.6}$$

The space is denoted as L_∞^m. More generally, the space L_p^m for $1 \leq p < \infty$ is defined as the set of piecewise continuous function $u : [0, \infty) \rightarrow R^m$ such that

$$\|u\|_{L_p} = \left(\int_0^\infty \|u(t)\|^p dt \right)^{1/p} < \infty \tag{9.7}$$

The subscript p in L_p^m refers to the type of p-norm used to define the space and the subscript m is the dimension of u. If $u \in L^m$ is considered a well-behaved input to the system, then y should be a well-behaved output for $y \in L^q$. The subscript q is the dimension of y. A system is stable for all well-behaved inputs $u \in L^m$ that generates well-behaved outputs $y \in L^q$. The definition of L_∞ stability is the familiar notion of bounded-input-bounded-output stability, i.e. if the system is L_∞ stable, then for every bounded input $u(t)$, the output $Gu(t)$ is bounded. The eventual end result of input–output stability analysis is the small gain theorem based on the early work of Zames (1966a, b). According to the gain theorem, the sufficient condition for the stability of a closed-loop system is given by

$$\|G\| \cdot \|H\| < 1 \tag{9.8}$$

where $\| \cdot \|$ denotes functional gain or norm of a system. Both classical and modern control system analysis use small gain conditions (Barreiro 1999). The first significant results from a direct application of input–output stability theory appeared in Espada and Barreiro (1994). French and Rogers (1998) applied input–output stability theory for designing direct neuro-fuzzy controllers.

From the small gain theorem, conicity criterion can also be derived, which is directly applicable to fuzzy control systems. Stability analysis of nonlinear control systems using conicity technique has a long history since Zames's work. The small-gain condition in (9.8) is a sufficient condition for stability of a system. If it does not hold, it does not conclude to the system's instability. In that case, a system engineer would prefer to increase the applicability of the small-gain condition by adding and subtracting a block C to the closed-loop system F shown in Fig. 9.2.

The transformed closed-loop system $T(F, C)$ is shown in Fig. 9.3. For a detail description on the transformed closed-loop form, interested readers are directed to Khalil (2002). In practice (under not very restrictive conditions), the transformation makes the stability conditions of the transformed closed-loop system $T(F, C)$ equivalent to the closed-loop system F, so that the small-gain condition can be

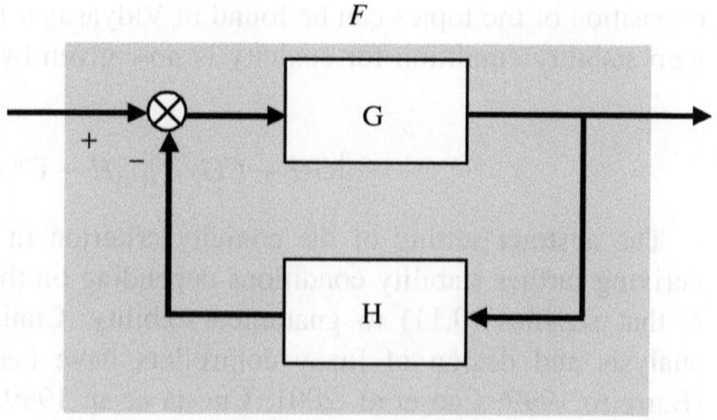

Fig. 9.2 Canonical closed-loop system

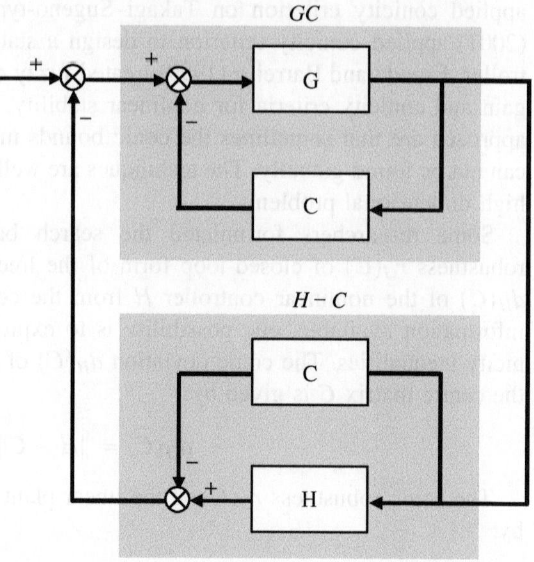

Fig. 9.3 Transformed closed-loop system

applied to $T(F, C)$ and the conicity criterion can be formulated. The closed-loop system F is stable if there is a linear operator C and a positive number $r > 0$ such that

$$\|H - C\| < r \tag{9.9}$$

$$\|G(I + CG)^{-1}\| \leq \frac{1}{r} \tag{9.10}$$

where $G(I + CG)^{-1}$ is the feedback configuration of the blocks G and C. The auxiliary elements C and r are called centre and radius respectively. The block C has to be linear to be combined with G and static to be combined with H. In other words, C has to be a constant matrix of appropriate dimensions. A detailed exposition of the topics can be found in Vidyasagar (1993). The equivalent small-gain stability condition for conicity is now given by:

$$\|G(I + CG)^{-1}\|.\|H - C\| < 1 \tag{9.11}$$

The abstract setting of the conicity criterion in (9.11) can now be used for deriving further stability conditions depending on the choice of the centre matrix C that satisfies (9.11) to guarantee stability. Conicity criterion based stability analysis and design of fuzzy controllers have been reported in the literature (Barreiro 1999; Cao et al. 2001; Cuesta et al. 1999; Espada and Barreiro 1999). Barreiro (1999) investigated the application of conicity criterion for the stability of a standard fuzzy controller with a high dimensional plant. Cuesta et al. (1999) also applied conicity criterion on Takagi–Sugeno-type fuzzy controller. Cao et al. (2001) applied conicity criterion to design a stable single-input PID fuzzy controller. Espada and Barreiro (1999) treated fuzzy controller design with the small-gain and conicity criteria for nonlinear stability. The main disadvantages of the approach are that sometimes the conic bounds may result very conservative and can not be found globally. The techniques are well-suited for open-loop stable and high dimensional problems.

Some researchers formulated the search based on two measures: conic robustness $r_G(C)$ of closed-loop form of the linear plant G and conic deviation $d_H(C)$ of the nonlinear controller H from the centre C. As there is no a priori information available, one possibility is to explore all centres verifying the conicity inequalities. The conic deviation $d_H(C)$ of the nonlinear controller H from the centre matrix C is given by:

$$d_H(C) = \|H - C\| \tag{9.12}$$

The conic robustness $r_G(C)$ of the linear plant G with the feedback C is given by:

$$r_G(C) = \frac{1}{\|G(I + C.G)^{-1}\|} \tag{9.13}$$

If C is chosen such that the closed-loop approaches instability, then the gain approaches infinity and $r_G(C)$ converges to zero. Using this rule of thumb, the search condition can be formulated as

$$\exists C, r : d_H(C) < r \leq r_G(C) \tag{9.14}$$

As mentioned earlier that the search for a suitable centre matrix C that satisfies the conditions of conicity can further lead to different stability criterion such as circle criterion.

Consider a linear system $G(s)$ with a feedback connection of a time-varying nonlinearity $h(.)$ and an external input $r = 0$ presented in Fig. 9.4. The unforced system can be described by:

Fig. 9.4 Feedback connection system

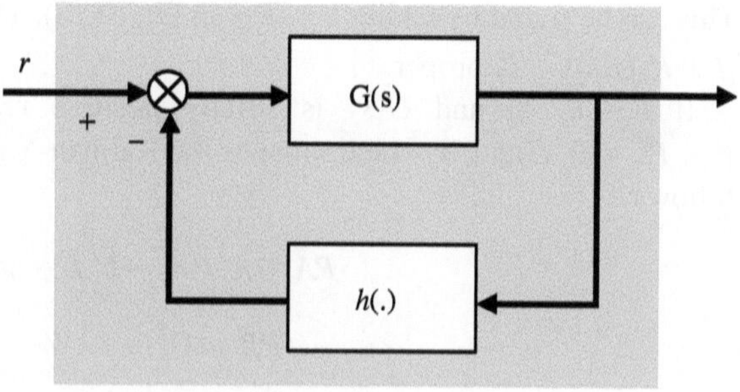

$$\dot{x} = Ax + Bu \tag{9.15}$$

$$y = Cx + Du \tag{9.16}$$

$$u = -h(t, y) \tag{9.17}$$

where $x \in R^n$, $u, y \in R^p$, (A, B) is controllable, (A, C) is observable. $h : [0, \infty) \times R^p \rightarrow R^p$ is time-varying nonlinearity and piecewise continuous in t and locally Lipschitz in y. By replacing y in Eq. (9.17), the feedback connection has a well-defined state model defined as

$$u = -h(t, Cx + Du) \tag{9.18}$$

The Eq. (9.18) has a unique solution for every (t, x) in the domain of interest when $D = 0$. The transfer function $G(s)$ of the linear system is given by:

$$G(s) = C(sI - A)^{-1}B + D \tag{9.19}$$

The transfer function matrix $G(s)$ in (9.19) is square and proper; also it ensures $\{A, B, C, D\}$ is a minimal realisation of $G(s)$. The nonlinearity is required to satisfy a sector condition for an equilibrium point of the system at origin $x = 0$ for all $y \in R^p$ or $y \in Y \subset R^p$. The use of loop transformations allows it to cover various sectors and Lyapunov function candidates leading to circle and Popov criteria (Khalil 2002).

The circle criterion states that the system described by (9.15)–(9.17) is absolutely stable if the following two conditions hold:

- $h \in [K_1, \infty]$ and $G(s)[I + K_1G(s)]^{-1}$ is strictly positive real or
- $h \in [K_1, K_2]$ with $K = K_2 - K_1 = K^T > 0$ and $[I + K_2G(s)][I + K_1G(s)]^{-1}$ is strictly positive real.

If the sector condition is satisfied only on a set $Y \subset R^p$, then the foregoing conditions ensure that the system is absolutely stable with a finite domain. A necessary condition for the Eq. (9.18) to have a unique solution u for every above criterion $h \in [K_1, \infty]$ or $h \in [K_1, K_2]$ is the nonsingularity of the matrix $(I + K_1D)$.

This can be found by taking $h = K_1 y$ in Eq. (9.18). Therefore, the transfer function $[I + K_1 G(s)]^{-1}$ is proper.

If $h \in [0, \infty]$ and $G(s)$ is strictly positive real, then there are matrices $P = P^T > 0$, L and W which satisfies the Kalman-Yakubovich-Popov equations as follows:

$$PA + A^T P = -L^T L - \varepsilon P \tag{9.20}$$

$$PB = C^T - L^T W \tag{9.20}$$

$$W^T W = D + D^T \tag{9.21}$$

where $\varepsilon > 0$ and $V(x) = (1/2)x^T Px$ is a storage function for the linear dynamical system.

The circle criterion was used by Ray et al. (1984) and Ray and Majumdar (1984), but their approach was limited to either single-input/single-output (SISO) systems, or decoupled multiple-input/multiple-output (MIMO) systems where a nonlinear plant was considered. Tso and Fung (1998) employed multivariable circle criterion approach to study the stability analysis of a fuzzy controlled double-integrator problem.

Consider a special case of the system (9.15)–(9.17) with $u_i = -h_i(y_i)$, $1 \le i \le p$ where $h_i : R \to R$ is a locally Lipschtz memoryless nonlinearity that belongs to the sector $[0, k_i]$. In this special case, the transfer function $G(s) = C(sI - A)^{-1}B$ is strictly proper and h is time invariant and decoupled.

The Popov criterion states that the system with the special case described above is absolutely stable if the following conditions hold for $1 \le i \le p$:

- $h_i \in [0, k_i]$, $0 < k_i \le \infty$ and
- There exists a constant $\gamma_i \ge 0$ with $(1 + \lambda_k \gamma_i) \ne 0$ for every eigenvalue λ_k of A such that $M + (I + s\Gamma)G(s)$ is strictly positive real.

where $\Gamma = diag(\gamma_1, \ldots, \gamma_p)$ and $M = diag(1/k_1, \ldots, 1/k_p)$. If the sector condition $h_i \in [0, k_i]$ is satisfied only on a set $Y \subset R^p$, then the foregoing conditions ensure that the system is absolutely stable with a finite domain.

The condition $(1 + \lambda_k \gamma_i) \ne 0$ implies that (A, C) is observable and $\{A, B, C, D\}$ is minimal realizable. If $M + (I + s\Gamma)G(s)$ is strictly positive real, then there exist matrices $P = P^T > 0$, L and W that satisfy the Kalman-Yakubovich-Popov equations as follows:

$$PA + A^T P = -L^T L - \varepsilon P \tag{9.22}$$

$$PB = (C + \Gamma CA)^T - L^T W \tag{9.23}$$

$$W^T W = 2M + \Gamma CB + B^T C^T \Gamma \tag{9.24}$$

The stability analysis of fuzzy logic controllers using Popov criterion has been widely reported in the literature (Choi et al. 2000; Lu et al. 2012; Wu et al. 2008).

Choi et al. (2000) demonstrated the stability analysis of a single-input fuzzy controller using Popov criterion. Wu et al. (2008) applied the parametric robust Popov criterion based on Lur's systems to stability analysis and design of fuzzy controllers. This is a valuable reference designing stable fuzzy controller. Lu et al. (2012) derived a sufficient condition for the T-S fuzzy control system to guarantee stability. The proposed method includes the circle criterion and Popov criterion-based stability conditions as special cases, which ensures the least conservative result. Yamashita et al. (1999) derived the absolute stability condition for a PI-type fuzzy controller using modified Yakubovich's (1967) method. The condition was represented by a similar condition to the usual Popov condition to analyse the stability of PI-type fuzzy controller.

Stability analysis of control systems can be carried out using any of the two broad categories of theories such as qualitative and general stability theory. There are two directions of stability analysis under the general stability theory. Sofonov (1980) established a conceptual framework for the two families of stability criteria, namely the Lyapunov and the input–output stability as shown in Fig. 9.5. This is considered as a general framework of stability analysis applicable to dynamical systems. Using the framework, one can derive a particular case for stability analysis for a fuzzy control system.

The general stability theory is well developed and widely applied to conventional mathematical model-based systems. The Takagi-Sugeo-type fuzzy system is taking the advantage of these well established methodologies as they let themselves represented as nominal linear model with uncertainties around the equilibrium of the system, which includes all the nonlinearities of the Takagi–Sugeno fuzzy model and then recast the control problem as a robust linear control problem with uncertainties (Feng 2001). Feng (2006) showed that Takagi–Sugeno fuzzy

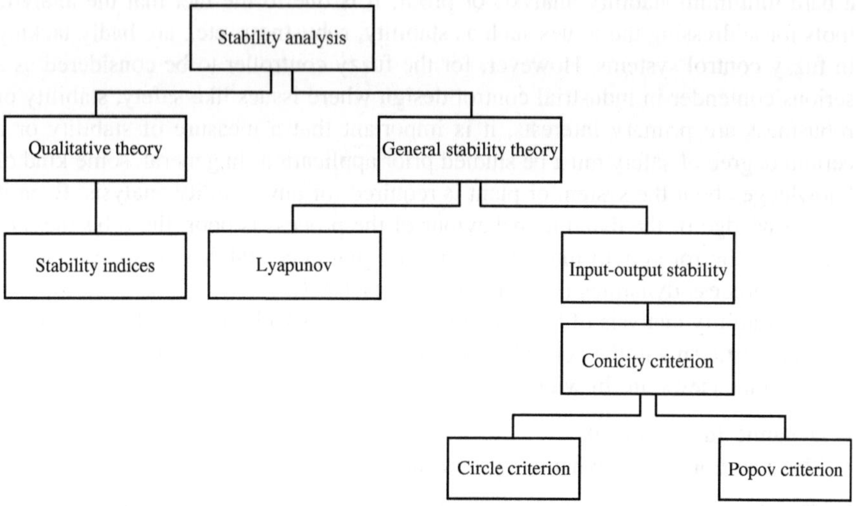

Fig. 9.5 General framework of different approaches to stability analysis

models can be used to represent complex MIMO systems with both fuzzy inference rules and local analytic linear dynamic models. In this way, general stability theory-based approaches to stability analysis can directly be applied to the Takagi–Sugeno fuzzy systems. A good number of research has been reported in the literature demonstrating the use of general stability analysis on Takagi–Sugeno-type fuzzy models (Ban et al. 2007a, b, 2010; Cao et al. 2011; Ding et al. 2003). Stability analysis of Takagi–Sugeno-type fuzzy systems has been pursued mainly based on Lyapunov stability theory but with different Lyapunov functions. Feng (2006) reported a good survey on different methods of stability analysis of Takagi–Sugeno fuzzy modes based on Lyapunov stability functions.

Though the performance of the Mamdani-type fuzzy controllers is satisfactory, they suffer from lack of systematic approaches for stability analysis. Unfortunately, the research community is somehow indifferent to stability analysis of Mamdani-type fuzzy systems. The possible reason may be the Mamdani-type fuzzy control systems are essentially heuristic and model free in the sense that the fuzzy control rules (i.e., If–Then rules) are based on an expert operator's control action or knowledge. Rigorous mathematical approaches of stability analysis such as Lyapunov, conicity, circle or Popov criteria are not easy to apply straight way. Therefore, the qualitative approaches of stability analysis deem useful for these types of fuzzy controllers.

9.3 Qualitative Stability Analysis of Fuzzy Controllers

Despite enormous success in a wide variety of nonlinear and industrial process control applications, the most fuzzy control systems in the literature do not provide a bare minimum stability analysis or proof. It is due to the fact that the analytic tools for addressing the issues such as stability, robustness, etc., are badly lacking in fuzzy control systems. However, for the fuzzy controller to be considered as a serious contender in industrial control design where issues like safety, stability or robustness are primary interests, it is important that a measure of stability or a certain degree of safety must be studied prior application. In general, some kind of knowledge about the system or plant is required for any stability analysis. Even if the knowledge of the dynamic behaviour of the process is poor, the robustness of the fuzzy control system must be studied to guarantee stability in spite of variations in process dynamics (French and Rogers 1998).

The stability analysis of fuzzy control systems, or lack of it, has been a subject of many criticisms in the control engineering literature. In fuzzy control literature, two distinct views are in wide use:

- Dynamic fuzzy systems
- Classical non-linear dynamic system theory.

The dynamic fuzzy system's view is associated with Zadeh's extension principle and so far is of theoretical interest. The classical non-linear dynamic system

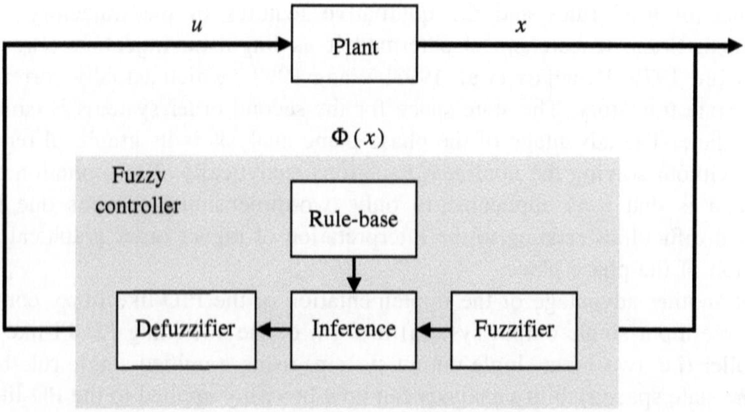

Fig. 9.6 Closed-loop fuzzy control system

theory approach is used to analyse fuzzy controllers where the closed-loop non-linear structure of a fuzzy controller can be represented by means of a non-linear function $u = \Phi(x)$ as shown in Fig. 9.6.

The closed-loop fuzzy control system in Fig. 9.6 is represented by the equations

$$\frac{dx}{dt} = f(x) + bu \tag{9.26}$$

$$u = \Phi(x) \tag{9.27}$$

where $x = [x_1, x_2]^T$ is the state, $f(x)$ is a non-linear function that represents the plant dynamics with $f(x) = 0$, x and b are vectors of dimension n ($n = 2$ here), u is the scalar control variable and $\Phi(x)$ is a non-linear function representing the fuzzy controller with $\Phi(x) = 0$. Suppose there are n_1 and n_2 fuzzy sets to cover the input domains x_1 and x_2 respectively, the rule-base of the fuzzy controller $\Phi(x)$ consists of $(n_1 \times n_2)$ rules. For the k-th rule in the rule-base

$$R^k : \text{ If } x_1 \text{ is } A_i^k \text{ and } x_2 \text{ is } B_j^k, \text{ Then } u \text{ is } C_l^k \tag{9.28}$$

where A_i is the fuzzy sets for input x_1 with $i = 1, 2, \ldots, n_1$, B_j is the fuzzy sets for input x_2 with $j = 1, 2, \ldots, n_2$, C_l is the fuzzy sets for output u with $l = 1, 2, \ldots, n_3$ and $k = 1, 2, \ldots, (n_1 \times n_2)$. The point (x_1, x_2) in the phase plane belongs to the subspace of the partition associated with the rule r, if it holds that

$$\mu_{A_1^r}(x_1) * \mu_{B_1^r}(x_2) \geq \mu_{A_1^k}(x_1) * \mu_{B_1^k}(x_2) \text{for} \quad \forall r \neq k \tag{9.29}$$

where * represents a t-norm.

A geometric interpretation of the state map based on vector field associated with the plant and the rule-base can be given by computing each point $\left(x_1^k, x_2^k\right)$ on the phase plane for each fired rule R^k. A trajectory can be constructed from the sequence of fired rules and the qualitative features of the trajectory can be examined. Some researchers also termed it as linguistic trajectory (Braae and Rutherford 1979; Driankov et al. 1993; Wang 1997), which actually corresponds to the state trajectory. The state space for the second-order systems is called the phase plane. The advantage of the phase plane analysis is its graphical representation without solving the nonlinear equations analytically. The limitation of this approach is that it is applicable to only two-dimensional systems due to the practical difficulties relating to the interpretation of higher order graphical representation of the phase plane.

Yet another advantage of the implementation of the PID-like fuzzy controller (i.e. three-input single-output system) in form of the switching PD-PI-like fuzzy controller (i.e. two-input single-output system) using a unified single rule-base is that the state space stability analysis can now be easily applied to the PD-like and PI-like fuzzy controllers on the same phase plane.

Figure 9.7 shows the partitioning of the input space $x = [x_1, x_2]$ determined by the above equation and the rule-base defined for a two-input single-output fuzzy controller. A closed-loop system trajectory can be mapped on the partition space, as a sequence of rules according to the order in which they are fired that form the so-called linguistic trajectory. Figure 9.8 shows an example of associated system trajectory. The state trajectory may seem converging slowly in the early stages. Modifying the associated rules and membership functions the trajectory can be corrected to converge faster. Figure 9.9a–d shows different forms of non-fired rules inside and outside of the partition space, which may have been caused by different modes of operation, working condition, improper partitioning of the input space, poor definition of membership functions and incomplete or inconsistent rules. In Fig. 9.9a, the rules centred around $x_2 = ZO$ are fired, which means inappropriate partitioning of input x_1. Similarly, Fig. 9.9b shows the fired rules centred around $x_1 = ZO$ indicating the inappropriate partitioning of the input x_2. In

Fig. 9.7 Partitioning of the input space $\{x_1, x_2\}$

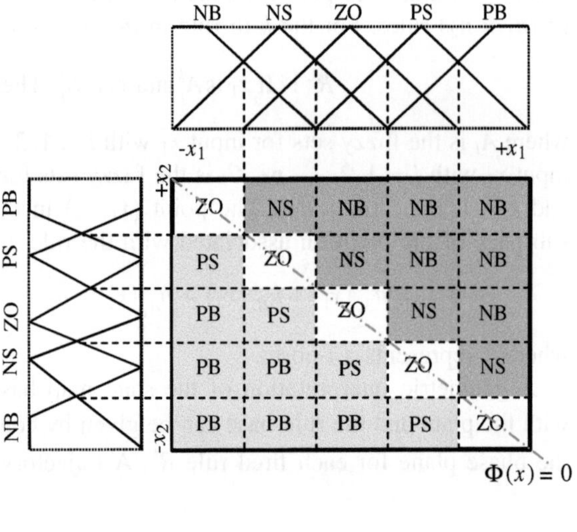

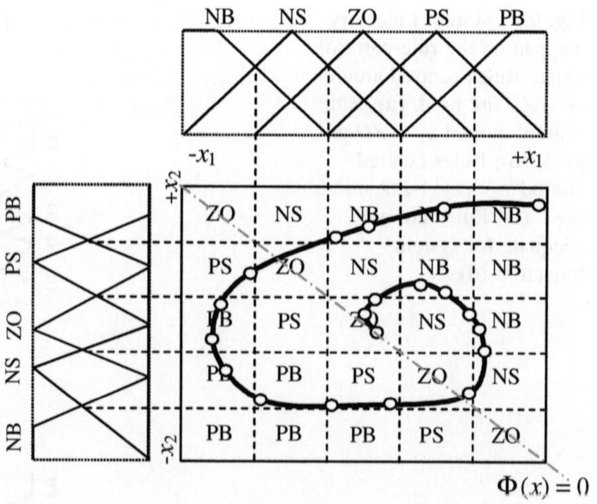

Fig. 9.8 System trajectory mapped on the partition space

Fig. 9.9c, the rules centred around $x_1 = x_2 = ZO$ are fired meaning the whole partitioning is improper causing so many non-operative rules. Figure 9.9d shows inadequate coverage of partition space. Non-fired rules in a given mode of operation or working conditions can easily be modified to force the system trajectory to go within the desired control space.

In general, the trajectory traverses from the edges of the input space (partitioned into cells representing the rules) to the centre for a stable system as demonstrated in Fig. 9.8. In the case of an unstable system, the tendency of the trajectory would be to go out of bound. From the design point of view, this provides interesting guidelines for analysis and modification of the design of fuzzy controllers. As far as stability is concerned, control systems have high potential to be unstable if the following are observed:

- Very fast rise time
- High over shoot
- Very long settling time.

Hence, to improve stability, a decrease in overshoot and in setting time is intended. These goals can be achieved by adjusting the membership functions, rule-base modification and tuning scaling factors.

The closed-loop behaviour will depend on the nature of $f(x)$ and $\Phi(x)$ of Eqs. (9.26)–(9.27). The direction of the vector field associated with fuzzy controller is determined by coefficients of b and the magnitude given by $b.\Phi(x)$. The vector field associated with $b.\Phi(x) = 0$ tends towards $\Phi(x) = 0$. The condition $\Phi(x) = 0$ defines a line on the phase plane shown in dotted line in Figs. 9.7, 9.8 and 9.9. The line is also called switching line and has some interesting features. The switching line $\Phi(x) = 0$ divides the phase plane into two subspaces: negative and positive as shown by a dotted line in Fig. 9.10. The negative subspace is shown in dark-shade and positive subspace is shown in light-shade in Fig. 9.10. When the state vector is

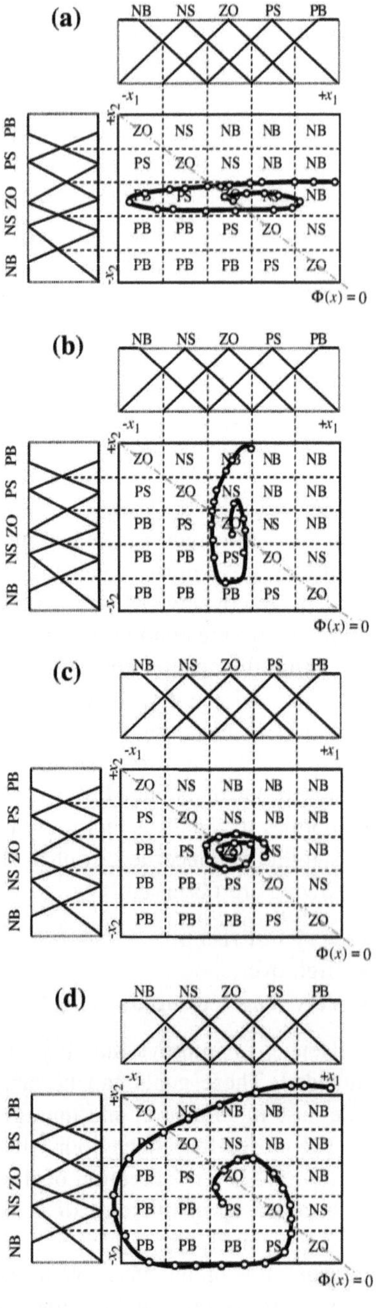

Fig. 9.9 System trajectory mapped on the rule-partition space. Rules centred around $x_2 = ZO$ are fired. (**a**); Rules centred around $x_1 = ZO$ are fired. (**b**); Rules centred around $x_1 = x_2 = ZO$ are fired. (**c**); Rules are not adequate for system trajectory (**d**)

far from the switching line $\Phi(x) = 0$, the control vector $b.\Phi(x)$ has greater influence on the closed-loop system. When the state vector gets closer to the switching line $\Phi(x) = 0$, the control vector $\Phi(x) = 0$ gets smaller and $f(x)$ has

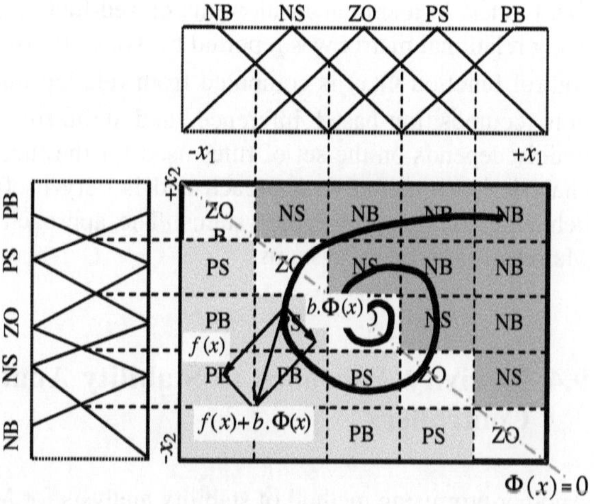

Fig. 9.10 Switching line and the components of vector field

greater influence on the closed-loop system. The characteristics of the vector fields $b.\Phi(x)$, $f(x)$ and the relationship between them determine the behaviour of the closed-loop system, which is illustrated in Fig. 9.10.

Considering different characteristics of the state space trajectories discussed above, a simple inspection approach can be developed to analyse stability and other dynamic phenomena of systems by using the relationship between $b.\Phi(x)$ and $f(x)$ (Aracil et al. 1988; Garcia-Cerezo et al. 1992; Driankov et al. 1993; Wang 1997).

Stable closed-loop systems: This is the case when the open-loop system $\dot{x} = f(x)$ is stable and the vector field $\Phi(x)$ tries to lead the system trajectories towards the switching line $\Phi(x) = 0$. When the trajectories approach this line (i.e. $\Phi(x) = 0$), the plant component of the vector field obtains a greater influence which makes the trajectories converge to the equilibrium point. Such an example case is shown in Fig. 9.10.

Limit cycles: This is the case when the open-loop system $\dot{x} = f(x)$is unstable and the vector field $\Phi(x)$ tries to stabilise the system. When the state vector is far from the switching line $\Phi(x) = 0$, the control vector $b.\Phi(x)$ has greater influence on the closed-loop system and tries to lead the trajectory towards the switching line $\Phi(x) = 0$. When the state vector gets closer to the switching line $\Phi(x) = 0$, the control vector $b.\Phi(x)$ gets smaller and the unstable system component $f(x)$ has greater influence on the closed-loop system and tries to diverge the trajectory away from the equilibrium point. The counter act between the vector field $b.\Phi(x)$ and plant component $f(x)$ makes the state oscillate around the equilibrium point creating a limit cycle.

The state space approach is one of the first methods used to analyse the stability of a closed-loop system and the concept of linguistic trajectory of closed-loop fuzzy control system was introduced by Braae and Rutherford (1979), and they established a relationship between the state space representation of system and fuzzy rules. In a similar fashion, the closed-loop analysis of fuzzy control system using relational matrix was reported by Tong (1976, 1980), in which the non-linear control function $\Phi(x)$ is computed from relation matrix $\tilde{R}$ by means of fuzzification, composition-based inference, and defuzzification. The relation matrix $\tilde{R}$ mainly depends on the set of rules used for the fuzzy controller. The phase plane analysis is a qualitative approach and is very useful for analysing the dynamic behaviour of fuzzy control systems. The approach is especially well suited for Mamdani-type fuzzy systems.

9.4 Passivity Approach to Stability Analysis of Fuzzy Controllers

Another promising method of stability analysis for Mamdani-type fuzzy controller is the passivity approach. Passivity provides with a useful tool for the nalysis of nonlinear control systems, which relates nicely to Lyapunov and L_2 stability (Khalil 2002). When the fuzzy controller $\Phi(x)$ is considered as a nonlinear mapping between the inputs and the output, the system's absolute stability can be derived based on the input–output dynamic characteristics (Shin and Xu 2009). The advantage of the passivity approach is that the method does not demand any explicit mathematical description of the control system.

A continuous time system with input $u(.) : R \rightarrow R$, output $y(.) : R \rightarrow R$ and the state vector $x \in R^n$ is said to be passive if there exists a continuous nonnegative real-valued storage function $V(x)$ with $V(0) = 0$ and a supply rate $s(u(\tau), y(\tau))$ such that the following inequality equation is satisfied.

$$V(x(t)) - V(x(0)) \leq \int_0^t s(u(\tau), y(\tau))d\tau \quad \forall x, u \tag{9.30}$$

The following passivity definitions are given from the relations between the supply rate and inputs and outputs.

- The system is said to be strictly input passive with respect to $s(u, y) = u^T y - \varepsilon u^T u$, $\varepsilon > 0$.
- The system is said to be strictly output passive with respect to $s(u, y) = u^T y - \varepsilon y^T y$, $\varepsilon > 0$.
- The system is said to be strictly input and output passive with respect to $s(u, y) = u^T y - \varepsilon_1 u^T u - \varepsilon_2 y^T y$, $\varepsilon_1 \varepsilon_2 > 0$.

The controllers discussed and studied in this monograph and most of the fuzzy controllers in the literature share some specific features. Ying (1994, 2005, 2006) observed that most of the fuzzy control applications have the following common features that are useful for the stability analysis:

- Two inputs e_1 and e_2 scaled to the same range $[-L, +L]$.
- The membership functions are sufficiently overlapped with adjacent membership functions. There is no assumption on the shape of the membership functions except that the trapezoidal shape around zero.
- In practice, the number of rules is quite small.
- The nonlinear control rules are symmetric with respect to the inputs.
- The control action (i.e. $u(t)$) corresponding to the central area of the rule-table is usually zero, i.e., $\Phi(0,0) = 0$ (output is null for null inputs).
- The control action increases gradually from left to right within a row and from bottom to top within a column.

The actual value of the controller output is obtained by an appropriate defuzzification algorithm, using the *min*- or *product*-inference based on fired rules and an eventually results in an output scalar value $u = \Phi(e_1, e_2)$ where e_1 and e_2 denote the two inputs to the fuzzy controller.

A fuzzy controller $u = \Phi(e)$ which posses above feature with respect to the inputs, rule-base and the output is called a sectorial fuzzy controller (SFC) (Calcev 1998; Shin and Xu 2009). This general class of fuzzy controller has specific sectorial properties of their input–output mapping, which are useful for analysis of fuzzy controllers (Calcev 1998). They are stated as follows:

- $\Phi(e_1, e_2)$ is globally Lipschitz continuous and bounded
- $\Phi(0,0) = 0$ is the steady state condition
- $\Phi(e_1, e_2) = -\Phi(-e_1, -e_2)$ is odd symmetry, and
- For every pair of (e_1, e_2) it satisfies

$$0 \le e_1(\Phi(e_1, e_2) - \Phi(0, e_2)) \le \lambda e_1^2$$

$$0 \le e_2(\Phi(e_1, e_2) - \Phi(e_1, 0)) \le \gamma e_2^2$$

The details of the proofs of these properties can be found in (Calcev 1998; Shin and Xu 2009; Xu and Shin 2005).

In order to obtain sufficient stability conditions for a fuzzy control structure, the first step is to derive the passivity properties of an SFC. For this purpose, the PD-like SFC is considered as a SISO nonlinear dynamical system, where $e_1(t)$ represents the error $e(t)$ and $e_2(t)$ represents change of error $\Delta e(t)$ and $u(t)$ is the control action $\Phi(e_1(t), e_2(t))$. The control system is defined as follows:

$$\dot{e}_1(t) = e_2(t) \tag{9.31}$$

$$u(t) = \Phi(e_1(t), e_2(t)) \tag{9.32}$$

From Eqs. (9.31)–(9.32), the PD-like fuzzy controller can be considered as nonlinear system with single-input $e_2(t)$, $e_1(t)$ as state and $u(t)$ as output. The fuzzy control system should have $u(0) = \Phi(e_1(0), e_2(0)) = 0$ as the equilibrium point. A proof of the passivity stability conditions for PD-like fuzzy controller is provided in (Shin and Xu (2009).

Similarly, to demonstrate the passivity of PI-like SFC, it can be considered as a SISO nonlinear dynamical system with $e_1(t)$ representing the error $e(t)$, $e_3(t)$ representing sum of error $\int e(t)(\Sigma e(k)$ in discrete case) and $u(t)$representing the control action $\Phi(e_1(t), e_3(t))$. The control system is defined as follows:

$$\dot{e}_3(t) = e_1(t) \tag{9.33}$$

$$u(t) = \Phi(e_1(t), e_3(t)) \tag{9.34}$$

From Eqs. (9.33)–(9.34), the PI-like fuzzy controller can be considered as nonlinear system with single-input $e_3(t)$, $e_1(t)$ as state and $u(t)$ as output. The fuzzy control system should have $u(0) = \Phi(e_1(0), e_3(0)) = 0$ as the equilibrium point. A proof of the passivity stability conditions for PI-like fuzzy controller is provided in (Shin and Xu 2009). The details of the proofs for discrete case for the PD- and PI-like fuzzy controllers are also provided in Shin and Xu (2009), Xu and Shin (2005).

In the next section, stability analysis of the Mamdani-type PD-PI-like fuzzy controller will be shown using the passivity and qualitative approach.

9.5 Stability Analysis of PD-PI-like Fuzzy Controller

The general theoretical analysis of the different approaches to the stability of fuzzy controllers has been made in the previous sections. There are practical limitations and disadvantages of each of the methods, especially when applying them to Mamdani-type fuzzy controllers. Among them, the passivity and state space methods are advantageous to be applied to Mamdani-type fuzzy controllers as they don't need any explicit mathematical descriptions of the control system.

The advantage of the PD-PI-like fuzzy controller developed in the previous chapters is that it is a two-input single-output Mamdani-type fuzzy controller consists of a PD-like and PI-like fuzzy controller executed in a sequence. The membership functions for change of error and sum of error are merged with a pursuit of unifying the two rule bases, i.e. the rule-base for the PD-like and the rule-base for PI-like fuzzy controller, to single rule-base carried out in Chap. 6. The corresponding rule-base has an odd symmetry and 'monotonicity' in linguistic terms only belongs to the row and column corresponding to zero inputs. The membership functions of the PD-PI-like fuzzy controller now satisfy the passivity conditions mentioned in Sect. 9.4. Therefore, the two-input and single-output Mamdani-type PD- or PI-like fuzzy controller can be written as a functional dependency as follows:

$$u(t) = \Phi(e_1(t), e_2(t)) \tag{9.35}$$

where e_1 and e_2 denote the two inputs to the fuzzy controller. The switching PD-PI-like controller executes two fuzzy controllers in a sequence to achieve the PID-like effect, i.e. executes PD-like controller first followed by PI-like controller.

Therefore, the functional dependency of the control action in Eq. (9.35) can be expressed as:

$$u_{t<t_s}(t) = \Phi(e_1(t), e_2(t)) = \Phi(e(t), \Delta e(t)) \quad \text{for PD controller} \tag{9.36}$$

$$u_{t\geq t_s}(t) = \Phi(e_1(t), e_2(t)) = \Phi(e(t), \Sigma e(t)) \quad \text{for PI controller} \tag{9.37}$$

where t_s is the time instant when the control mechanism switches from PD-like to PI-like controller. The stability of the PD-like and PI-like fuzzy controller has been analysed in Sect. 9.4. Therefore, when the two fuzzy controllers are executed in a sequence, the PD-PI-like fuzzy controller will remain passive stable.

The advantage of the state space stability analysis is that it can be shown graphically for a two-input and single-output control systems. The obvious advantage of the PD-PI-like fuzzy controller is that it uses a single rule-base and the control action is given by (9.35). The primary objectives of the switching PD-PI-like fuzzy controller were to achieve fast rise time, minimum overshoot and minimum steady state error or shorter settling time. The secondary objective was a design simplification by merging the universes of discourse for change of error and sum of error and a unified single rule-base. The design simplification of the PD-PI-like fuzzy controller eventually followed the guidelines of the stability criteria discussed in Sect. 9.3. That is, as far as stability is concerned, control systems have high potential to be unstable if the following situations occur during the execution of the controller:

- A very fast rise time that also causes very high overshoot
- High overshoot, which also causes oscillation and prolongs settling time
- Very long settling time, which also means oscillation.

Hence, to improve stability, a decrease in overshoot is ensured by executing the PD-like fuzzy controller first and as a consequence setting time is shortened as the PI-like fuzzy controller is executed after the switching point, i.e. around the point of maximum overshoot. These goals have been achieved by rule-base modification carried out in Chap. 6, by adjusting the membership functions carried out in Chap. 6 and by tuning the scaling factors carried out in Chap. 7 and 8. From the definition and discussion in Sect. 9.3, the stability of the PD-PI-like fuzzy controller can be analysed using the state trajectory. The objective is to drive the state variables error and change of error or sum of error to the equilibrium point.

Figure 9.11 shows the system trajectory of the switching PD-PI-like fuzzy controller. As noted the trajectory shows oscillatory behaviour, which may have been caused by the membership functions. The big sudden change of the trajectory around 10 of the variable error in the direction of change of error is possibly due to switching from PD to PI fuzzy controller, which took place during this time. From a good designer's point of view, a smooth system trajectory is desired. An adjustment of the membership functions may improve the oscillatory behaviour of the system trajectory. Genetic algorithm was used to adjust the membership function with the hope that the oscillation in the system trajectory will diminish with an improvement in the overshoot and settling time as well. The system

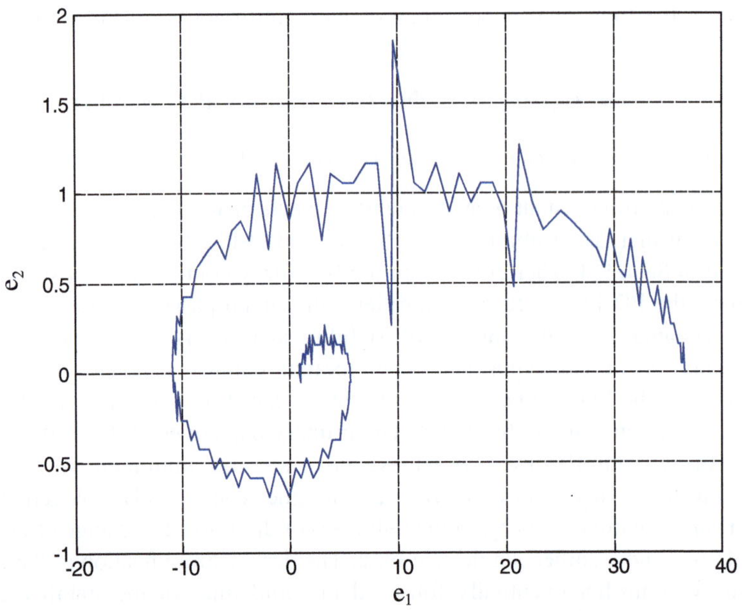

Fig. 9.11 State trajectory of the PD-PI-like fuzzy controller

performance has improved in respect of overshoot and settling time, as shown in Fig. 6.19 in Chap. 6, but surprisingly the system trajectory became more oscillatory, which is revealed by investigating the system trajectory shown in Fig. 9.12. The option that was open to improve stability was to tune the scaling parameters of the PD-PI-like fuzzy controller. A single neuron network was used to tune the scaling parameters k_d' and k_i' instead of adjusting the membership functions.

Figure 9.13 shows the system trajectory with the neuro-fuzzy controller. At this stage of the development, the performance of the system has improved significantly and the system trajectory smoothed.

Inclusion of the sigmoidal activation function and training of the neural network using genetic algorithm seemed not promising in comparison with neuro-fuzzy controller with a linear activation function but the system trajectory with GA-based neuro-fuzzy controller, shown in Fig. 9.14, reveals that the rules do cover all system trajectory.

9.6 Summary

Consistent analysis of fuzzy controllers has been a painful part of fuzzy systems theory for a long time. Fuzzy control was accused of being unreliable approximate engineering approach, which uses experience, intuition and rules of thumb instead

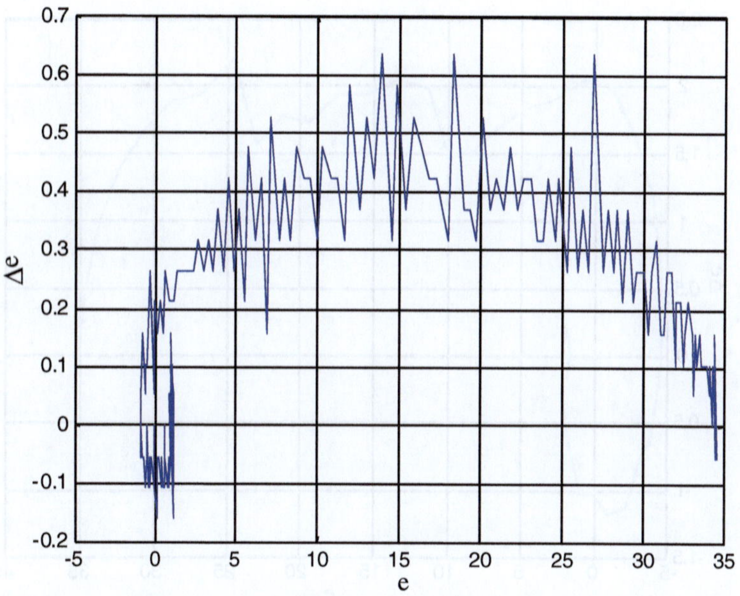

Fig. 9.12 State trajectory of the GA-Fuzzy controller

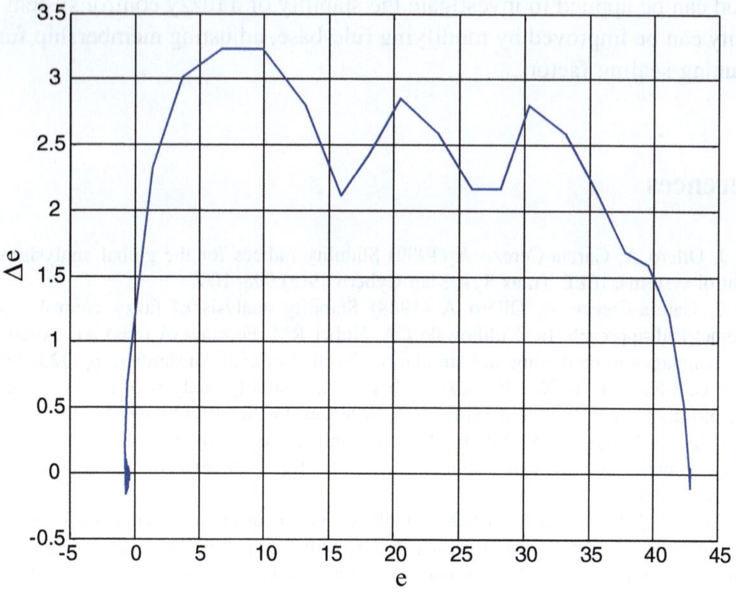

Fig. 9.13 State trajectory of the neuro-fuzzy controller

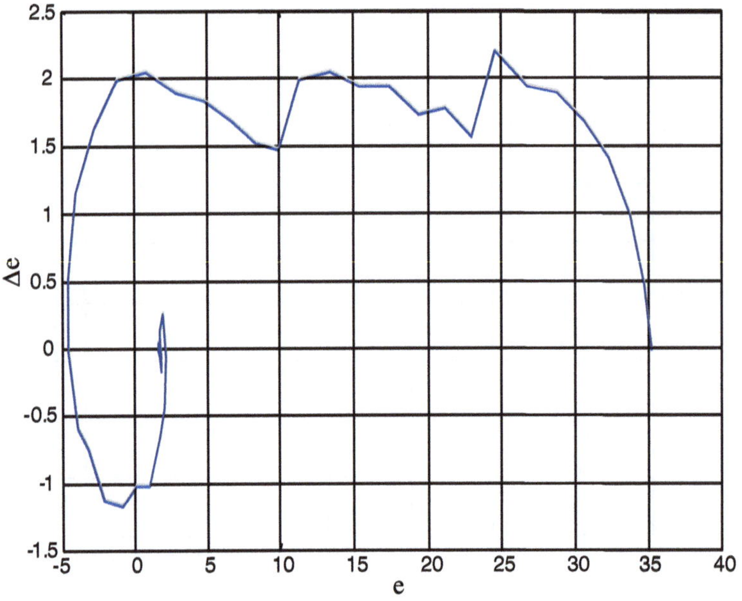

Fig. 9.14 State trajectory of the GA-based neuro-fuzzy controller

of consistent firm analytical theory. The study reveals that the system trajectory method can be applied to investigate the stability of a fuzzy control system and its stability can be improved by modifying rule-base, adjusting membership functions and tuning scaling factors.

References

Aracil J, Ollero A, Garcia-Cerezo A (1989) Stability indices for the global analysis of expert control systems. IEEE Trans Syst Man Cybern 19(5):998–1007

Aracil J, Garcia-Cerezo A, Ollero A (1988) Stability analysis of fuzzy control systems: a geometrical approach. In: Kulikowski CA, Huber RM, Ferrate GA (eds) AI, expert systems and languages in modelling and simulation. North-Holland, Amsterdam, pp 323–330

Ban XJ, Gao XZ, Huang XL, Vasilakos AV (2007a) Stability analysis of the simplest Takagi-Sugeno fuzzy control system using circle criterion. Inf Sci 177(20):4387–4409

Ban XJ, Gao XZ, Huang XL, Yin H (2007b) Stability analysis of the simplest Takagi-Sugeno fuzzy control system using Popov criterion. Int J Innovative Comput, Inf Control 3(5):1087–1096

Ban XJ, Gao XZ, Huang XL (2010) L_2 -Stability analysis of proportional Takagi–Sugeno fuzzy controller based control systems. Intel Autom Soft Comput 16(1):61–71

Barreiro A (1999) Stability of fuzzy control systems by conicity. In: Aracil J, Gordillo F (eds) Stability issues in fuzzy control. Physica-Verlag, Wurzburg

Braae M, Rutherford DA (1979) Selection of parameters for a fuzzy logic controller. Fuzzy Sets Syst 2:185–199

Calcev G (1998) Some remarks on the stability of mamdani fuzzy control systems. IEEE Trans Fuzzy Syst 6(3):436–442

Cao YD, Hu BG, Gao DJ (2001) Genetic-based robust optimal design for one-input fuzzy PID controllers. In: IEEE international conference on systems, man and cybernetics, pp 2263–2268

Cao KR, Gao XZ, Huang XL, Ban XJ (2011) Stability analysis of a type of Takagi–Sugeno PI fuzzy control systems using circle criterion. Int J Comput Intel Syst 4(2):196–207

Choi B-J, Kwak S-W, Kim BK (2000) Design and stability analysis of single-input fuzzy logic controller. IEEE Trans Syst Man Cybern B Cybern 30(2):303–309

Cuesta F, Gordillo F, Aracil J, Ollero A (1999) Stability analysis of nonlinear multivariable Takagi–Sugeno fuzzy control systems. IEEE Trans Fuzzy Syst 7(5):508–520

Ding Y, Ying H, Shao S (2003) Typical Takagi-Sugeno PI and PD fuzzy controllers: analytical structures and stability analysis. Inf Sci 151:245–262

Driankov D, Hellendoorn H, Reinfrank M (1993) An introduction to fuzzy control. Springer, Berlin

Espada A, Barreiro A (1994) Robust design of fuzzy controllers based on small gain conditions. In: Proceedings of AIRTC'94, Valencia, Spain

Espada A, Barreiro A (1999) Robust stability of fuzzy control systems based on conicity conditions. Automatica 35(4):643–654

Feng G (2001) Approaches to quadratic stabilization of uncertain fuzzy dynamic systems. IEEE Trans Circuits Syst I 48(6):760–769

Feng G (2003) Controller synthesis of fuzzy dynamic systems based on piecewise Lyapunov functions. IEEE Trans Fuzzy Syst 11(5):605–612

Feng G (2004) Stability analysis of discrete-time fuzzy dynamic systems based on piecewise Lyapunov functions. IEEE Trans Fuzzy Syst 12(1):22–28

Feng G (2006) A survey on analysis and design of model-based fuzzy control systems. IEEE Trans Fuzzy Syst 14(5):676–697

French M, Rogers E (1998) Input/output stability theory for direct neuro-fuzzy controllers. IEEE Trans Fuzzy Syst 6(3):331–345

Fuh C-C, Tung P-C (1997) Robust stability analysis of fuzzy control systems. Fuzzy Sets Syst 88:289–298

Garcia-Cerezo A, Ollero A, Aracil J (1992) Stability of fuzzy control systems by using nonlinear systems theory. In: Proceedings of the IFAC symposium on AI in real-time control, delft, The Netherlands, pp 171–176

Khalil HK (2002) Nonlinear systems, 3rd edn. Prentice-Hall, Upper Saddle River

Kikert WJM, Mamdani EH (1978) Analysis of a fuzzy logic controllers. Fuzzy Sets Syst 1(1):29–44

Kluska J (1999) Absolute stability of continuous fuzzy control systems. In: Aracil J, Gordillo F (eds) Stability issues in fuzzy control. Physica-Verlag, Wurzburg

Lam HK, Leung FHF (2007) Fuzzy controller with stability and performance rules for nonlinear systems. Fuzzy Sets Syst 158(2):147–163

Li W, Chang X, Farrellt J (2001) Stability and performance analysis of an enhanced hybrid fuzzy P + ID controller. In: Proceedings of the American control conference, Arlington, VA June 25–27, pp 3855–3860

Lu H, Cao K, Ban X, Huang X (2012) Stability analysis for a type of T-S fuzzy control systems using integral quadratic constraints. In: 9th International conference on fuzzy systems and knowledge discovery (FSKD 2012), pp 376–380

Lyapunov AM (1992) The general problem of stability of motion (100th Anniversary). Taylor & Francis, London

Martynnyuk AA (2003) Advances in stability theories at the end of the 20th Century, Edited. Taylor & Francis, London

Merkin DR (1997) Introduction to the theory of stability. Springer, New York

Ohtake H, Tanaka K, Wang HO (2003) Fuzzy modelling via sector nonlinearity concept. Integr Comput-Aided Eng 10(4):333–341

Ohtake H, Tanaka K, Wang HO (2006) Switching fuzzy design based on switching Lyapunov function for a class of non-linear systems. IEEE Trans Syst Man Cybern B Cybern 36(1):13–23

Ohtake H, Tanaka K, Wang HO (2007) Piecewise fuzzy model construction and controller design based on piecewise Lyapunov function. In: Proceedings of American control conference, New York, pp 259–262

Ray KS, Majumdar DD (1984) Application of circle criteria for stability analysis of linear SISO and MIMO system associated with fuzzy logic controller. IEEE Trans Syst Man Cybern 14(2):345–349

Ray KS, Ghosh AM, Majumdar DD (1984) L_2-stability and the related design concept for SISO linear system associated with fuzzy logic controller. IEEE Trans Syst Man Cybern 14(6):932–939

Sofonov MG (1980) Stability and robustness of multivariable feedback systems. The MIT Press, Cambridge

Shin YC, Xu C (2009) Intelligent systems—modeling, optimization and control. CRC Press, Taylor & Francis Group

Tanaka K, Wang HO (2001) Fuzzy control systems design and analysis: a linear matrix inequality approach. Wiley, New York

Tanaka K, Sugeno M (1992) Stability analysis and design of fuzzy control systems. Fuzzy Sets Syst 45(2):135–156

Tong RM (1976) Analysis of fuzzy control algorithms using the relational matrix. Int J Man Mach Stud 8:679–686

Tong RM (1980) The evaluation of fuzzy models derived from experimental data. Fuzzy Sets Syst 4:1–12

Tso SK, Fung YH (1998) Synthesis and stability analysis of linguistic fuzzy controlled systems, 1998 IEEE World Congress on computational intelligence. In: Proceedings of the international conference on fuzzy systems, 4–9 May, AK, pp 366–371

Vidyasagar M (1993) Nonlinear systems analysis. Prentice-Hall Inc, London

Wang L-X (1997) A course in fuzzy systems and control. Prentice-Hall Inc, Englewood Cliffs

Wang HO, Tanaka K, Griffin M (1995) An analytical framework of fuzzy modelling and control of nonlinear systems: stability and design issues. In: Proceedings of the American control conference, Seattle, Washington, June 1995, pp 2272–2276

Wu BF, Ma LS, Perng JW, Chin HI (2008) Absolute stability analysis in uncertain static fuzzy control systems with the parametric robust popov criterion. In: IEEE international conference on fuzzy systems (FUZZ 2008), pp 1325–1330

Xu C, Shin YC (2005) Design of multi-level fuzzy controller for nonlinear systems and stability analysis. IEEE Trans Fuzzy Syst 13(6):761–778

Yakubovich VA (1967) Frequency conditions for the absolute stability of the control systems with several nonlinear or linear non-stationary blocks. Autom Remote Control 28:857–880

Yamashita T, Katoh R, Kohama T (1999) Stability of a control system with PI type fuzzy controller. In: Aracil J, Gordillo F (eds) Stability Issues in Fuzzy Control. Physica-Verlag, Wurzburg

Ying H (1994) Practical design of nonlinear fuzzy controllers with stability analysis of regulating process with unknown mathematical models. Automatica 30:1185–1195

Ying H (2005) Structure and stability analysis of general Mamdani fuzzy dynamic models. Int J Intell Syst 20:103–125

Ying H (2006) Deriving analytical input-output relations for fuzzy controllers using arbitrary input fuzzy sets and Zadeh fuzzy AND operator. IEEE Trans Fuzzy Syst 14(5):654–662

Zames G (1966a) On the Input-Output Stability of Nonlinear Time-varying Feedback Systems, Part-I, IEEE Trans. Automatic Control, AC-11, pp 228–238

Zames G (1966b) On the Input-Output Stability of Nonlinear Time-varying Feedback Systems, Part-II, IEEE Trans. Automatic Control, AC-11, pp 465–477

Zhang H, Feng G (2008) Stability analysis and $H\infty$ controller design of discrete-time fuzzy large-scale systems based on Piecewise Lyapunov functions
Zhang HB, Li CG, Liao XF (2006) Stability analysis and $H\infty$ controller design of fuzzy large-scale systems based on piecewise Lyapunov functions. IEEE Trans Syst Man Cybern B Cybern 36(3):685–698

Chapter 10
Future Work

10.1 Epilogue

In this research monograph, intelligent control techniques have been explored with theoretical exposition and practical application to flexible robotic arm. Although fuzzy control and neuro-control have been well established, covering a wide spectrum of applications in the literature, the application to controlling a flexible arm is very limited.

Fuzzy PD-, PI-like controllers have the same characteristics as the conventional PD- and PI-like controllers. The fuzzy PID-like controllers have better performance over the fuzzy PD- or PI-like controllers at the expense of a huge time consuming rule-base processing. To overcome the drawback, a variety of modularisations have been investigated and a switching PD-PI-like fuzzy controller was presented in this book, which achieves a faster rise time, smaller overshoot, smaller settling time and reduced steady state error. The run-time rule base of PD-PI-like fuzzy controller is reduced to $n \times n = n^2$, i.e., for $n = 5$ there will be only 25 rules whereas in a PID-like fuzzy controller it is 125. A rule-base reduction of 80 % can be achieved by employing this switching type PD-PI fuzzy controller.

Initially, the membership functions have been constructed using heuristic rules. Then an optimisation/learning of the membership functions has been due because of merging of the two universes of discourse namely change of error and sum of error. Genetic algorithm has been used to optimise the membership functions for the inputs and output as well.

It is still believed that the membership functions for change of error and sum of error are not the same although it is optimised by genetic algorithm. A further adjustment of the membership functions can enhance the performance of the PD-PI-like fuzzy controller. Adjustment of membership functions requires several parameters to tune and hence tuning of the scaling factors is chosen. Tuning scaling factors is simpler task than adjusting membership functions and yields the same result. A mechanism has been sought to tune the scaling factors of the PD-PI fuzzy controller by using a neural network where an online adaptation is also an important issue. A neural network with multiple layers and a number of neurons in

N. Siddique, *Intelligent Control*, Studies in Computational Intelligence 517,
DOI: 10.1007/978-3-319-02135-5_10,

the hidden layer can best do the approximation of the non-linear behaviour of the system, but a significant amount of time will be consumed in calculating the updated scaling factors. In order to minimise the calculation time, a single neuron network has been used for an online updating of the scaling factors. Firstly backpropagation learning algorithm has been derived and used in training the network.

Experimentations in Chaps. 7 and 8 show that the nonlinearity can be represented by the sigmoidal function and its shape. There are several algorithms like backpropagation that learn the weights and biases of neural network but very few algorithms are reported that learn shape of the sigmoidal function. Genetic algorithm has been chosen to adjust the weights, biases and shape of the sigmoidal function simultaneously.

The stability of control system is a basic requirement for the design of practical systems, especially in real-time control of nonlinear systems. There are several well-established methods existing for Sugeno-type fuzzy controllers but unfortunately very few methods are reported for Mamdani-type fuzzy controllers. Among them are linguistic trajectory method is widely appreciated and used. Since GA has some characteristic of random search, some search points may cause practical learning process unstable. Similarly, for some search points of weight updates of the neural network learning the system can be unstable as well. Stability analysis of the PD-PI-fuzzy, GA-fuzzy and Neuro-fuzzy controller has been investigated using the common linguistic trajectory method in Chap. 9.

10.2 Future Research Directions

The aim of this work was to develop an intelligent controller incorporating fuzzy logic, neural network and genetic algorithms. Of course there were constraints and time limitation of this research work due to which many other ideas could not be implemented. These ideas can be investigated and implemented in a future research. Some of the proposed methods and ideas are described in this chapter.

In recent years, learning-based neural-network controllers have emerged as an alternative to adaptive control. The most popular neural networks in neuro-control are the multilayer perceptron (MLP) network, the functional layer network, associative memory network such as cerebral model articulation control (CMAC), the radial basis function (RBF) network and the B-Spline network. Feedforward neural-network is essentially global in nature and slow since all the weights are updated during each learning cycle. The CMAC and B-Spline networks are examples of local generalisation networks, in which learning interference is minimised and learning is relatively fast owing to the minimum number of weights required to update for each training pair. These are discussed in the following sections.

10.3 Adaptive Neural Network Control

The widely used structures of neural network based control systems are similar to those employed in adaptive control, where a neural network is used to estimate the unknown nonlinear system and the controller is then formulated using the estimation results. The estimation uses the measured input and output from the system and can be realised via various types of neural networks, such as MLP networks, RBF networks and B-Spline networks. A variety of neural control schemes are available in the literature such as NN-based direct control (Park et al. 2005), NN-based indirect control (Sen et al. 1998), Backpropagation through time control (Werbos 1990b; Omatu et al. 1995), NN-based direct inverse control (Werbos 1990a; Sen et al. 1998), NN-based model predictive control (Akesson and Toivonen 2006; Yuzgec et al. 2008), NN-based adaptive control (Narendra 1996; Narendra and Mukhopadhyay 1997), NARMA-L2 (Feedback Linearization) Control (Narendra and Mukhopadhyay 1997; Pukrittayakame et al. 2002).

Figure 10.1 shows a direct neuro-controller, which can be applied to flexible-link arm. The derivation of a learning algorithm such as backpropagation is straightforward and simple. The number of neurons in the hidden layer and number of hidden layers are to be optimised so that the network can be employed in a real time manner.

10.3.1 Adaptive Neuro-Fuzzy Controller

Most of the fuzzy logic controllers for flexible arm reported up to now are Mamdani-type rule-based fuzzy logic controllers. Mamdani-type controllers

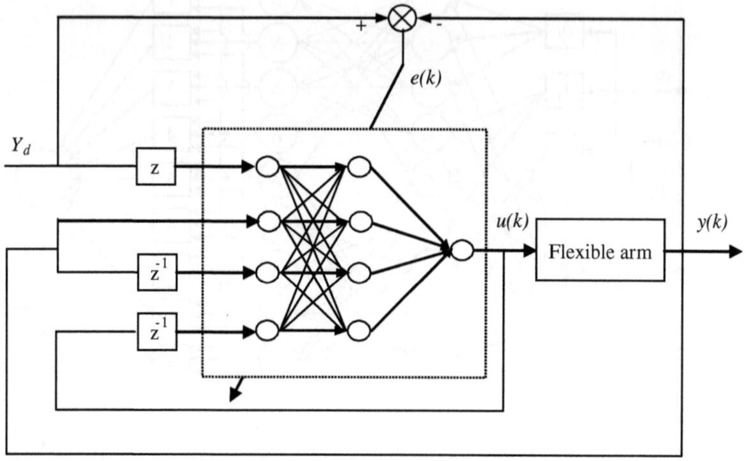

Fig. 10.1 Direct neuro-controller

require processing of the rule base of dimension $n \times m$, n and m are the number of primary fuzzy sets of the inputs in PD- or PI-type controllers. Most of the rule processing time is spent in the consequent part in calculating the control output by centre of gravity method of defuzzification. The processing time of the rule base can be minimised by avoiding defuzzification procedure suggesting a Sugeno-type fuzzy controller that avoids methods of defuzzification and uses linear functions for the consequent part.

The idea of a neuro-fuzzy system is to find the parameters of a fuzzy system by means of learning methods obtained from neural networks. A common way to apply a learning algorithm to a fuzzy system is to represent it in a special neural-network-like architecture. Then a learning algorithm, such as backpropagation, is used to train the system. A variety of neuro-fuzzy systems have been proposed in the literature such as Fuzzy Adaptive Learning Control Network (FALCON) (Lin and Lee 1991, 1994), Adaptive Neuro-Fuzzy Inferencing Systems (ANFIS) (Jang 1993; Jang et al. 1997), Neuro-Fuzzy Controller (NEFCON) (Nauck et al. 1997) and Coactive Neuro-Fuzzy Inference System CANFIS (Mizutani et al. 1994; Mizutani and Jang 1995).

An adaptive neuro-fuzzy controller with two inputs, each with three membership functions, and one output is shown in Fig. 10.2.

Layer 1: Every node i in this layer is an adaptive node with membership functions A_k and B_j where e and Δe are angle error and change of error. These nodes calculate the membership grade of the inputs.

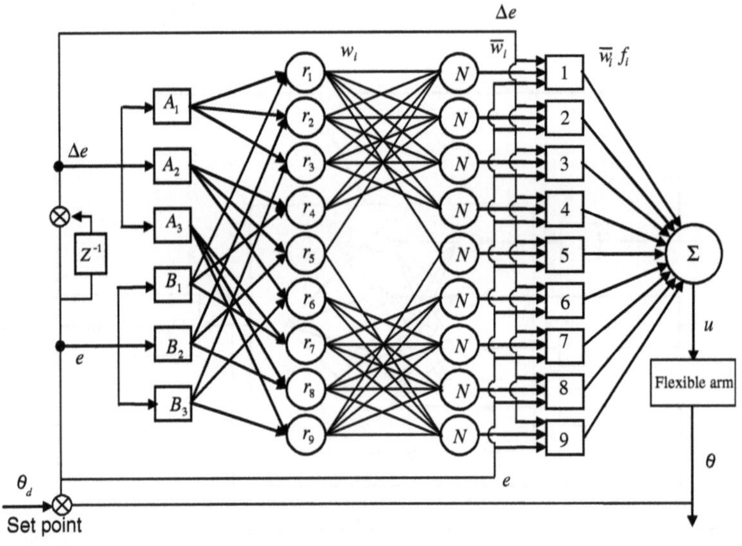

Fig. 10.2 Adaptive neuro-fuzzy control architecture

$$O_{1,k} = \mu_{A\,k}(e), \quad for\ k = 1,2,3 \tag{10.1}$$

$$O_{1,j} = \mu_{B\,j}(\Delta e), \quad for\ j = 1,2,3 \tag{10.2}$$

Layer 2: Every node in this layer is a fixed node representing 9 rules labeled $r_1 \ldots r_9$. Each node determines the firing strength of a rule as

$$w_i = \mu_{A_k}(e).\mu_{B_j}(\Delta e), \quad k, j = 1,2,3 \tag{10.3}$$

Layer 3: Every node in this layer is fixed node labeled N. Each node calculates the normalized firing strength.

$$\overline{w}_i = \frac{w_i}{\sum_{i=1}^{9} w_i}, \quad i = 1,2,\ldots 9 \tag{10.4}$$

Layer 4: Every node in this layer is an adaptive node with a linear function defined by

$$f_i = a_i.e + b_i.\Delta e + c_i, \quad i = 1,2,\ldots 9 \tag{10.5}$$

where a_i, b_i and c_i, $i = 1, 2, \ldots, 9$ are the parameters of the consequent part of the rule base.

Layer 5: The single node in this layer produces the control output by aggregating all the fired rule values.

$$u = \sum_i \overline{w}_i.f_i, \quad i = 1,2,\ldots 9 \tag{10.6}$$

Learning Antecedent and Consequent Parameters

Antecedent parameters: Different consequents of fuzzy rules describe the behaviour within the region via various constituents and result in different fuzzy inference systems but their antecedents are always the same. Antecedent parameters are the parameters of the membership functions. In fact, any continuous and piecewise differentiable function, such as commonly used Gaussian, bell-shaped, trapezoidal or triangular-shaped membership functions can be used.

Consequent Parameters: When the values of the antecedent parameters are fixed, the overall output can be expressed as a linear combination of the consequent parameters as

$$u = \overline{w}_1 f_1 + \overline{w}_2 f_2 + \cdots + \overline{w}_9 f_9 \tag{10.7}$$

where the consequent functions $f_i = a_i.e + b_i.\Delta e + c_i, \quad i = 1, 2, \ldots 9$. The parameters a_i, b_i, c_i, $i = 1, 2, 3, \ldots 9$, are the consequent parameters.

Jang proposed a hybrid-learning algorithm. In the forward pass of the algorithm, functional signals go up to layer 4 and the consequent parameters are identified by the least squares estimate. In the backward pass, the error rates propagate backward and the antecedent parameters are updated by the gradient descent (Jang 1992, 1993).

10.3.2 B-Spline Neural Network

B-Spline neural network (BSNN) is intended for use in the area of on-line adaptive modelling and control, as well as for static off-line design. A BSNN is constructed from a linear combination of basis functions, which are piecewise polynomials of order k such that the output of the network is

$$y = a^T w \tag{10.8}$$

where a is the basis function output vector and w is a weight vector. The network output is a piecewise polynomial of order k defined by the designer.

B-spline interpolants can be used like static fuzzy logic, for non-adaptive control, such as real time motion planning (Harris et al. 1993).

10.3.3 CMAC Network

The CMAC has a similar structure to a three layer neural network with association cells playing the role of hidden layer units (Albus 1975a, b). Mathematically, CMAC may be described as consisting of a series of mappings: $X \to A \to U$ where A is an N dimensional cell space. A fixed mapping $X \to A$ transforms each $\mu \in X$ into an N-dimensional binary association vector $a(\mu)$ in which only N_L elements have the values of one, where $N_L < N$ is referred to as generalisation width. In other words, each μ activates precisely N_L association cells or geometrically each μ is associated with a neighbourhood in which N_L cells may be included. The structure of a CMAC network is shown in Fig. 10.3.

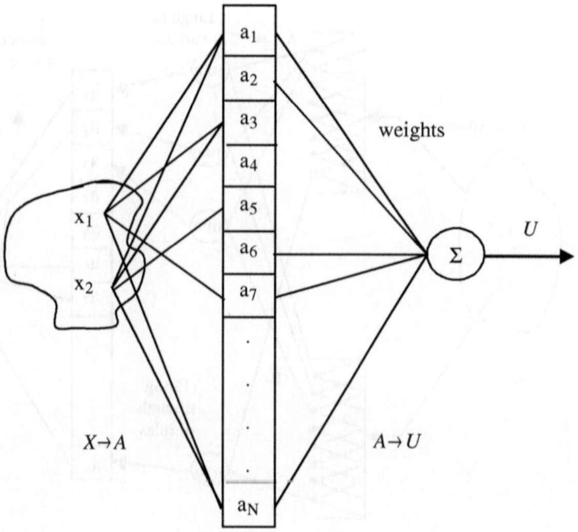

Fig. 10.3 Structure of CMAC

An important property of CMAC is local generalisation derived from the fact that nearby input vectors μ_i and μ_j have some overlapping neighbourhood and therefore share some common association cells. The degree to which the neighbourhoods of μ_i and μ_j are overlapping depends on the Hamming distance H_{ij} between μ_i and μ_j. If H_{ij} is small, the intersection of μ_i and μ_j should be large and vice versa.

According to the above principle, Albus developed a mapping algorithm consisting of two sequential mappings: $X \rightarrow M \rightarrow A$, which perform a content addressing task (Albus 1975a). The n components of μ are first mapped into N dimensional vectors and these vectors are then concatenated into a binary association vector a with only N_L elements being one.

The mapping $A \rightarrow U$ is simply a procedure of summing the weights of the association cells excited by the input vector μ to produce the output. More specifically, each component u_k is given by

$$u_k = \sum a^i(\mu) w_k^i \tag{10.9}$$

where w_k^i denotes the weight connecting the jth association cell to the kth output.

A CMAC network needs training for determining appropriate values of its weights to represent a given nonlinear function. A supervised learning scheme can be used on observed data pairs $\{x, y\}$. The learning rule of CMAC can be given as linear adaptation in multidimensional space as

$$w(t+1) = w(t) + \frac{\eta a(y_d - y)}{q} \tag{10.10}$$

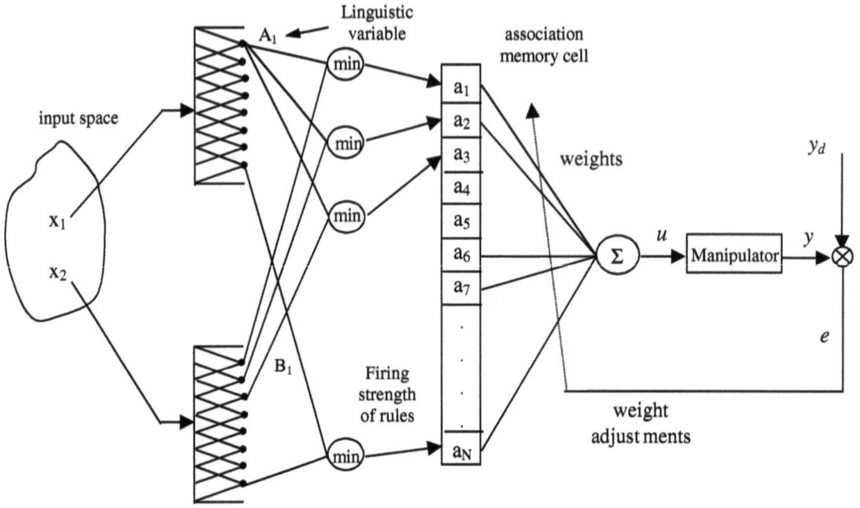

Fig. 10.4 Fuzzified CMAC for control

where q is the total number of weights involved in the input, y_d is the desired output and η is the learning rate. A schematic diagram of fuzzified CMAC controller is shown in Fig. 10.4.

The network's n-dimensional input vector is denoted by x, and the network's sparse internal representation is denoted by the ρ-dimensional vector a; this vector is called the transformed input vector or the basis function output vector. The transformed input vector, a, has, as elements, the outputs of the basis functions in the hidden layer, and the output, y, of the CMAC network is formed from a linear combination of these basis functions. The network output is therefore given by:

$$y(t) = \sum a_i(t)w_i(t-1) = \sum a_{ad(i)}(t)w_{ad(i)}(t-1) \tag{10.11}$$

where $ad(i)$ is a function which returns the address of the ith non-zero basis function.

10.3.4 Binary Neural Network-Based Fuzzy Controller

The basic idea of the binary neural network (BNN)-based fuzzy logic controller is that the inference mechanism of a rule-based fuzzy controller is implemented by using a BNN. All linguistic variables of both antecedent and consequent part are translated into fuzzy sets described by membership functions and the fuzzy sets are then converted into fuzzy numbers or numeric values. The BNN network can be trained off-line by presenting all rules sequentially to the network. Once the BNN is trained, the network can then be inserted in the control loop for on-line operation. The inputs come directly from measured inputs, fuzzified into fuzzy numbers, presented to the network and the control input is calculated using defuzzification.

Training of BNN Using Linguistic Variables

The rule-base of a Mamdani-type fuzzy logic controller with two inputs and a single output, namely the error and change of error or sum of error, consisting of the following rules is used to train the BNN.

$$\text{IF } X_1 \text{ is } A_i \text{ AND } X_2 \text{ is } B_j \text{ THEN } U \text{ is } C_k \tag{10.12}$$

where $X_1 \hat{=}$ error, $X_2 \hat{=}$ change of error, $U \hat{=}$ control input, A_i, $i = 1, 2, \ldots, N$, B_j, $j = 1, 2, \cdots, M$ are the input linguistic variables and C_k, $k = 1, 2, \cdots, L$ are the output linguistic variables. N, M and L are the maximum number of linguistic variables (primary fuzzy sets).

The training procedure of the BNN, using these fuzzy numbers, is shown in Fig. 10.5. Training the BNN is equivalent to rule-base construction in a Mamdani-type fuzzy controller, where the user has to decide the number of fuzzy sets (linguistic labels) for each input and output (Linkens and Nie 1994). The generic rule-base of such Mamdani-type fuzzy controller is shown in Table 10.1, where $C_r \in \{C_1, C_2, \cdots, C_L\}$ are the output fuzzy sets to be learnt. The neural network

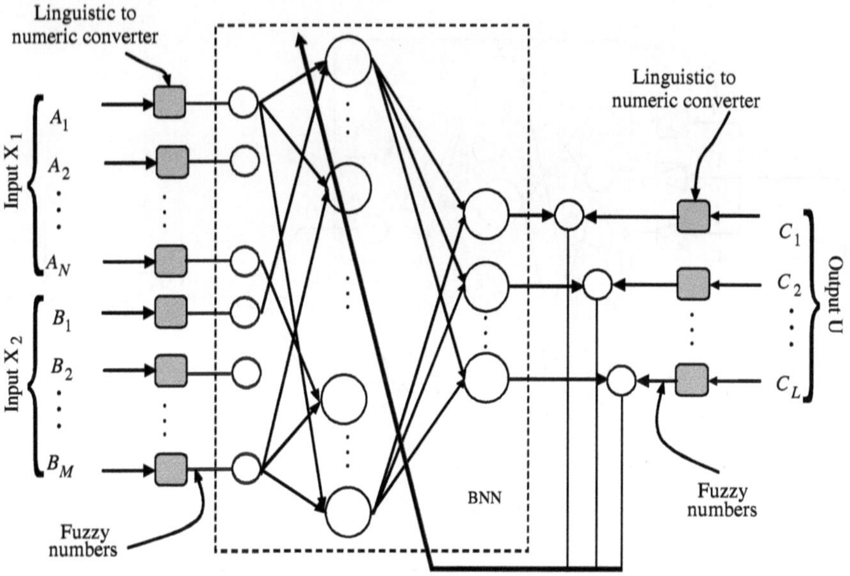

Fig. 10.5 Training of BNN using linguistic description

Table 10.1 Rule-base of FLC

	Input X_2				
	U	B_1	B_2	$\cdots$	B_M
Input X_1	A_1	C_r	C_r	$\cdots$	C_r
	A_2	C_r	C_r	$\cdots$	C_r
	$\vdots$	$\vdots$	$\vdots$	$\ddots$	$\vdots$
	A_N	C_r	C_r	$\cdots$	C_r

will, eventually, learn the rule-base with fixed predefined linguistic variables defined by membership functions.

A linguistic label, such as large, medium or small, is typically a fuzzy set described by a membership function. The linguistis labels used in the fuzzy rules in (10.12) is defined by membership functions of any type e.g. triangular or Gaussian and converted into fuzzy numbers. The fuzzy numbers are typically characterised by a central value with an interval around the center (Dubois and Prade 1987a, b). The width of the associated interval determines the degree of fuzziness. For example, $\mu_{A_1}(x) = \Gamma\{a_i, m_i, b_i\}$ is a fuzzy number with a triangular converter function $\Gamma\{a_i, m_i, b_i\}$ where m denotes the central value, $\{a, b\}$ denotes left and right values of a triangular membership function.

The BNN-based fuzzy controller architecture with two inputs, error (e) and change of error (Δe), and one output, torque input (u), thus can be used for the

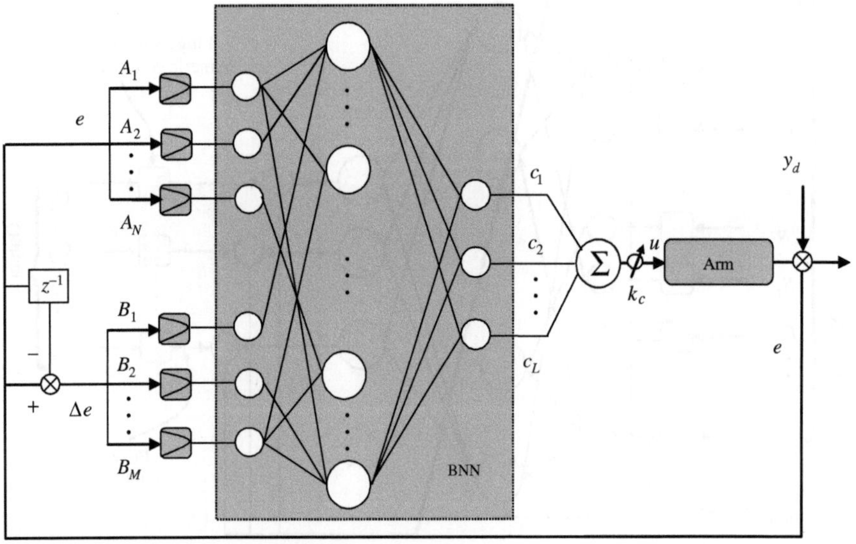

Fig. 10.6 BNN-based FLC in operation

fuzzy controller for the flexible arm as shown in Fig. 10.6. The definition of e and Δe is obvious from the Figure. The control input u to the flexible arm can be obtained using defuzzification procedure in the BNN-FLC, which is defined as

$$u = k_c \sum_{i=1}^{L} c_i = k_c(c_1 + c_2 + \cdots + c_L) \tag{10.13}$$

where k_c is the scaling factor and c_i is the output of the BNN comparable to predefuzzified values of the linguistic labels in Table 10.1. The scaling factor k_c can be broken down into components $k_c = [\,k_{c1} \quad k_{c2} \quad \cdots \quad k_{cL}\,]$ and Eq. (10.13) is rewritten as

$$u = [\,k_{c1} \quad k_{c2} \quad \cdots \quad k_{cL}\,] \begin{bmatrix} c_1 \\ c_2 \\ \vdots \\ c_L \end{bmatrix} = \sum_{i=1}^{L} k_{ci} \cdot c_i \tag{10.14}$$

From Eqs. (10.13)–(10.14), it is established that the tuning of the scaling factor k_c is equivalent to tuning the vector of scaling factors $[\,k_{c1} \quad k_{c2} \quad \cdots \quad k_{cL}\,]$.

The proposed configuration of the fuzzy controller is a PD-like controller. In a similar fashion, a PI-like controller can also be developed.

10.4 Summary

Fuzzy controllers have been extensively applied to many engineering and industrial problems. There still many problems associated with the construction and processing membership functions, rule-base and defuzzification. This chapter firstly, highlights the salient features of the PD-PI-like fuzzy controllers in combination with neural networks and genetic algorithms and secondly provides few future research directions that can be associated with the current research. The ideas of direct NN, adaptive NN, B-spline NN, CMAC and BNN are very promising and require further investigation.

References

Akesson BM, Toivonen HT (2006) A neural network model predictive controller. J Process Control 16:937–946

Albus JS (1975a) A new approach to manipulator control: the cerebeller model articulation controller (CMAC). Trans ASME J Dyn Syst Measur Control 63(3):220–227

Albus JS (1975b) Data storage in the cerebeller model articulation controller (CMAC). Trans ASME J Dyn Syst Measur Control 63(3):228–233

Dubois D, Prade H (1987a) Fuzzy numbers: an overview, in analysis of fuzzy information 1. CRC Press, Boca Raton, pp. 3–39
Dubois D, Prade H (1987b) Mean value of a fuzzy number. Fuzzy Sets Syst 24(3):279–300
Harris CJ, Moore CG, Brown M (1993) Intelligent control: aspects of fuzzy logic and neural nets. World scientific series vol. 6, World Scientific, Singapore
Jang J-SR (1992) Self-learning fuzzy controller based on temporal back-propagation. IEEE Trans Neural Netw 3(5):714–723
Jang J-SR (1993) ANFIS: adaptive-network-based fuzzy inference system. IEEE Trans Syst Man Cybern 23(3):665–685
Jang J-SR, Sun C-T, Mizutani E (1997) Neuro-fuzzy and soft computing, Prentice Hall, New Jeresy
Lin TC, Lee CS (1991) Neural network based fuzzy logic control and decision system. IEEE Trans Comput 40(12):1320–1336
Lin TC, Lee CS (1994) Neural fuzzy control systems with structure and parameter learning. World Scientific, Singapore
Linkens DA, Nie J (1994) Backpropagation neural network based fuzzy controller with a self-learning teacher. Int J Control 60(1):17–39
Mizutani E, Jang J-SR (1995) Coactive neural fuzzy modelling. In: Proceedings of the international conference on neural networks, pp 760–765
Mizutani E, Jang J-SR, Nishio K, Takagi H, Auslander DM (1994) Coactive neural networks wit adjustable fuzzy membership functions and their applications. In: Proceedings of the international conference on fuzzy logic and neural networks, Japan, pp 581–582
Narendra KS (1996) Neural networks for control: theory and practice. In: Proceedings of the IEEE, vol 84. pp 1385–1406
Narendra KS, Mukhopadhyay S (1997) Adaptive control using neural networks and approximate models. IEEE Trans Neural Netw 8(1997):475–485
Nauck D, Klawonn F, Kruse R (1997) Foundations of neuro-fuzzy systems. Wiley, Chichester
Omatu S, Khalid M, Yusof R (1995) Neuro-control and its applications. Springer, London
Park J-H, Huh S-H, Kim S-H, Seo S-J, Park G-T (2005) Direct adaptive controller for non-affine non-linear systems using self-structuring neural networks. IEEE Trans Neural Netw 16(2):414–422
Pukrittayakame A, De Jesus O, Hagan MT (2002) Smoothing the control action for NARMA-L2 controllers. In: Proceedings of the 2002 45th midwest symposium on circuit and systems (MWSCS-2002), vol 3. pp 37–40
Sen P, Hearn GE, Zhang Y (1998) Adaptive neural controller. In: Leondes CT (ed) Neural network systems techniques and applications, industrial and manufacturing systems, vol 4. Academic Press, Massachusetts, pp 273–344
Werbos PJ (1990a) Neuro-control and related techniques. In: Maren A, Harston C, Pap R (eds) Handbook of neural computing applications. Academic Press, San Diego, pp 345–380
Werbos PJ (1990b) Overview of designs and capabilities. In: Miller WT III, Sutton RS, Werbos PJ (eds) Neural networks for control. The MIT Press, Cambridge, pp 59–65
Yuzgec U, Becerikli Y, Turker M (2008) Dynamic neural network based model predictive control of an industrial baker's yeast process. IEEE Trans Neural Netw 19(7):1231–1242

Index

N. Siddique, *Intelligent Control*, Studies in Computational Intelligence 517,
DOI: 10.1007/978-3-319-02135-5,

MIX
Papier aus verantwortungsvollen Quellen
Paper from responsible sources
FSC® C105338

If you have any concerns about our products,
you can contact us on
ProductSafety@springernature.com

In case Publisher is established outside the EU,
the EU authorized representative is:
Springer Nature Customer Service Center GmbH
Europaplatz 3, 69115 Heidelberg, Germany

Printed by Libri Plureos GmbH
in Hamburg, Germany